RED
BLOOD CELL
AGGREGATION

RED BLOOD CELL AGGREGATION

Oguz Baskurt
Björn Neu
Herbert J. Meiselman

CRC Press
Taylor & Francis Group
Boca Raton London New York

CRC Press is an imprint of the
Taylor & Francis Group, an **informa** business

Cover image provided by Dr. Jonathan K. Armstrong

CRC Press
Taylor & Francis Group
6000 Broken Sound Parkway NW, Suite 300
Boca Raton, FL 33487-2742

First issued in paperback 2019

© 2012 by Taylor & Francis Group, LLC
CRC Press is an imprint of Taylor & Francis Group, an Informa business

No claim to original U.S. Government works

ISBN-13: 978-1-4398-4180-8 (hbk)
ISBN-13: 978-0-367-38231-5 (pbk)

Visit the Taylor & Francis Web site at
http://www.taylorandfrancis.com

and the CRC Press Web site at
http://www.crcpress.com

Contents

Foreword

One's first reaction to the publication of a book on red blood cell (RBC) aggregation might be to ask, "Why write a book on a subject that gets 160,000 results on Google, and about which an enormous literature has accumulated over the last 150 years with papers in journals of anatomy, histology, hematology, pathology, physiology, cardiology, biophysics, rheology, and biomedical and biomechanical engineering?" If you seek an answer, turn to the first page of Chapter I of this book and look at the beautiful photo of a network of rouleaux of human biconcave red cells lying on a microscope slide. You might ask yourself these questions: How does such an extraordinary phenomenon occur? Why are most of the cells so nicely packed face to face? What is the underlying mechanism that can produce such a pattern? How is it measured and does this really happen in the circulation, and if so, does rouleau formation have a useful function? Does it make extra work for the heart in pumping blood through the vessels? Is it more prominent in disease? Do the red cells of other species exhibit this phenomenon, and if not, why not? These are the questions dealt with in this book. While vast numbers of papers dealing with various aspects of RBC aggregation have been published and continue to be published, and the potential mechanisms of RBC aggregation and its physiological significance have been discussed in books on hemorheology and hemodynamics, this book is the first to assemble the subject in all its aspects in one place.

After an introductory chapter on the history of observations related to RBC aggregation, the determinants of such aggregation—the biconcave shape, age, hematocrit, and above all the presence of macromolecules in the suspending medium—are outlined. Chapter 3 presents the two hypotheses that provide potential mechanisms for adhesive forces that lead to the regular packing of the cells in rouleaux: the bridging model of adsorbed fibrinogen and other macromolecules linking adjacent cell surfaces, as opposed to the depletion interaction model in which aggregation occurs due to lower macromolecule concentration near the cell surface than in the bulk suspension. Chapter 4 gives a very thorough review of all methods used, classical (e.g., the erythrocyte sedimentation rate test) and modern (e.g., image analysis techniques), to quantify RBC aggregation in vitro and their importance in clinical practice. Chapter 5 deals with the effect of RBC aggregation on the in vitro rheology of blood, essentially its viscometric properties, and is followed by Chapter 6 on the effect of aggregation on flow through tubes—the famous Fåhraeus and Fåhraeus-Lindqvist effects. We then arrive at what actually happens in the circulation when red cells aggregate in Chapter 7, perhaps the most interesting one in the book, and Chapter 8, dealing with variations in RBC aggregation due to physiological and pathophysiological challenges. The authors leave us with a fascinating short chapter on comparative RBC aggregation in mammals. Imagine obtaining blood samples from kangaroos, koala bears, Tasmanian devils, and Weddell seals separated by 8,000 to 10,000 miles from Bowhead whales.

The authors have succeeded in producing a well-written and very informative state-of-the-art book that will be invaluable for investigators and students of mammalian red blood cells.

Giles R. Cokelet, Sc.D.
Montana State University, Bozeman, Montana

Harry L. Goldsmith, Ph.D.
McGill University, Montreal, Canada

Michael W. Rampling, Ph.D.
Imperial College, London, United Kingdom

Preface

Red blood cell (RBC) aggregation and the phenomena associated with aggregate formation have been of interest since the very early work by Hippocrates (5th century BCE), Aristotle (384–322 BCE) and Galen (131–201 CE). In general, these studies are associated with other scientific areas such as blood rheology, cardiovascular physiology, clinical medicine and direct observations of blood flow in living systems. Within the last several decades, there has been a growing interest in the basic science and clinical aspects of aggregation; such interest has been the foundation for the establishment of research groups and numerous peer review publications. Currently, there are no texts or handbooks devoted to RBC aggregation.

The concept of a book focused on RBC aggregation arose because of the long-time association among the authors: Baskurt spent one year in Meiselman's laboratory and this interaction has continued for over 20 years, and Neu was a fellow and a faculty member with Meiselman for six years (1999-2005) and this interaction has also continued. More importantly, the authors share a common interest in RBC aggregation and continue to carry out cooperative research and to publish jointly. While all areas of aggregation are of interest, each author tends to explore different aspects: in vivo circulatory effects (Baskurt), biophysics of cell adhesion (Neu), blood rheology and cellular determinants of aggregation (Meiselman). Based on these areas of common interest, we decided that a book should be written and, while there are many experts in the field, that just the three of us would write it.

In organizing the book, the goal was to cover several topics including historical aspects, measurement techniques, biophysics and models for aggregation, effects on in vitro and in vivo blood flow, cellular factors that affect aggregation (i.e., RBC "aggregability") and comparative aspects. Each author was assigned specific chapters, but in order to avoid embarrassment and attack by those who disagree with a section or statement, no author's name is associated with the chapters.

We hope we have been successful in reaching our objectives for this book, and that it will be of value to basic and clinical scientists engaged in relevant areas. In particular, we trust that this book will serve to foster cooperation among researchers. As will be evident upon reading the book, many aspects of RBC aggregation raise substantial questions that will require broad-based input in order to be answered. These further efforts will be mutually beneficial, and will, we trust, lead to improved health, especially for individuals with various clinical conditions.

The authors wish to thank their wives and families for their patience and support during the many months of writing and the almost countless emails between Los

Angeles, Istanbul and Singapore. We also wish to extend our sincere gratitude to Ms. Roaslinda B. Wenby who read and re-read versions of the text in order to correct the many mistakes made by the authors. This book would have been markedly delayed, and probably not as good, without her help.

O.K. Baskurt
Björn Neu
H.J. Meiselman

Authors

Oguz Baskurt graduated from Hacettepe Medical Faculty at Ankara, Turkey in 1982 and completed his PhD in physiology in the same institution in 1988. He currently serves as a professor of physiology at Koc University School of Medicine, Istanbul, Turkey. Dr. Baskurt's research is focused on the role of hemorheological factors in *in vivo* flow dynamics of blood. He conducted research on the mechanisms of red blood cell aggregation and hemorheological instrumentation. His scientific interest also includes comparative aspects of circulatory physiology and hemorheology in a wide variety of mammalian species. He is among the editors of *Handbook of Hemorheology and Hemodynamics* published in 2007 and the international journal entitled *Clinical Hemorheology and Microcirculation*. He served as the president of The International Society of Clinical Hemorheology for two terms between 1999 and 2005. Dr. Baskurt has published more than 150 peer-reviewed papers.

Björn Neu received his doctorate in biophysics in 1999 at the Humboldt University in Berlin. He then did post-doctoral research in the area of hemorheology at the Keck School of Medicine, University of Southern California in Los Angeles. Currently he is a professor of bioengineering at the Nanyang Technological University in Singapore. His research interests include cell interactions, polymers at bio-interfaces and the rheological behavior of blood.

Herbert J. Meiselman received his doctorate in chemical engineering at the Massachusetts Institute of Technology, Cambridge. His doctoral work focused on the effects of polymers on blood rheology. He then did post-doctoral research in the areas of microvascular blood flow and hemorheology at the California Institute of Technology, Pasadena. He joined the faculty of the Keck School of Medicine, University of Southern California in 1972 and has continued research in the field of blood rheology in health and disease, including comparative studies of various mammalian species; he is currently professor and vice-chair of physiology and biophysics. He is among the editors of the *Handbook of Hemorheology and Hemodynamics* (IOS Press, 2007) and the international journals *Biorheology* and *Clinical Hemorheology and Microcirculation*. He has published over 250 peer-reviewed papers, has received the Fåhraeus Award and the Poiseuille Gold Medal, and is currently president of the International Society for Clinical Hemorheology.

1 Introduction

1.1 PHENOMENON OF RED BLOOD CELL AGGREGATION

Human red blood cells (RBC), as well as RBC from most mammals, have a tendency to form aggregates that initially consist of face-to-face linear structures that resemble a stack or roll of coins (Chien and Jan 1973; Chien and Sung 1987; Fåhraeus 1929). Such individual linear structures are often termed *rouleau*, with rouleaux being the plural. Figure 1.1 shows a representative assortment of rouleaux for normal human RBC in autologous plasma, where it is obvious that (1) the number of RBC per rouleau can vary widely, and (2) side-to-side or side-to-end branching can occur. While Figure 1.1 represents normal aggregation in a thin two-dimensional geometry (i.e., between a microscope slide and cover slip), rouleaux can form three-dimensional structures under appropriate conditions (Branemark 1971; Cokelet and Goldsmith 1991). As anticipated, the formation of these larger structures is affected by factors such as available space and the level of cell–cell attractive forces.

The aggregates shown in Figure 1.1 were photographed under static, no-flow conditions after sufficient time had elapsed for the aggregates to reach a stable configuration. However, as will be detailed in other sections of this monograph, RBC aggregation involves relatively weak attractive forces (Neu and Meiselman 2002; Neu et al. 2003; Skalak and Zhu 1990). Hence, it is possible to disperse aggregates into smaller structures or individual cells by the application of forces resulting from fluid flow or mechanical shear (Rampling 1990; Schmid-Schönbein et al. 1969). It is important to recognize that RBC aggregation is a reversible process: aggregates will reform when external forces are reduced or eliminated. The reversible nature of RBC aggregation is different from blood coagulation in which the soluble plasma protein fibrinogen forms a network of insoluble strands and in which RBC and blood platelets can become enmeshed. The distinction between aggregation and blood coagulation can be initially unclear because fibrinogen is a major determinant of RBC aggregation and also participates in coagulation (Meiselman 2009; Rampling 1990). Unlike coagulation, fibrinogen remains soluble during aggregation.

1.2 DEFINITIONS

Given the large number of investigators who have studied or currently are studying RBC aggregation, the use of common terminology is deemed essential. Recent publications (Baskurt et al. 2000; Baskurt and Meiselman 2009) have dealt explicitly with two terms that are briefly detailed below.

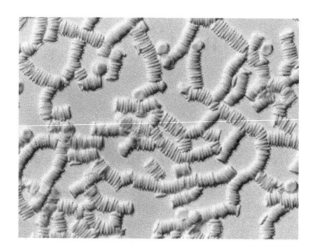

FIGURE 1.1   Red blood cell aggregates. Image provided by Dr. Jonathan K. Armstrong.

1.2.1   RED BLOOD CELL AGGREGATION

RBC aggregation is the process of RBC forming linear and three-dimensional aggregates. This process is reversible such that aggregates will reform at stasis upon the removal of external forces. Various indices of aggregation for blood or RBC suspensions can be measured using several techniques (see Chapter 4). Such indices are specific to the cell suspension being studied, with their magnitude affected by the properties of the cells and the suspending medium.

1.2.2   RED BLOOD CELL AGGREGABILITY

RBC aggregability reflects the intrinsic tendency of RBC to form aggregates, and is thus a cellular property that can be affected by several physiochemical factors. The determination of RBC aggregability requires a comparison of cell populations in the same suspending medium: cells exhibiting greater aggregation in the same medium have enhanced aggregability. Note that merely measuring aggregation, even in a defined suspending medium, does not by itself provide a basis for judging aggregability; comparisons using the same medium are essential. The difference between aggregation and aggregability is further discussed in Chapter 2 and Chapter 4, Section 4.3.

1.3   HISTORICAL ASPECTS

1.3.1   ANTIQUITY

RBC aggregation and the phenomena associated with aggregation have been of interest for millennia, with almost all studies intertwined with other scientific areas such as circulatory physiology, blood rheology, clinical medicine, and *in vivo* observations. Hippocrates (fifth century BCE), Aristotle (384–322 BCE) and Galen (131–201 CE) were all involved with studies of the heart, lungs, and blood vessels and hence, in some way, with blood (Copley 1985; Rampling 2007). According to the theories of Hippocrates and Galen, health is conceived as being dependent on the normal mixture of four fluids that were believed to be the contents of blood vessels: yellow bile (*cholera*), the serum that separates from a blood clot; mucus (*phlegma*), which is now known as the buffy coat and is composed of leukocytes, platelets, and polymerized fibrinogen (i.e., fibrin); blood (*sanguis*), the bright red oxygenated packed red blood cells; and black bile (*melancholia*), red blood cells that failed to become oxygenated when in the body. To analyze the relative proportions of each fluid, native blood without an anticoagulant was allowed to settle, clot, and separate in a cylindrical container; at equilibrium, the amount of each fluid was determined and used as a diagnostic tool (Figure 1.2). Although not recognized at that time, it now seems obvious that the rate at which the blood layers settled and separated was markedly affected by the rate of RBC settling or sedimentation. Also now obvious is the association between RBC sedimentation and RBC aggregation—larger aggregates settle faster (Fåhraeus 1921; Fåhraeus 1929).

In 1665, Malpighi (1628–1694) was the first to demonstrate the existence of RBC as one of the formed elements of blood. Interestingly, Malpighi mistook these cells as globules of fat. This error was corrected about ten years later in 1674 by Anthony van Leeuwenhoek (1618–1723), who gave the first accurate description of red cells (Copley 1985). In a historical review, Rampling (1988, p. 49) provides two interesting quotes regarding RBC aggregation. The first is from a 1786 lecture by John Hunter: "In all inflammatory dispositions ... blood has an increased disposition to separate into its component parts, the red globules become less uniformly diffused and their attraction to one another becomes stronger, so that the blood ... when spread over any surface, it appears mottled, the red blood attracting itself and forming spots of red." The second is by Lister in 1858: "If a drop of blood just shed is placed between two plates of glass and examined with a microscope, the red corpuscles are seen to be applied to one another by their flat surfaces, so as to form long cylindrical masses

FIGURE 1.2 Hematoscopy was used as an important diagnostic tool for many centuries. See the text for explanation. Reproduced from Schmid-Schönbein, H., G. Grunau, H. Brauer H. 1980. *Exempla hämorheologica "Das strömende Organ Blut."* Wiesbaden, Germany: Albert-Roussel Pharma GmbH with permission.

like piles of money, as first observed in 1827 by my father and Dr. Hodgkin: the terminal corpuscles of each 'rouleau' adhering to other rouleaux, a network is produced with intervals of colorless liquid sanguinis."

1.3.2 ROBIN FÅHRAEUS AND MELVIN KNISELY

During the early and middle portions of the twentieth century, two individuals published several seminal papers that served as a stimulus to the current interest in RBC aggregation.

Robin Fåhraeus (1888–1968) was a pathologist at the University of Uppsala in Sweden and was interested in the "suspension stability" of blood and in blood rheology (see Figure 1.3). His very earliest observations dealt with the increased settling velocity of RBC during normal pregnancy. His doctoral thesis, The Suspension-Stability of the Blood, was published in 1921 (Fåhraeus 1921, p. 3) and in 1929 he prepared a manuscript version of this work (Fåhraeus 1929). In the Introduction of the 1921 thesis he states, "[T]he subsiding speed of the blood corpuscles differs under various physiological and pathological conditions to a very great extent—in other words the suspension-stability of the blood is subjected to considerable and constitutional changes." Since the sedimentation speed, often termed settling or sinking or subsiding speed, of red blood cells is enhanced by increased particle size and hence by RBC aggregate formation, Fåhraeus concluded that it is the varying degree of aggregation that determines the sinking velocity.

FIGURE 1.3 Robin Fåhraeus, as illustrated on the Fåhraeus Medal, which is presented to a scientist working in the fields of blood rheology and/or microcirculatory blood flow.

Fåhraeus made careful observations of aggregate behavior in blood from both healthy and diseased subjects. He reports that in healthy blood, aggregates (i.e., rouleaux) contain relatively few RBC, are somewhat irregular in shape, and appear evenly distributed in a microscopic field. Conversely, in diseased blood, rouleaux contain many more RBC and are clustered with cell-free plasma between clusters. He tested the rheological behavior of aggregates contained between glass slides: "In normal blood the rouleaux fall apart at the slightest movement in the suspending phase, whilst the aggregates in pathological blood are not disturbed until very violent currents are produced." He thus concludes that aggregates in pathological blood have a decidedly more "solid" structure. He also evaluated the effects of plasma composition and appears to be the first to prove "a pronounced parallelism between the amount of fibrinogen or serum globulin and sinking velocity" (Fåhraeus 1929, p. 255).

Based in part on the *in vitro* observations of Fåhraeus, Melvin Knisely, and coworkers carried out extensive studies of the *in vivo* circulation in both animals and man (Knisely et al. 1947). Their initial studies focused on the surface of solid organs (e.g., brain, kidney, spleen) and on some transparent tissues. With transparent tissues, it was possible to view both the top and side of vessels and hence they were able to detect any gravitational effects (e.g., RBC settling, two-phase flow). Subsequently, they employed binocular oblique illumination of the bulbar conjunctiva in living, unanesthetized animals, healthy human subjects, and patients with a wide variety of pathologic conditions and diseases (Harding and Knisely 1958; Knisely 1961; Malcom et al. 1972).

Seminal observations by these investigators included the following: (1) Under normal blood flow conditions, circulating RBC were not agglutinated (i.e., aggregated) and did not form rouleaux but rather flowed as individual cells. (2) In animals with experimentally induced enhanced aggregation and in many pathologies, RBC were markedly aggregated into masses of cells that were often separated by clear plasma and were resistant to dispersion by fluid forces. (3) In vessels with slow flow or stasis, RBC sedimentation occurred leaving a cell-poor region above the cell pack. Blood exhibiting such large masses was termed sludged blood and was associated with conditions ranging from normal uncomplicated pregnancy to sepsis and severe trauma. As outlined in the following text, the associations between sludged blood, clinical states, and impaired tissue perfusion led to attempts to reduce RBC aggregation and improve blood flow.

1.3.3 DEXTRANS AND RED BLOOD CELL AGGREGATION

Studies starting in about the mid-1940s began to explore the use of the polyglucose dextran as a blood or plasma substitute (Bicher 1972; Derrick and Guest 1971; Eiseman and Bosomworth 1963; Gronwall 1957; Segal 1964; Squire et al. 1955). These early studies used solutions of several molecular mass fractions (e.g., 40, 70, 150 kDa) of this polymer, initially in animal models of hemorrhage and circulatory shock; such studies were then expanded to include humans. Several investigators, primarily from Sweden, evaluated the effects of dextran on colloid osmotic pressure, RBC aggregation, and RBC life span in the circulation. With time, these studies focused on the 70 kDa fraction as a plasma expander; Pharmacia AB in Uppsala, Sweden marketed Macrodex™, a sterile solution of this polymer intended for human use.

Given the continued interest in RBC aggregation and sludged blood as determinants of *in vivo* blood flow, the twenty-year period from mid-1950 was marked by an increasing interest in the 40 kDa fraction of dextran; Pharmacia AB in Uppsala, Sweden marketed Rheomacrodex™, a sterile solution of this polymer intended for human use. Literature at that time contained numerous references to RBC aggregation and to the ability of Rheomacrodex™ to reduce RBC aggregate formation. That is, Rheomacrodex™ was proposed to decrease the degree of blood sludging and was a "flow improver." Many national and international conferences focused on the effects of this 40-kDa polymer on blood flow and Pharmacia AB distributed bound volumes of reference citations regarding this agent. Lars Eric Gelin, a Swedish surgeon and an active researcher studying dextrans indicated: "Increase in erythrocyte aggregation induced by high molecular mass dextran was countered by low molecular mass dextran" (Gelin 1961). In 1968 Pharmacia AB stated, "When intravascular aggregation is reversed by the infusion of Rheomacrodex™ with an average molecular mass of 40 kDa, oxygen consumption increases and acidosis is allayed" (Pharmacia AB 1968, p. 289).

Interestingly, the antiaggregating ability of the 40 kDa fraction was challenged in 1967 based upon *in vitro* viscometric data, with the *in vivo* benefits suggested as not being due to a specific effect but rather to blood dilution, reduced hematocrit and blood viscosity, and thus higher intravascular shear rates dispersing the aggregates (Meiselman et al. 1967). More recent studies have confirmed that 40 kDa dextran neither reduces nor promotes aggregation for human RBC in plasma (Armstrong et al. 2004) and hence is neutral with respect to aggregate formation. Similar findings have been reported for 40 kDa dextran added to cat blood (Eliasson and Samelius-Broberg 2009).

1.4   RECENT HISTORY AND FUTURE DIRECTIONS

As discussed in the following chapters, RBC aggregation is of current basic science and clinical interest. Work since the mid-1980s has focused on two general areas: (1) mechanisms of aggregation and the role of cellular factors in aggregate formation, and (2) *in vivo* significance of RBC aggregation and its role as a determinant of blood flow and vascular resistance. There are three scientific societies with a strong interest in aggregation (i.e., International Society for Clinical Hemorheology, International Society for Biorheology, European Society for Clinical Hemorheology and Microcirculation) and two scientific journals that frequently publish relevant papers (*Biorheology,* and *Clinical Hemorheology and Microcirculation*, IOS Press, Netherlands). All three societies organize international scientific meetings on a regular basis in order to exchange information and further the understanding of RBC aggregation.

LITERATURE CITED

Armstrong, J. K., R. Wenby, H. J. Meiselman, and T. C. Fisher. 2004. "The hydrodynamic radii of macromolecules and their effect on red blood cell aggregation." *Biophysical Journal* 87:4259-4270.

Baskurt, O. K., M. Bor-Kucukatay, O. Yalcin, H. J. Meiselman, and J. K. Armstrong. 2000. "Standard aggregating media to test the "aggregability" of rat red blood cells." *Clinical Hemorheology and Microcirculation* 22:161-166.

Baskurt, O. K. and H. J. Meiselman. 2009. "Red blood cell aggregability." *Clinical Hemorheology and Microcirculation* 43:353-354.

Bicher, H. I. 1972. *"Blood Cell Aggregation in Thrombotic Processes."* Springfield, IL: Charles C. Thomas.

Branemark, P. I. 1971. *"Intravascular Anatomy of Blood Cells in Man."* Basel, Switzerland: S. Karger.

Chien, S. and K. M. Jan. 1973. "Ultrastructural basis of the mechanism of rouleaux formation." *Microvascular Research* 5:155-166.

Chien, S. and L. A. Sung. 1987. "Physicochemical basis and clinical implications of red cell aggregation." *Clinical Hemorheology* 7:71-91.

Cokelet, G. R. and H. L. Goldsmith. 1991. "Decreased hydrodynamic resistance in the two-phase flow of blood through small vertical tubes at low flow rates." *Circulation Research* 68:1-17.

Copley, A. L. 1985. "The History of Clinical Hemorheology." Clinical Hemorheology 5:765-811.

Derrick, J. R. and M. M. Guest. 1971. *"Dextrans: Current Concepts of Basic Actions and Clinical Applications."* Springfield, IL: Charles C. Thomas.

Eiseman, B. and P. Bosomworth. 1963. *"Evaluation of Low Molecular Weight Dextran in Shock."* Washington, DC: National Academy of Sciences.

Eliasson, R. and U. Samelius-Broberg. 2009. "The Effect of Various Dextran Fractions on the Suspension Stability of the Blood after Intravenous Injection in Cats." *Acta Physiologica Scandinavica* 58:211-215.

Fåhraeus, R. 1929. "The suspension stability of the blood." Physiological Reviews 9:241-274.

Fåhraeus, R. 1921. "The suspension stability of the blood." *Acta Medica Scandinavica* 55:1-228.

Gelin, L. E. 1961. "Hemorheological Disturbances in Surgery." *Acta Clinica Scandinavica* 122:287-293.

Gronwall, A. 1957. *"Dextran and its use in colloidal infusion solutions."* Stockholm, Sweden: Almqvist and Wiksell.

Harding, H. F. and M. H. Knisely. 1958. "Settling of Sludge in Human Patients; a Contribution to the Biophysics of Disease." *Angiology* 9:317-341.

Knisely, M. H. 1961. "The Settling of Sludge During Life. First Observations, Evidences and Significances. A contribution to the Biophysics of Disease." *Acta Anatomica* 44:1-64.

Knisely, M. H., E. H. Bloch, T. S. Eliot, and L. Warner. 1947. "Sludged blood." *Science* 106:431-440.

Malcom, R., H. I. Bicher, M. H. Knisely, and R. C. Duncan. 1972. "Behavioral Effects of Erythrocyte Aggregation." *Microvascular Research* 4, no. 1:94-&.

Meiselman, H. J. 2009. "Red blood cell aggregation: 45 years being curious." *Biorheology* 46:1-19.

Meiselman, H. J., E. W. Merrill, E. W. Salzman, E. R. Gilliland, and G. A. Pelletier. 1967. "The Effect of Dextran on the Rheology of Human Blood: Low Shear Viscosity." *Journal of Applied Physiology* 22:480-486.

Neu, B. and H. J. Meiselman. 2002. "Depletion-mediated red blood cell aggregation in polymer solutions." *Biophysical Journal* 83:2482-2490.

Neu, B., S. O. Sowemimo-Coker, and H. J. Meiselman. 2003. "Cell-cell affinity of senescent human erythrocytes." *Biophysical Journal* 85:75-84.

Pharmacia, AB. 1968. *"Blood Flow Improvement: Clinical and Experimental Data on Rheomacrodex."* Uppsala, Sweden: Beckman Hanson AB.

Rampling, M. W. 1988. "Red cell aggregation and yield stress." In *Clinical blood rheology*, ed. Lowe, G. D. O., 45-64. Boca Raton, FL: CRC Press, Inc.

Rampling, M. W. 1990. "Reuleaux formation -its causes, estimation and consequences." *Doga Turkish Journal of Medical Sciences* 14:447-453.

Rampling, M. W. 2007. "History of Hemorheology." In *Handbook of Hemorheology and Hemodynamics*, ds. Baskurt, O. K., M. R. Hardeman, M. W. Rampling, and H. J. Meiselman, 3-17. Amsterdam, The Netherlands: IOS Press.

Schmid-Schönbein, H., R. Wells, and R. Schildkraut. 1969. Microscopy and Viscometry of Blood Flowing under Uniform Shear Rate (Rheoscopy). *Journal of Applied Physiology* 26:674-678.

Schmid-Schönbein, H., G. Grunau, H. Brauer H. 1980. Exempla hämorheologica "Das strömende Organ Blut." Wiesbaden, Germany: Albert-Roussel Pharma GmbH.

Segal, A. 1964. *"The Clinical Use of Dextran Solutions."* New York, NY: Grune and Stratton.

Skalak, R. and C. Zhu. 1990. "Rheological aspects of red blood cell aggregation." *Biorheology* 27:309-325.

Squire, J. R., J. P. Bull, W. A. Maycock, and C. R. Ricketts. 1955. *"Dextran: Its Properties and Use in Medicine."* Springfield, IL: Charles.

2 Determinants of Red Blood Cell Aggregation

As outlined in the previous chapter, red blood cells (RBC) in static human blood form loose aggregates with a characteristic face-to-face morphology, similar to a stack of coins, termed *rouleaux*. Such aggregation is caused by the presence of various macromolecules, such as fibrinogen, in the plasma. Similar aggregation can also be caused if RBC are suspended in solutions of high molecular mass water-soluble polymers such as dextran, poly(ethylene glycol) abbreviated PEG, or polyvinylpyrrolidone (PVP). However, RBC aggregation does not occur in simple salt solutions (e.g., phosphate-buffered saline, PBS). Figure 2.1 shows human RBC from the same donor suspended at a low hematocrit in PBS (panel a) and in autologous plasma (panel b). RBC suspended in plasma form rouleaux, while no aggregation occurs for RBC in PBS, thus demonstrating the important role of plasma factors.

Although RBC aggregation requires the presence of an aggregant in the suspending medium, it is also known that cell-specific factors can influence RBC aggregation. For example, RBC in certain pathological conditions have a greater tendency to form aggregates in a standard polymer-containing medium (Rampling et al. 2004). Thus, in order to distinguish between the effects of the suspending medium (e.g., polymer or protein type and concentration) and those intrinsic to red blood cells, the term *aggregability* has been coined to indicate the intrinsic tendency of RBC to form aggregates. It is important to distinguish between these two definitions: *RBC aggregation* refers to the measured rate, extent, or strength of RBC aggregation in any medium and *RBC aggregability* to the measured rate, extent, or strength of RBC aggregation in a defined medium (see also Chapter 1, Section 1.2.2). Aggregability is thus used to compare RBC populations or subpopulations in the same medium: greater aggregation for one cell population versus another indicates greater RBC aggregability. This chapter deals with the determinants of RBC aggregation, starting with macromolecules as promoters and inhibitors of RBC aggregation followed by an outline of factors and conditions that influence red cell aggregability.

2.1 FACTORS AFFECTING RED BLOOD CELL AGGREGATION

The extent of aggregation is determined by the balance between the forces promoting aggregation and those forces opposing aggregation (i.e., a force balance determines aggregation). Although disaggregating forces are more straightforward to define, there is not yet a consensus on the nature of aggregating forces (see Chapter 3).

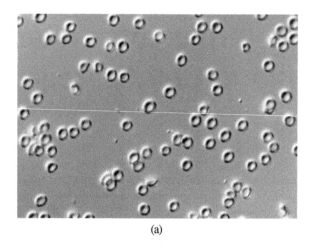

(a)

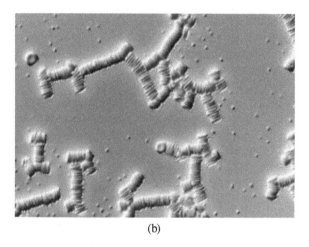

(b)

FIGURE 2.1   Red blood cells (RBC) suspended in phosphate-buffered saline (a) or autologous plasma (b) RBC were isolated from the same blood sample and suspended in either medium at low hematocrit. Note the rouleaux formation in plasma, while no RBC aggregation is observed in PBS. Small particles in both panels are blood platelets that were not entirely removed during preparation of the suspensions.

Disaggregating forces include the following:

A. *Shear forces*: These arise due to mechanical forces such as shear in a rotating viscometer or during flow through a tube. Shear forces prevent rouleaux formation or disperse existing aggregates, and thus there is an inverse relationship between the magnitude of shear forces and the size of RBC aggregates. In normal, nonpathologic blood, aggregates usually are dispersed when they are subjected to low shear rates of 20–40 s^{-1}. However, greater cell–cell attractive forces can exist in pathologic blood or for RBC suspended in media containing large polymers, and therefore higher shear forces are required for disaggregation. See Chapter 5 for a detailed discussion regarding the influence of shear forces on RBC aggregation.

B. *Surface charge density*: RBC possess a net negative surface charge, mainly due to the sialic acid residues on the membrane surface. Therefore, there is a repulsive electrostatic force pushing individual RBC apart from each other. This force clearly opposes rouleaux formation. See Section 2.3.5.1 for further discussion.

C. *Membrane strain*: RBC need to deform to allow an area of close contact between cells and hence to allow aggregation; membrane strain opposes this deformation. It is generally accepted that impaired RBC deformability leads to decreased RBC aggregation (Chien and Sung 1987; Kobuchi et al. 1988; Meiselman 1993), and experimental alterations of RBC deformability can cause changes in aggregability (see Section 2.3.5.2 of this chapter).

It is clear from the previous discussion that RBC aggregation in a given red cell suspension is modulated by both suspending phase composition and RBC properties; the influence of suspending phase properties and cellular properties are discussed in Sections 2.2 and 2.3 of this chapter.

2.1.1 BICONCAVE-DISCOID SHAPE IS A PREREQUISITE FOR AGGREGATION

The typical biconcave-discoid shape of mammalian RBC is an important feature for rouleaux formation, and any divergence from this shape leads to deviations from normal aggregation behavior. Oval RBC of camelids have almost no aggregation (Windberger and Baskurt 2007; Johnn et al. 1992), and even slight changes in shape due to osmotic pressure (i.e., increase or decrease of cell volume) or free-radical attack may significantly affect RBC aggregation (Cicha et al. 2003). Isovolumic shape changes from a biconcave-disc toward a crenated form (i.e., echinocytes) can markedly reduce or abolish aggregation (Meiselman 2009).

2.1.2 HEMATOCRIT EFFECT

The volume fraction of RBC in a suspension, termed the *hematocrit*, which can be expressed as either volume percent (%) or as a fraction (liter/liter, l/l), is an important determinant of the aggregation process (Agosti et al. 1988) inasmuch as cells must come in contact to form aggregates. Clearly, too low a hematocrit will impede cell contact and hence aggregate formation. As detailed in Chapter 3, aggregation begins when the random movement of RBC brings them into close proximity (Kim et al. 2005; Kim et al. 2007), and obviously the frequency of close contacts increases with the number of RBC per unit volume (i.e., hematocrit). The time course of RBC aggregation becomes faster as the hematocrit is increased (Baskurt et al. 1998; Hardeman et al. 2001; Shin et al. 2009); hematocrit is also an important determinant of *in vivo* aggregation (Kim et al. 2005; Kim et al. 2007).

See Chapter 4, Section 4.1.6.7 for further discussion regarding the effects of hematocrit on aggregation parameters, and Chapter 5 for the effects of hematocrit and RBC aggregation on the *in vitro* viscometry of blood and RBC suspensions.

2.2 MACROMOLECULES AS DETERMINANTS OF RED BLOOD CELL AGGREGATION

2.2.1 PLASMA PROTEINS

Fibrinogen is generally accepted to be the most potent plasma protein for inducing RBC aggregation. It is a glycoprotein with a molecular mass of 340 kDa that is synthesized by the liver and composed of three pairs of nonidentical polypeptide chains denoted Aα, Bβ, and γ; it has a hydrated diameter in the range of ~40 nm. The physiological plasma concentration is ~150–300 mg/dl but higher levels are possible; it is an *acute phase* protein in that its production and plasma levels are enhanced in several diseases such as cardiovascular disease or any form of inflammation (see Chapter 8). Fibrinogen affects essentially all aspects of RBC aggregation (e.g., aggregate size, yield stress, viscosity of red cell suspensions, and erythrocyte sedimentation rate) (Skalak et al. 1981; Maeda and Shiga 1986; Maeda et al. 1990; Lowe 1988).

In addition to fibrinogen, there are several other large plasma proteins known to induce RBC aggregation, including immunoglobulin G (IgG), which accounts for about 20% of the plasma proteins. Reports regarding the effects of IgG on RBC aggregation are not consistent, with some studies suggesting that IgG causes RBC aggregation (Maeda and Shiga 1986) and others indicating no effect (Madl et al. 1993). Another protein is albumin, a small globular and anionic plasma protein with a molecular mass of 66 kDa. Usually albumin is considered to have no impact on RBC aggregation (Rampling and Martin 1992), yet some studies have indicated that it can inhibit aggregation (Lacombe et al. 1988; Zimmermann et al. 1996). It has been reported that IgG-induced RBC aggregation can be inhibited by albumin (Maeda and Shiga 1986) whereas other studies have reported the opposite effect (Weng et al. 1998). Thus, even though the impact of some plasma proteins has been clearly identified (e.g., aggregation promotion by fibrinogen), there are several apparently contradicting reports for others such as albumin. Such contradictions suggest a complex interplay between the various plasma constituents and the RBC membrane in determining if a molecule induces or prevents aggregation or if it has no effect at all (Table 2.1).

2.2.2 DEXTRAN

The aggregation of erythrocytes is not only induced by several plasma proteins but can also be affected by various polymers and macromolecules that are not naturally found in plasma. One of these is dextran, a neutral polyglucose that, above a molecular mass of 40 kDa, induces RBC aggregation (Chien and Jan 1973). However, if the molecular mass is below this threshold, dextran can either have no effect on RBC aggregation or can reduce or inhibit rouleau formation in the presence of other aggregating agents

(Neu et al. 2001). This biphasic effect plus the availability of several molecular mass fractions has made dextran one of the most widely studied aggregants.

Numerous investigators have described the effects of polymer concentration and molecular mass on RBC aggregation (Boynard and Lelievre 1990; Brooks 1988; Chien and Sung 1987; Nash et al. 1987). The experimental data usually reflect two typical aspects of dextran-induced RBC aggregation: (1) biphasic, bell-shaped response to the polymer concentration, and (2) the extent or strength of aggregation increases with molecular mass.

Dextran-induced aggregation has been evaluated over a wide range of molecular mass, including fractions of up to 28 MDa (Neu et al. 2008). In contrast to previous reports indicating increasing aggregation with increasing molecular mass (Barshtein et al. 1998; Chien 1975; Pribush et al. 2007), this study indicated an optimal or maximum aggregation at about 500 kDa. Figure 2.2 gives a schematic summary of the dependence of RBC aggregation on dextran molecular mass and concentration. Note, however, that although dextrans with molecular masses below 2 MDa have been studied extensively, only one study to date has utilized a molecular mass of ~28 MDa (Neu et al. 2008). Thus, even though the degree of aggregation at 28 MDa is negligible (Figure 2.2), no threshold for negligible aggregation has been determined experimentally nor has the area between 2 MDa and 28 MDa been explored.

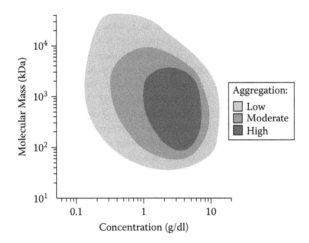

FIGURE 2.2 Estimated degree of aggregation of normal red blood cells induced by dextran as a function of molecular mass and concentration; higher density (i.e., darker) points indicate a higher degree of aggregation. Dextran with a molecular mass below 2 MDa has been studied extensively, whereas there is only one study to date involving a molecular mass of ~28 MDa (Chien, S. and K. M. Jan. 1973. "Ultrastructural Basis of the Mechanism of Rouleaux Formation." *Microvascular Research* 5:155–166, and Neu, B., R. Wenby, and H. J. Meiselman. 2008. "Effects of Dextran Molecular Weight on Red Blood Cell Aggregation." *Biophysical Journal* 95:3059–3065). Thus, even though the degree of aggregation at 28 MDa is negligible, no threshold for lack of aggregation has been determined nor has the molecular mass range between 2 MDa and 28 MDa been explored.

TABLE 2.1

Effects of Various Plasma Proteins on RBC Aggregation[a]

Plasma Protein	M_m (kDa)	Plasma concentration (mg/ml)	Effects on Aggregation
Fibrinogen	340	1.5–3.0	↑
Immunoglobulin G	150	8	↑, no effect
Immunoglobulin M	900	1.5	↑
C-reactive protein	25	<0.01	↑
Transferrin	80	2.0–3.6	no effect
Haptoglobin	38	0.5–2.5	↑, no effect
Ceruloplasmin	151	0.2–0.4	no effect
α_1-acid glycoprotein	40	0.5–1.0	↑, no effect
Albumin	66	20–50	↓,↑, no effect

[a] Data summarized from these literature reports: Madl, C., R. Koppensteiner, B. Wendelin, K. Lenz, L. Kramer, G. Grimm, A. Kranz, B. Schneeweiss, and H. Ehringer. 1993. "Effect of Immunoglobulin Administration on Blood Rheology in Patients with Septic Shock." Circulation and Shock 40:264–267; Lacombe, C., C. Bucherer, J. Ladjouzi, and J. C. Lelievre. 1988. "Competitive Role between Fibrinogen and Albumin on Thixotropy of Red Cell Suspensions." Biorheology 25:349–354; Zimmermann, J., L. Schramm, C. Wanner, E. Mulzer, H. A. Henrich, R. Langer, and E. Heidbreder. 1996. "Hemorheology, Plasma Protein Composition and von Willebrand Factor in Type I Diabetic Nephropathy." Clinical Nephrology 46:230–236; Maeda, N., and T. Shiga. 1986. "Opposite Effect of Albumin on the Erythrocyte Aggregation Induced by Immunoglobulin G and Fibrinogen." Biochimica et Biophysica Acta 855:127–135; and Weng, X. D., G. O. Roederer, R. Beaulieu, and G. Cloutier. 1998. "Contribution of Acute-Phase Proteins and Cardiovascular Risk Factors to Erythrocyte Aggregation in Normolipidemic and Hyperlipidemic Individuals." Thrombosis and Haemostasis 80:903–908. Plasma concentrations are normal, nonpathological concentrations; ↑: promotes aggregation; ↓: inhibits aggregation. Note that for some proteins different effects on aggregation have been reported.

2.2.3 POLYMER HYDRODYNAMIC RADIUS AS A DETERMINANT OF RED BLOOD CELL AGGREGATION

A wide range of macromolecules is capable of inducing RBC aggregation, including polyhemoglobin, gelatin, dextran, ficoll, starches, poly(ethylene glycol) (PEG), polyvinylpyrrolidone (PVP), and various poloxamers (Chien and Sung 1987; Armstrong et al. 1999; Brooks 1988; Neu and Meiselman 2001; Stoltz 1994). Most of these polymers have very little in common except for observations that (1) molecules that induce aggregation are those with a relatively high molecular mass and (2) low molecular mass species of the same polymer often have no effect on aggregation or even inhibit aggregation. For example, fibrinogen (360 kDa) and tetrameric (poly)albumin (260 kDa) both promote RBC aggregation, while native albumin (65 kDa) usually does not (Chien 1975; Rampling and Martin 1992; Lowe 1988; Stoltz 1994). However, this molecular mass–aggregation relation is not always valid: Both 60 kDa dextran and 18.5 kDa PEG strongly aggregate RBC, whereas 40 kDa dextran or 7.5 kDa PEG inhibit RBC aggregation when added to RBC in plasma (Armstrong et al. 1999; Chien 1975).

Armstrong et al. (2004) have demonstrated that the hydrodynamic size of polymers is a better criterion to distinguish polymers that induce aggregation from those that inhibit aggregation or have no effect. They studied the impact of various uncharged, nonionic polymers on human RBC aggregation in plasma and in isotonic polymer solutions. The hydrodynamic radii of various fractions of dextran, PVP, and PEG over a wide range of molecular mass (1.5–2,000 kDa) were calculated from their intrinsic viscosities using the Einstein relation and directly by quasi-elastic light scattering. RBC aggregation was quantified via a Myrenne aggregometer and by the ratio of low- to high-shear blood viscosity (see Chapter 4). Their results clearly demonstrated that despite the different structures for the three polymers, all fractions with a hydrodynamic radius less than 4 nm were found to inhibit aggregation while those with a hydrodynamic radius greater than 4 nm enhanced aggregation (Figure 2.3). For polymers with a radius less than 4 nm, inhibition of RBC aggregation increased with radius and was maximal at about 3 nm.

2.2.4 INHIBITION OF RED BLOOD CELL AGGREGATION BY SMALL POLYMERS

As mentioned previously, some macromolecules also prevent RBC aggregation. Studies in which two different fractions of the same polymer were tested together, or in which the effects of two different polymer species were examined, have been described: (1) employing dextran homologs, 40 kDa fractions have been shown to reduce low-shear viscosity and mean aggregate size for RBC suspended in dextran with a molecular mass of 70 kDa (Jan et al. 1982; Razavian et al. 1991); (2) dextrans with molecular masses of 10 and 20 kDa, as well as simple glucose, have similar effects in reducing the velocity of RBC aggregation for cells suspended in dextran with a molecular mass of either 70 kDa or 500 kDa (Maeda and Shiga 1985); (3) dextran 10 kDa and 40 kDa have been shown to reduce RBC aggregation for cells suspended in autologous plasma (Pribush et al. 2000; Singh and Joseph 1987); (4) PEG with a molecular mass below 18.5 kDa reduces aggregation for cells in autologous plasma or in 3 g/dl 70 kDa dextran; (5) the 8.4 kDa nonionic triblock copolymer, poloxamer 188 markedly reduces plasma-induced RBC aggregation (Toth et al. 1997) and aggregation induced by several water-soluble polymers such as dextran, polyvinylpyrrolidone, and poly-l-glutamic acid (Toth et al. 2000). In these studies, poloxamer 188 inhibited both the extent and strength of aggregation in a dose-dependent manner.

In another study, RBC were washed and resuspended in buffer containing either 73 kDa dextran or 35 kDa PEG, and then tested for aggregation in these solutions with or without various concentrations of smaller dextrans (10.5 and 18.1 kDa) or PEGs (3.35, 7.5, and 10 kDa). This study (Neu et al. 2001) indicated that heterogeneous mixtures had a greater effect on the reduction of aggregation (Figure 2.4) for both small PEGs added to the 73 kDa dextran or small dextrans added to the 35 kDa PEG. For cells in the dextran, aggregation decreased with increasing molecular

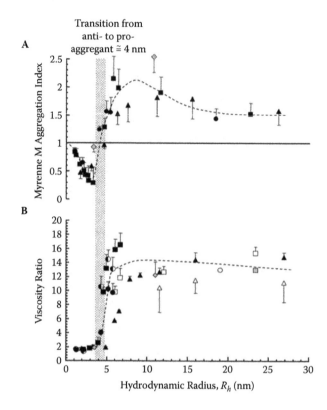

FIGURE 2.3 Effect of polymer hydrodynamic radius (R_h) on RBC aggregation in plasma (A) and blood viscosity (B). Cells were suspended in dextran (triangles), PEG (boxes), and PVP (circles) at polymer concentrations of 3% w/v (half-solid symbols), 1.5% w/v (solid symbols), 0.5% w/v (open symbols), and 0.25% w/v (shaded symbols). The effects of albumin and fibrinogen (shaded diamond) are shown for comparison. The shaded bar shows the transition from an anti- to a pro-aggregating system occurring at a hydrodynamic radius of around 4 nm for all macromolecules studied. Copied with permission from Armstrong, J. K., R. B. Wenby, H. J. Meiselman, and T. C. Fisher. 2004. "The Hydrodynamic Radii of Macromolecules and Their Effect on Red Blood Cell Aggregation." *Biophysical Journal* 87:4259–4270.

mass and concentration of the small dextrans or PEGs, and for cells in the 35 kDa PEG solution small dextrans decreased aggregation with increasing molecular mass and concentration; small PEGs had minimal effects or even further promoted RBC aggregation for cells in the larger PEG (Figure 2.4). Thus, in general, heterogeneous binary polymer systems (e.g., dextran + PEG) yield greater inhibition of aggregation than homogenous ones (e.g., dextran + dextran or PEG + PEG).

2.3 CELLULAR FACTORS DETERMINING THE EXTENT OF RED BLOOD CELL AGGREGATION

Most studies related to RBC aggregation have primarily focused on the effects of protein or polymer type and concentration, with less attention directed toward the

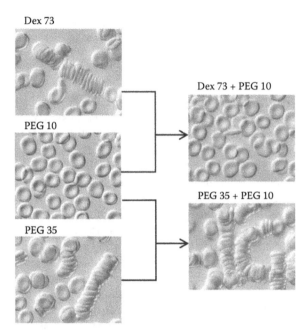

FIGURE 2.4 Photomicrographs illustrating the effect of homogenous and heterogeneous polymer mixtures on RBC aggregation. RBC were suspended in isotonic solutions of dextran 73 kDa (2 g/dl), PEG 10 kDa (0.6 g/dl), and PEG 35 kDa (0.35 g/dl) and in mixtures of these solutions. Left column: both dextran 73 kDa and PEG 35 kDa clearly induce aggregation, whereas PEG 10 kDa has no effect; right column: adding PEG 10 to dextran 73 completely abolishes RBC aggregation whereas mixing the two PEGs enhances aggregation.

effects of cellular factors. One of the first reports in this area appears to be the observation that density-separated (i.e., age-separated) RBC exhibit different degrees of aggregation when suspended in autologous plasma (Nordt 1983): older, denser cells exhibited greater aggregation than younger, less-dense RBC. Subsequent reports have observed large variations in the aggregating tendency of cells from different normal, healthy subjects (Nash et al. 1987; Sowemimo-Coker et al. 1989). These results clearly indicated that there are intrinsic cell-specific factors that control the aggregability of red cells and that these factors differ greatly among individuals. Quite interestingly, the observations regarding intrinsic cell-specific factors were initially rather surprising since it was assumed that all normal adult red cells had the same aggregating tendency. Any differences of RBC aggregation between different subjects for cells in plasma were assumed to be due only to differences in plasma concentrations of various aggregating agents. Further, differences in RBC aggregation between different donors for cells in the same polymer solution were, if considered at all, ascribed to technical problems (e.g., measurement errors). Observations that cell-specific factors are also a determinant of RBC aggregation is an important development in that it provides a better understanding of the physicochemical aspects of RBC aggregation as well as its significance in health and disease. In addition, identifying cellular factors that determine RBC aggregability may possibly aid in the development of new diagnostic or therapeutic tools.

2.3.1 Donor-Specific Effects

Differences in RBC aggregability can be found between various healthy donors, with Sowemimo-Coker and coworkers (1989) the first to systematically study this effect. They compared the aggregation of RBC from different adult subjects when suspended in either autologous plasma or in a phosphate-buffered saline solution containing dextran 70 kDa at a concentration of 3 g/dl. All suspensions were made to the same hematocrit and the degree of aggregation was measured using a Myrenne erythrocyte aggregometer (see Chapter 4, Section 4.1.6.9.1). The extent of the aggregation in plasma versus the extent in dextran for each subject is shown in Figure 2.5.

As is evident from Figure 2.5, differences in aggregation between subjects were large, with a twofold range in the dextran solution and a fivefold range in autologous plasma. Based upon the dextran data it is obvious that these ranges must be due to subject-to-subject variations of cell-specific factors that affect RBC aggregation (i.e., aggregability) since the properties of the suspending medium were constant. On the other hand, the plasma data do not allow distinguishing between the role of aggregability and changes of the suspending media properties since the level of pro-aggregant plasma proteins (e.g., fibrinogen) can vary between donors. Plasma protein levels were not determined in this study and therefore it was concluded that both cell and medium properties contribute to the large range observed for RBC aggregation in autologous plasma (Meiselman 1993).

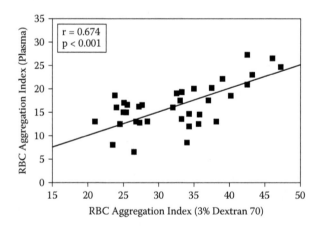

FIGURE 2.5   RBC aggregation in autologous plasma versus aggregation for the same cells in a polymer solution containing 3 g/dl 70 kDa dextran. Note the wide range of aggregation in dextran and the linear correlation between aggregation in plasma and dextran; the correlation and the wide range of aggregation in dextran both suggest a wide range of RBC aggregability. (Meiselman, H. J. 1993. "Red-Blood-Cell Role in RBC Aggregation." *Clinical Hemorheology Microcirculation* 13:575–592. Meiselman, H.J. 2009. "Red Blood Cell Aggregation." *Biorheology* 46: 1-19.)

2.3.2 Effects of *in Vivo* Cell Age

As discussed in greater detail in Chapter 8, Section 8.1.6, human RBC have a life span of about 120 days following which they are removed from the circulation (Berlin 1964). During this aging process, RBC undergo several physicochemical changes, including an increase of cytoplasmic hemoglobin concentration and cell density (Linderkamp et al. 1983; Muller et al. 1992). This age-related increase in density allows RBC to be age-separated via high-speed centrifugation: in normal individuals, denser cells are older whereas less-dense cells are younger (Murphy 1973). In one of the first reports related to RBC aggregability, Nordt (1983) tested age-separated cells in autologous plasma and observed that the 10% densest cells had an aggregation index twice as large as the 10% least-dense cells, and that cells with an intermediate density exhibited aggregation between these two limits.

2.3.3 Clinical Conditions

Literature reports dealing with red blood cell aggregation in clinical states usually employ either whole blood or RBC-plasma suspensions, and most indicate enhanced aggregation for RBC in autologous plasma (Lowe 1988; Stoltz et al. 1999; Baskurt et al. 2007). See Chapter 8 for a discussion of clinical conditions associated with altered RBC aggregation. Studies designed to evaluate altered RBC aggregability in clinical states seem to be less common (Rampling et al. 2004), although there are a few reports indicating correlations between RBC aggregation in plasma and in polymer solutions; cell-specific changes induced by clinical conditions have not yet been fully explored. Table 2.2 summarizes several literature reports specifically dealing with changes of RBC aggregation in polymer solutions and hence with RBC aggregability. Note that alterations of RBC aggregability vary widely from condition to condition, and are subject to variations for the same disease. For example, studies in subjects with type 2 diabetes indicate either no alteration of aggregability (Bauersachs et al. 1989) or significantly greater aggregation in dextran solutions (Chong-Martinez et al. 2003); the latter report showed decreased aggregability with improved glycemic control.

2.3.4 Impact of the Aggregant on Red Blood Cell Aggregability

Another interesting aspect of RBC aggregability is shown in Figure 2.5 where the aggregation in 70 kDa dextran is linearly correlated with the aggregation of cells from the same donor suspended in autologous plasma. This finding suggests that cell-specific factors that result in changes of aggregability in one medium may be of similar importance for cells in other aggregating media. Experimental evidence supporting this suggestion has been presented by Whittingstall and coworkers (Whittingstall et al. 1994). They investigated RBC aggregation induced by isotonic solutions of poly-vinylpyrrolidone (PVP), poly-l-glutamic acid (P-L-Glu), and sodium heparin as well as dextran and autologous plasma. For each of the four polymers, the extent of aggre-gation was correlated with aggregation in plasma (Figure 2.6). These results clearly indicate that regardless of differences in polymer size and charges (i.e., dextran and

TABLE 2.2
Literature Reports of RBC Aggregability in Various Clinical States

Condition	Aggregability[a]	References
Angina	↑	(Ben Ami et al. 2001)
Myocardial Infarction	No change	(Ben Ami et al. 2001)
Bacterial Infection	↑	(Ben Ami et al. 2001)
Beta-Thalassemia	↑	(Chen et al. 1996)
Cardiac Syndrome X	↑	(Lee et al. 2008)
Diabetes: Type 2	No change	(Bauersachs et al. 1989)
Diabetes: Type 2	↑	(Chong-Martinez et al. 2003)
Gaucher's Disease	↑	(Adar et al. 2006)
Pregnancy-Induced Hypertension	↑	(Gamzu et al. 2002)
Leprosy	↑	(Meiselman 1993)
Morbid Obesity	↑	(Samocha-Bonet et al. 2004)
Nephrotic Syndrome	No change	(Ozanne et al. 1983)
Normal Tension Glaucoma	↑	(Vetrugno et al. 2004)
Sickle cell Disease	↑	(Obiefuna and Photiades 1990)

[a] ↑ indicates an increase of RBC aggregability.

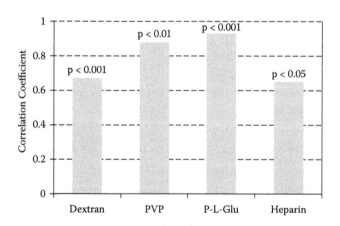

FIGURE 2.6 Correlations between plasma- and polymer-induced RBC aggregation. Dextran = 70 kDa, 3% solution; PVP = polyvinylpyrrolidone 360 kDa, 0.5% solution; P-L-Glu = poly-l-glutamic acid 61.2 kDa, 0.6% solution; Heparin = 17 kDa, 6% solution. Significance values are for linear regressions tested versus a slope of zero. (Whittingstall, P., K. Toth, R. B. Wenby, and H. J. Meiselman. 1994. "Cellular Factors in RBC Aggregation: Effects of Autologous Plasma and Various Polymers," in *Hemorheologie et Aggregation Erythrocytaire*, ed. J. F. Stoltz. Paris: Editions Medicales Internationales.)

PVP are neutral, P-L-Glu and heparin are negative), aggregation induced by each aggregant exhibits a significant correlation with that induced by plasma. In addition, significant correlations were found between aggregation induced by any two polymers (Whittingstall et al. 1994). This latter finding indicates that aggregability, as tested by these four polymers, shows minimal dependence on the nature of the aggregant. However, this independence may not be observed when the aggregant produces either extremely weak or extremely intense aggregation since sensitivity to subtle cellular changes is reduced.

2.3.5   MODIFYING AGGREGABILITY

While many alterations of RBC aggregability follow from current knowledge about RBC aggregation mechanisms, some remain unexpected and/or unexplained (Neu et al. 2003). During the past few decades, substantial effort has been directed toward identifying cellular factors responsible for altered RBC aggregability. One reason for this effort relates to *in vivo* studies of abnormal or altered aggregation in which polymers that modify aggregation, or RBC suspended in a polymer solution, are infused (see Chapter 7, Section 7.3.2). A problem associated with such infusions is that they also involve concomitant alterations of plasma viscosity plus possible adverse effects of the dissolved polymers. Consequently, an alternative approach is to alter RBC aggregability via changing RBC physicochemical properties. In the material below, several strategies aimed at modifying RBC aggregability are summarized.

2.3.5.1   Enzyme Treatment of Red Blood Cells

One approach to alter cellular factors that affect RBC aggregability is to modify the cell surface via specific enzymes such as neuraminidase. Jan and Chien (1973a) reported that treating RBC with neuraminidase to remove membrane-bound sialic acid, and thus to reduce membrane charge density, greatly increased the aggregability of human erythrocytes with similar results being reported by Maeda and coworkers (1984). Nash et al. (1987) extended this work by using neuraminidase treatment on age-separated cells. Their results clearly demonstrated that neuraminidase treatment slightly reduced aggregation in plasma but markedly increased aggregation in dextran 70 kDa at 3.0 g/dl. Further, they found that neuraminidase treatment did not significantly alter the aggregation ratio for young and old cells (i.e., old>>young) in either plasma or dextran solutions. The enhanced aggregation of neuraminidase-treated RBC in dextran was consistent with earlier reports (Jan and Chien 1973b), whereas the reasons for the lack of neuraminidase treatment to affect RBC aggregation in autologous plasma as reported by Nash et al. (1987) remains unclear.

Other proteolytic enzymes can also have a marked impact on RBC aggregability as demonstrated by Pearson (1996), who compared the effects of digesting the RBC glycocalyx with a variety of proteolytic enzymes including chymotrypsin, trypsin, and bromelain and with neuraminidase that removes sialic acid residues. These results (Figure 2.7) indicated an increased aggregation for cells in dextran with a molecular mass of 40 and 70 kDa. Bromelain was clearly the most effective proteolytic enzyme; neuraminidase markedly increased aggregation for cells in both 40 and 70 kDa dextran solutions. Conversely, chymotrypsin and trypsin had minimal

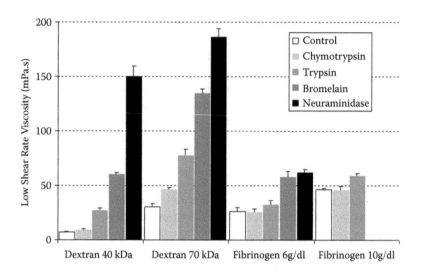

FIGURE 2.7 Low-shear-rate viscosity of enzyme-treated RBC suspended in isotonic solutions of 40 or 70 kDa dextran or fibrinogen. Data are mean ± SEM viscosity values for 0.45 l/l hematocrit suspensions; viscosity values obtained at 0.28 s^{-1} and have units of mPa.s. Proteolytic enzymes were used at 5 mg per ml cells and neuraminidase at 33 U/ml. Data from Pearson, M. J. 1996. "An Investigation into the Mechanism of Rouleaux Formation and the Development of Improved Methods for Its Quantitation." PhD diss., London University.

effects for cells in fibrinogen solutions, and neuraminidase only increased aggregation modestly. These results thus indicate that the effects of cellular factors on RBC aggregation may, under some circumstances, be specific to the aggregant.

2.3.5.2   Heat Treatment and Aldehyde Fixation of Red Blood Cells

Heating RBC suspensions at 48°C for 8–9 minutes results in an increase of the membrane elastic shear modulus and membrane viscosity and hence reduced cellular deformability, but does not affect RBC morphology (Nash and Meiselman 1985). For short time periods, the effects of heating are minimal or nonexistent but can lead to significant changes if the cells are treated for longer periods. Therefore, heat-treated RBC have been used in several studies to explore the dependence of aggregability on RBC mechanical properties. Rakow et al. (1981) investigated the changes of aggregation in dextran and plasma. For RBC that had been heat treated for 2 minutes, there were no significant changes of aggregability in dextran, whereas prolonged treatment for 9 minutes led to a decrease of the extent and rate of aggregation in plasma and an increase of the extent and rate in dextran 70 kDa. In another study, Reinhart and Singh (1990) reported significant time effects for cells in plasma and dextran 70 kDa: heat treatment for 2 and 5 minutes resulted in a slight increase of aggregation in both media followed by progressive decreases if the cells were treated over the range of 10 to 80 minutes.

    It is also possible to alter RBC aggregability by changing the mechanical properties of RBC with aldehydes such as formaldehyde or glutaraldehyde. This treatment also results in RBC with normal morphology but with decreased cellular deformability (Corry and Meiselman 1978). Overnight fixation with 2% formaldehyde decreased aggregation in dextran with a molecular mass of 70 kDa (Knox et al. 1977). Using the bifunctional fixative glutaraldehyde (GA) at and above 0.05%, Reinhart and Singh indicate markedly reduced aggregation in 70 kDa dextran as judged by two-hour Erythrocyte Sedimentation Rate (ESR) values (Reinhart and Singh 1990). Decreased aggregation (i.e., lower M and M1 Myrenne indices) in autologous plasma has also been reported at and above 0.01% GA (Baskurt and Meiselman 1997). Interestingly, results obtained at low GA levels are somewhat paradoxical: (1) Reinhart and Singh (1990) show increased aggregation (i.e., greater ESR) at 0.01 and 0.02% GA for cells in 70 kDa dextran, but (2) GA at 0.005% reduces RBC deformability yet does not alter aggregation (i.e., M and M1 indices) in autologous plasma (Baskurt and Meiselman 1997). The reason for decreased aggregation at high GA concentrations is clear: reduced RBC deformability inhibits the formation of close contacts between adjacent cells. It is tempting to speculate that the results at low GA levels seem to reflect a "balance" between altered glycocalyx properties and cell rigidity. That is, if polymer or protein penetration of the glycocalyx is reduced, then greater depletion occurs and hence aggregation is favored, whereas any reduction of RBC deformability would tend to oppose cell–cell contact (see Chapter 3). Note that the effects of GA treatment at very low concentrations may depend on specific experimental conditions, such as the ratio of RBC number to volume of GA solution, the time and temperature during treatment, and the manner in which cells are stirred while being exposed to GA.

2.3.5.3 Macromolecular Binding to Reduce Aggregation

Another possibility to change the aggregability of RBC is to covalently attach polymers to the RBC membrane. Stabilizing RBC suspensions via a reduction of aggregation was first reported by Armstrong and coworkers (Armstrong et al. 1997). A simple method to covalently bind PEG to the RBC was developed: monomethoxy-PEG of 5 kDa molecular mass activated with cyanuric chloride was added to washed RBC in buffer and the suspension gently mixed at room temperature for one hour, following which the RBC were washed and resuspended in autologous plasma. This process yielded RBC with normal morphology and deformability, and with PEG covalently attached to the cell membrane. Rheological testing of such PEG-coated red blood cells revealed a 93% reduction of the aggregation index at stasis in plasma, a 75% reduction of low-shear blood viscosity, and nearly Newtonian flow behavior (Armstrong et al. 1997).

2.3.5.4 Macromolecular Binding to Enhance Aggregation

The previous studies describing PEG-coated RBC clearly demonstrated that RBC aggregation could be decreased by various degrees subsequent to the attachment of PEG to the cell surface (Armstrong et al. 1999). However, in some applications, it is also of interest to increase RBC aggregation or to destabilize blood flow without the addition of polymer to the suspending medium; this approach prevents altering the physicochemical properties of the medium (e.g., plasma).

One possibility to enhance RBC aggregation withouth adding polymers to a suspension or infusing polymers into an *in vivo* system is to covalently attach nonionic poly(ethylene glycol)-containing block copolymers, often termed poloxamers, to the RBC surface (Armstrong et al. 2001). One characteristic of these copolymers is that they exhibit a critical micellization temperature (CMT) above which a phase transition occurs from predominately single, fully hydrated copolymer chains to micelle-like structures. The CMT is a function of the polymer concentration and its molecular mass (Alexandridis and Hatton 2006). Thus, if cells are coated with an appropriate poloxamer, these cells will form aggregates above the CMT since polymers from different cells would start to form micelles, thus causing cell aggregation.

In their studies, Armstrong et al. (2001) employed reactive derivatives of poloxamers containing 80% PEG and molecular masses of 8.4, 11.4, 13.0, and 14.6 kDa covalently attached to the cell surface. In order to avoid the formation of permanently covalently cross-linked aggregates, the attachment was done at temperatures below the CMT. RBC coated with poloxamer were then resuspended in native, unaltered autologous plasma and tested for rheological properties via viscometry (Figure 2.8). Both poloxamer molecular mass and temperature affected rheological behavior; (1) At both 25 and 37°C, cells coated with the 8.4-kDa poloxamer, which has a CMT of 48°C, showed markedly reduced low-shear viscosity and nearly Newtonian behavior. (2) At 25°C, cells coated with the 11.4-kDa poloxamer, which has a CMT of 37°C, demonstrated a flow behavior similar to that of the control cells, thereby indicating a degree of RBC aggregation similar to the untreated control sample. However, at 37°C, the low-shear viscosity was significantly above that of the control cells, thus

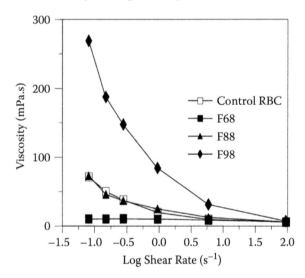

FIGURE 2.8 Viscosity–shear rate data for poloxamer-coated RBC in buffer at 25°C. F68 is an 8.4 kDa poloxamer, F88 is an 11.4 kDa poloxamer, and F98 is a 13.0 kDa poloxamer. See text for details. Redrawn with permission from Armstrong, J. K., H. J. Meiselman, R. B. Wenby, and T. C. Fisher. 2001. "Modulation of Red Blood Cell Aggregation and Blood Viscosity by the Covalent Attachment of Pluronic Copolymers." *Biorheology* 38:239–247.

indicating a significant enhancement of aggregation; (3) Cells coated with either 13.0 kDa (CMT = 30°C) or 14.6 kDa (CMT = 23°C) showed a greatly increased low-shear viscosity at 37°C.

2.3.6 MISCELLANEOUS

The composition of the phospholipid bilayer of RBC varies between the inner and outer leaflet and the specific levels of the various components can be changed via several approaches. It has been demonstrated that, for example, the ratio of phosphatidylcholine (PC) to phosphatidylserine (PS) in the outer leaflet can affect aggregation in solutions containing dextran: the incorporation of additional PC and hence an increasing PC/PS ratio promotes aggregation whereas the incorporation of PS has no effect on aggregability (Evans and Kukan 1983; Othmane et al. 1990). 4,4'-diisothiocyano-2,2'-stillbenedisulfonic acid, DIDS, which is an inhibitor of anion transport across the RBC membrane, inhibits aggregation at 100–200 µM levels (Fabry 1987; Norris et al. 1996) but the underlying mechanism of this reduction has not yet been resolved. However, it has been suggested that a change of RBC morphology from a biconcave disc to a crenated sphere may be involved. Procaine hydrochloride, a cationic local anesthetic, is also known to affect aggregation in dextran. It has been demonstrated that procaine hydrochloride inhibits aggregation at low concentrations (80 to 800 µM) and that it leads to a markedly increased aggregation at higher levels (Sowemimo-Coker and Meiselman 1989), or to always inhibit aggregation at levels of 4 µM and greater (Ramakrishnan et al. 1999). Blood storage also affects aggregability in 0.5% dextran 500 kDa, with an eightfold increase of the aggregation over a 42-day period (Hovav et al. 1999).

LITERATURE CITED

Adar, T., R. Ben-Ami, D. Elstein, A. Zimran, S. Berliner, S. Yedgar, and G. Barshtein. 2006. "Aggregation of Red Blood Cells in Patients with Gaucher Disease." *British Journal of Haematology* 134:432–437.

Agosti, R., A. Clivati, M. Dettorre, F. Ferrarini, R. Somazzi, and E. Longhini. 1988. "Hematocrit Dependence of Erythrocyte Aggregation." *Clinical Hemorheology* 8:913–924.

Alexandridis, P., and T. A. Hatton. 2006. "PEG-PPO-PEG Block Copolymer Surfactants in Aqueous Solutions and at Interfaces—Thermodynamics, Structures, Dynamics, and Modeling." *Colloids and Surfaces* 96:1–46.

Armstrong, J. K., H. J. Meiselman, and T. C. Fisher. 1997. "Covalent Binding of Poly(Ethylene Glycol) (PEG) to the Surface of Red Blood Cells Inhibits Aggregation and Reduces Low Shear Blood Viscosity." *American Journal of Hematology* 56:26–28.

Armstrong, J. K., H. J. Meiselman, and T. C. Fisher. 1999. "Evidence against Macromolecular Bridging as the Mechanism of Red Blood Cell Aggregation Induced by Nonionic Polymers." *Biorheology* 36:433–437.

Armstrong, J. K., H. J. Meiselman, R. B. Wenby, and T. C. Fisher. 2001. "Modulation of Red Blood Cell Aggregation and Blood Viscosity by the Covalent Attachment of Pluronic Copolymers." *Biorheology* 38:239–247.

Armstrong, J. K., R. B. Wenby, H. J. Meiselman, and T. C. Fisher. 2004. "The Hydrodynamic Radii of Macromolecules and Their Effect on Red Blood Cell Aggregation." *Biophysical Journal* 87:4259–4270.

Barshtein, G., I. Tamir, and S. Yedgar. 1998. "Red Blood Cell Rouleaux Formation in Dextran Solution: Dependence on Polymer Conformation." *European Biophysics Journal*, 27:177–181.

Baskurt, O.K. and H.J. Meiselman. 1997. "Cellular Determinants of Low-Shear Blood Viscosity" *Biorheology*. 34: 235:247.

Baskurt, O. K., H. J. Meiselman, and E. Kayar. 1998. "Measurement of Red Blood Cell Aggregation in a "Plate-Plate" Shearing System by Analysis of Light Transmission." *Clinical Hemorheology Microcirculation* 19:307–314.

Baskurt, O. K., M. R. Hardeman, H. J. Meiselman, and M. W. Rampling, eds. 2007. *Handbook of Hemorheology and Hemodynamics*. Vol. 69, *Biomedical and Health Research*. Amsterdam: IOS Press.

Bauersachs, R. M., S. J. Shaw, A. Zeidler, and H. J. Meiselman. 1989. "Red Blood Cell Aggregation and Blood Viscoelasticity in Poorly Controlled Type 2 Diabetes Mellitus." *Clinical Hemorheology* 9:935–952.

Ben Ami, R., G. Barshtein, D. Zeltser, Y. Goldberg, I. Shapira, A. Roth, G. Keren, H. Meller, V. Prochorov, A. Eldor, S. Berliner, and S. Yedgar. 2001. "Parameters of Red Blood Cell Aggregation as Correlates of the Inflammatory State." *American Journal Physiology Heart and Circulatory Physiology* 280:H1982–H1988.

Berlin, N. I. 1964. "Life Span of the Red Cell" in *The Red Blood Cell: A Comprehensive Treatise*, ed. C. Bishop and D. M. Surgenor. New York: Academic Press; 423–450.

Boynard, M. And J.C. Lelievre. 1990. "Size determination of red blood cell aggregates induced by dextron using ultrasound backscattering phenomenon" Biorheology. 27: 39-46.

Brooks, D. E. 1988. "Mechanism of Red Cell Aggregation" in *Blood Cells, Rheology, and Aging*, ed. D. Platt. Berlin: Springer Verlag; 158–162.

Chen, S., A. Eldor, G. Barshtein, S. Zhang, A. Goldfarb, E. Rachmilewitz, and S. Yedgar. 1996. "Enhanced Aggregability of Red Blood Cells of B-Thalassemia Major Patient." *American Journal of Physiology* 270:H1951–H1956.

Chien, S. 1975. "Biophysical Behavior of Red Cells in Suspensions" in *The Red Blood Cell*, ed. D. M. Surgenor. New York: Academic Press; 1031–1033.

Chien, S, and K. M. Jan. 1973. "Ultrastructural Basis of the Mechanism of Rouleaux Formation." *Microvascular Research* 5:155–166.

Chien, S., and L. A. Sung. 1987. "Physicochemical Basis and Clinical Implications of Red Cell Aggregation." *Clinical Hemorheology* 7:71–91.

Chong-Martinez, B., T. A. Buchanan, R. Wenby, and H. J. Meiselman. 2003. "Decreased Red Blood Cell Aggregation Subsequent to Improved Glycemic Control in Type 2 Diabetes Mellitus." *Diabetic Medicine* 20:301–306.

Cicha, I., Y. Suzuki, N. Tateishi, and N. Maeda. 2003. "Changes of RBC Aggregation in Oxygenation-Deoxygenation: pH Dependency and Cell Morphology." *American Journal of Physiology—Heart and Circulatory Physiology* 284:H2335–H2342.

Corry, W. D., and H. J. Meiselman. 1978. "Modification of Erythrocyte Physicochemical Properties by Millimolar Concentrations of Glutaraldehyde." *Blood Cells* 4:465–480.

Evans, E., and B. Kukan. 1983. "Free Energy Potential for Aggregation of Erythrocytes." *Biophysical Journal* 44:255–260.

Fabry, T. L. 1987. "Mechanism of Erythrocyte Aggregation and Sedimentation." *Blood* 70:1572–1576.

Gamzu, R., G. Barshtein, F. Tsipis, J. B. Lessing, A. S. Berliner, M. J. Kupferminc, A. Eldor, and S. Yedgar. 2002. "Pregnancy-Induced Hypertension Is Associated with Elevation of Aggregability of Red Blood Cells." *Clinical Hemorheology and Microcirculation* 27:163–169.

Hardeman, M. R., J. G. G. Dobbe, and C. Ince. 2001. "The Laser-Assisted Optical Rotational Cell Analyzer (LORCA) as Red Blood Cell Aggregometer." *Clinical Hemorheology and Microcirculation* 25:1–11.

Hovav, T., S. Yedgar, N. Manny, and G. Barshtein. 1999. "Alteration of Red Cell Aggregability and Shape during Blood Storage." *Transfusion* 39:277–281.

Jan, K. M., and S. Chien. 1973a. "Role of Surface Electric Charge in Red Blood Cell Interactions." *Journal of General Physiology* 61:638–54.

Jan, K. M., and S. Chien. 1973b. "Role of the Electrostatic Repulsive Force in Red Cell Interactions." *Biblioteca Anatomica* 11:281–8.

Jan, K. M., S. Usami, and S. Chien. 1982. "The Disaggregation Effect of Dextran 40 on Red Cell Aggregation in Macromolecular Suspensions." *Biorheology* 19:543–554.

Johnn, H., C. Phipps, S. C. Gascoyne, C. Hawkey, and M. W. Rampling. 1992. "A Comparison of the Viscometric Properties of the Blood from a Wide Range of Mammals." *Clinical Hemorheology* 12:639–647.

Kim, S., A. S. Popel, M. Intaglietta, and P. C. Johnson. 2005. "Aggregate Formation of Erythrocytes in Postcapillary Venules." *American Journal of Physiology—Heart and Circulatory Physiology* 288:H584–H590.

Kim, S., J. Zhen, A. S. Popel, M. Intaglietta, and P. C. Johnson. 2007. "Contributions of Collision Rate and Collision Efficiency to Erythrocyte Aggregation in Postcapillary Venules at Low Flow Rates." *American Journal of Physiology—Heart and Circulatory Physiology* 293:H1947–H1954.

Knox, R. J., F. J. Nordt, G. V. F. Seaman, and D. E. Brooks. 1977. "Rheology of Erythrocytes Suspensions: Dextran-Mediated Aggregation of Deformable and Non-Deformable Cells." *Biorheology* 14:75–84.

Kobuchi, Y., I. Tadanao, and A. Ogiwara. 1988. "A Model for Rouleaux Pattern Formation of Red Blood Cells." *Journal of Theoretical Biology* 130:129–145.

Lacombe, C., C. Bucherer, J. Ladjouzi, and J. C. Lelievre. 1988. "Competitive Role between Fibrinogen and Albumin on Thixotropy of Red Cell Suspensions." *Biorheology* 25:349–354.

Lee, B. K., A. Durairaj, A. Mehra, R. B. Wenby, H. J. Meiselman, and T. Alexy. 2008. "Microcirculatory Dysfunction in Cardiac Syndrome X: Role of Abnormal Blood Rheology." *Microcirculation* 15:451–459.

Linderkamp, O., P. Y. Wu, and H. J. Meiselman. 1983. "Geometry of Neonatal and Adult Red Blood Cells." *Pediatric Research* 17:250–253.

Lowe, G. D. O. 1988. *Clinical Blood Rheology*. Boca Raton, FL: CRC Press.

Madl, C., R. Koppensteiner, B. Wendelin, K. Lenz, L. Kramer, G. Grimm, A. Kranz, B. Schneeweiss, and H. Ehringer. 1993. "Effect of Immunoglobulin Administration on Blood Rheology in Patients with Septic Shock." *Circulation and Shock* 40:264–267.

Maeda, N., K. Imaizumi, M. Sekiya, and T. Shiga. 1984. "Rheological Characteristics of Desialyated Erythrocytes in Relation to Fibrinogen-Induced Aggregation." *Biochimica et Biophysica Acta* 776:151–158.

Maeda, N., M. Seike, T. Nakajima, Y. Izumida, M. Sekiya, and T. Shiga. 1990. "Contribution of Glycoproteins to Fibrinogen-Induced Aggregation of Erythrocytes." *Biochimica et Biophysica Acta* 1022:72–8.

Maeda, N, and T Shiga. 1985. "Inhibition and Acceleration of Erythrocyte Aggregation Induced by Small Macromolecules." *Biochimica et Biophysica Acta* 843:128–136.

Maeda, N., and T. Shiga. 1986. "Opposite Effect of Albumin on the Erythrocyte Aggregation Induced by Immunoglobulin G and Fibrinogen." *Biochimica et Biophysica Acta* 855:127–135.

Meiselman, H. J. 1993. "Red-Blood-Cell Role in RBC Aggregation." *Clinical Hemorheology Microcirculation* 13:575–592.

Meiselman, H. J. 2009. "Red Blood Cell Aggregation." *Biorheology* 46:1–19.

Muller, G. H., H. Schmid-Schönbein, and H. J. Meiselman. 1992. "Development of Viscoelasticity in Heated Hemoglobin Solutions." *Biorheology.* 29:203–216.

Murphy, J. R. 1973. "Influence of Temperature and Method of Centrifugation on the Separation of Erythrocytes." *Journal of Laboratory and Clinical Medicine* 82:334–341.

Nash, G. B., and H. J. Meiselman. 1985. "Alteration of Red Cell Membrane Viscoelasticity by Heat Treatment: Effect on Cell Deformability and Suspension Viscosity." *Biorheology* 22:73–84.

Nash, G. B., R. B. Wenby, S. O. Sowemimo-Coker, and H. J. Meiselman. 1987. "Influence of Cellular Properties on Red Cell Aggregation." *Clinical Hemorheolology and Microcirculation* 7:93–108.

Neu, B., J. K. Armstrong, T. C. Fisher, and H. J. Meiselman. 2001. "Aggregation of Human RBC in Binary Dextran-PEG Polymer Mixtures." *Biorheology* 38:53–68.

Neu, B., and H. J. Meiselman. 2001. "Sedimentation and Electrophoretic Mobility Behavior of Human Red Blood Cells in Various Dextran Solutions." *Langmuir* 17:7973–7975.

Neu, B., S. O. Sowemimo-Coker, and H. J. Meiselman. 2003. "Cell-Cell Affinity of Senescent Human Erythrocytes." *Biophysical Journal* 85:75–84.

Neu, B., R. Wenby, and H. J. Meiselman. 2008. "Effects of Dextran Molecular Weight on Red Blood Cell Aggregation." *Biophysical Journal* 95:3059–3065.

Nordt, F. J. 1983. "Hemorheology in Cerebrovascular Diseases: Approaches to Drug Development." *Annals of the New York Academy of Science* 416:651–661.

Norris, S. S., D. D. Allen, T. P. Neff, and S. L. Wilkinson. 1996. "Evaluation of DIDS in the Inhibition of Rouleaux Formation." *Transfusion* 36:109–112.

Obiefuna, P. C. M., and D. P. Photiades. 1990. "Sickle Discocytes Form More Rouleaux *In Vitro* Than Normal Erythrocytes." *Journal of Tropical Medicine and Hygiene* 93:210–214.

Othmane, A., M. Bitbol, P. Snabre, and P. Mills. 1990. "Influence of Altered Phospholipid Composition of the Membrane Outer Layer on Red Blood Cell Aggregation." *European Biophysics Journal* 18:93–99.

Ozanne, P., R. B. Francis, and H. J. Meiselman. 1983. "Red Blood Cell Aggregation in Nephrotic Syndrome." *Kidney International.* 23:519–525.

Pearson, M. J. 1996. "An Investigation into the Mechanism of Rouleaux Formation and the Development of Improved Methods for its Quantitation," Ph.D. diss., London University.

Pribush, A., D. Mankuta, H. J. Meiselman, D. Meyerstein, T. Silberstein, M. Katz, and N. Meyerstein. 2000. "The Effect of Low-Molecular Weight Dextran on Erythrocyte Aggregation in Normal and Preeclamptic Pregnancy. *Clinical Hemorheology and Microcirculation* 22:143–152.

Pribush, A., D. Zilberman-Kravits, and N. Meyerstein. 2007. "The Mechanism of the Dextran-Induced Red Blood Cell Aggregation." *European Biophysics Journal with Biophysics Letters* 36:85–94.

Rakow, A., S. Simchon, L. A. Sung, and S. Chien. 1981. "Aggregation of Red Cells with Membrane Altered by Heat Treatment." *Biorheology* 18:3–8.

Ramakrishnan, S., R. Grebe, M. Singh, and H. Schmid-Schönbein. 1999. "Influence of Local Anesthetics on Aggregation and Deformability of Erythrocytes." *Clinical Hemorheology and Microcirculation* 20:21–26.

Rampling, M. W., and G. Martin. 1992. "Albumin and Rouleaux Formation." *Clinical Hemorheology* 12 (5):761–765.

Rampling, M. W., H. J. Meiselman, B. Neu, and O. K. Baskurt. 2004. "Influence of Cell-Specific Factors on Red Blood Cell Aggregation." *Biorheology* 41:91–112.

Razavian, S. M., M. T. Guillemin, R. Guillet, Y. Beuzard, and M. Boynard. 1991. "Assessment of Red-Blood-Cell Aggregation with Dextran by Ultrasonic Interferometry." *Biorheology* 28:89–97.

Reinhart, W. H., and A. Singh. 1990. "Erythrocyte Aggregation: The Roles of Cell Deformability and Geometry." *European Journal of Clinical Investigation.* 20:458–62.

Samocha-Bonet, D., R. Ben-Ami, I. Shapira, G. Shenkerman, S. Abu-Abeid, N. Stern, T. Mardi, T. Tulchinski, V. Deutsch, S. Yedgar, G. Barshtein, and S. Berliner. 2004. "Flow-Resistant Red Blood Cell Aggregation in Morbid Obesity." *International Journal of Obesity* 28:1528–1534.

Shin, S., Y. Yang, and J. S. Suh. 2009. "Measurement of Erythrocyte Aggregation in a Microchip-Based Stirring System by Light Transmission." *Clinical Hemorheology and Microcirculation* 41:197–207.

Singh, M., and K. P. Joseph. 1987. "Erythrocytes Sedimentation Profiles under Gravitational Field as Determined by He-Ne Laser. VII. Influence of Dextrans, Albumin, and Saline on Cellular Aggregation and Sedimentation Rate." *Biorheology* 24:53–61.

Skalak, R., P. R. Zarda, K. M. Jan, and S. Chien. 1981. "Mechanics of Rouleau Formation." *Biophysical Journal* 35:771–81.

Sowemimo-Coker, S. O., and H. J. Meiselman. 1989. "Effect of Procaine Hydrochloride on the Electrophoretic Mobility of Human Red Blood Cells." *Cell Biophysics* 15:235–248.

Sowemimo-Coker, S. O., P. Whittingstall, L. Pietsch, R. M. Bauersachs, R. B. Wenby, and H. J. Meiselman. 1989. "Effects of Cellular Factors on the Aggregation Behavior of Human, Rat, and Bovine Erythrocytes." *Clinical Hemorheology and Microcirculation* 9:723–737.

Stoltz, J. F. 1994. *Hemorheologie et Aggregation Erythrocytaire*. Cachan, France: Editions Medicales Internationales.

Stoltz, J. F., M. Singh, and P. Riha. 1999. *Hemorheology in Practice*. Amsterdam: IOS Press.

Toth, K., L. Bogar, I. Juricskay, M. Keltai, S. Yusuf, L. J. Haywood, and H. J. Meiselman. 1997. "The Effect of RheothRx Injection on the Hemorheological Parameters in Patients with Acute Myocardial Infarction (CORE Trial Substudy)." *Clinical Hemorheology and Microcirculation* 17:117–125.

Toth, K., R. B. Wenby, and H. J. Meiselman. 2000. "Inhibition of Polymer-Induced Red Blood Cell Aggregation by Poloxamer 188." *Biorheology* 37:301–12.

Vetrugno, M., G. Cicco, F. Cantatore, L. Arnese, N. Delle Noci, and C. Sborgia. 2004. "Red Blood Cell Deformability, Aggregability, and Cytosolic Calcium Concentration in Normal Tension Glaucoma." *Clinical Hemorheology and Microcirculation* 31:295–302.

Weng, X. D., G. O. Roederer, R. Beaulieu, and G. Cloutier. 1998. "Contribution of Acute-Phase Proteins and Cardiovascular Risk Factors to Erythrocyte Aggregation in Normolipidemic and Hyperlipidemic Individuals." *Thrombosis and Haemostasis* 80:903–908.

Whittingstall, P., K. Toth, R. B. Wenby, and H. J. Meiselman. 1994. "Cellular Factors in RBC Aggregation: Effects of Autologous Plasma and Various Polymers" in *Hemorheologie et Aggregation Erythrocytaire*, ed. J. F. Stoltz. Paris: Editions Medicales Internationales.

Windberger, V. And O.K. Baskurt. 2007. "Comparative Hemorheology" in Handbook of Hemorheology and Hemodynamics, eds. Baskurt. I.K., M.R. Hardeman, M.W. Rampling, and H. J. Meiselman. Amsterdam, Berlin, Oxford, Tokyo, Washington, D.C.: IOS Press. 267-285.

Zimmermann, J., L. Schramm, C. Wanner, E. Mulzer, H. A. Henrich, R. Langer, and E. Heidbreder. 1996. "Hemorheology, Plasma Protein Composition, and von Willebrand Factor in Type I Diabetic Nephropathy." *Clinical Nephrology* 46:230–236.

3 Mechanism of Red Blood Cell Aggregation

As outlined in the previous chapter, there is general agreement regarding correlations between elevated levels of fibrinogen or other large plasma proteins and enhanced red blood cell (RBC) aggregation, as well as the effects of molecular mass and concentration for neutral polymers such as dextran or poly(ethylene glycol) (PEG) (Chien and Lang 1987). It has also been shown that cellular factors affect RBC aggregation via changing the intrinsic tendency of RBC to form aggregates (i.e., RBC aggregability, see Chapters 1 and 2) (Rampling et al. 2004), and that specific cellular factors can be linked to the observed changes in aggregability.

3.1 CURRENT MODELS OF RBC AGGREGATION

To date the mechanisms involved in RBC aggregation have not been fully elucidated. Currently, there are two coexisting but almost mutually exclusive models for RBC aggregation: the bridging model and the depletion model. In the bridging model, RBC is proposed to occur due to large macromolecules such as plasma proteins cross-linking binding sites on adjacent cells (Brooks 1988). In contrast, the depletion model proposes quite the opposite. In this model, RBC aggregation occurs because of a lower localized protein or polymer concentration near the cell surface compared to the suspending medium (i.e., relative depletion near the cell surface). This exclusion of macromolecules near the cell surface leads to an osmotic gradient, depletion interaction, and attractive forces (Bäumler et al. 1996; Bäumler et al. 2001). The following material briefly outlines these two models, followed by a section describing the depletion theory in more detail.

3.1.1 BRIDGING HYPOTHESIS

The bridging model proposes that RBC aggregation occurs when bridging forces due to the adsorption of macromolecules onto adjacent cell surfaces exceed disaggregating forces due to electrostatic repulsion, membrane strain, and mechanical shearing (Brooks 1973b; Brooks 1988; Chien 1975; Chien et al. 1973; Chien and Lang 1987; Snabre and Mills 1985). Studies related to the intercellular distance between adjacent cells (Chien and Jan 1973) have demonstrated that separations between adjacent cells are less than the size of the hydrated molecules, leading to the model illustrated in Figure 3.1. This model suggests that the terminal portions of the flexible polymers are adsorbed onto the surfaces of adjacent cells, resulting in a cell–cell separation that increases with increasing polymer size but is still smaller than the diameter of the hydrated polymer (Chien 1975).

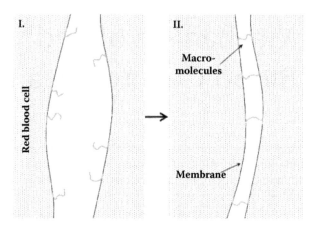

FIGURE 3.1 Schematic illustration of how macromolecular bridging brings adjacent cells into intimate contact: I. Macromolecules are adsorbed onto adjacent RBC surfaces. II. Aggregation of RBC occurs due to macromolecular bridging. Modified from Chien, S. 1975. "Biophysical Behavior of Red Cells in Suspensions," in *The Red Blood Cell*, ed. D. M. Surgenor. New York: Academic Press. 1031–1033.

The bridging model also leads to speculations regarding the dynamics of rouleau formation as shown in Figure 3.2 (Chien 1975). In this approach, it is assumed that when two cells are suspended in a solution containing weakly adsorbing macromolecules and are brought in close contact via external forces, macromolecules already adsorbed on one surface can adsorb onto the adjacent cell and thereby cross-link the two RBC. Once the cells are in close proximity via these bridges, further bridging will occur. Due to the high deformability of RBC, it is suggested that rotation of the red blood cells then makes it possible to form multiple bridges, thereby maximizing the area in close contact, eventually leading to rouleau formation.

This proposed mechanism is similar to other cell interactions such as agglutination, with the main difference being that the proposed adsorption energy of the macromolecules is much smaller in order to be consistent with the relative weakness of these forces. Several reports have been published in support of the bridging model. Chien and Jan (1973) studied the effects of dextran of various molecular masses on RBC aggregation for cells suspended in isotonic solutions. They found that dextran with a molecular mass equal to or larger than 40 kDa induces RBC aggregation and that the molar efficiency increased with molar mass. They also investigated RBC aggregates via electron microscopy to determine the distance between the membranes of adjacent cells and found that the cell–cell separation of RBC aggregates increases with increasing molecular mass of dextran: 19 nm for dextran 40 kDa to 32 nm for dextran 2,000 kDa.

These observations can be easily interpreted in light of the bridging hypothesis, since it can be expected that larger dextran molecules would cross-link the cells at greater separations. In addition, larger cell-separation would result in lower electrostatic repulsion, thereby requiring fewer molecules to cross-link the cells. On the other hand, below a critical cell–cell distance, the electrostatic

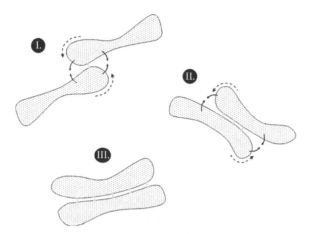

FIGURE 3.2 Schematic illustration of the postulated dynamics of rouleau formation in the bridging model. When two cells are suspended in a solution containing weakly adsorbing macromolecules and they are brought in close contact via external forces, macromolecules already adsorbed on one surface can adsorb to the adjacent cell and thereby (I) cross-link the two RBC. Once the cells are brought in close proximity via these bridges further bridging will occur (II). Due to the high deformability of RBC, it is suggested that the rotation of the red blood cells makes it then possible to form multiple bridges, thereby maximizing the surface area (II) in close contact, eventually leading to rouleau formation (III). Modified from Chien, S. 1975. "Biophysical Behavior of Red Cells in Suspensions," in *The Red Blood Cell*, ed. D. M. Surgenor. New York: Academic Press; 1031–1033. Chien, S., and K. M. Jan. 1973. "Ultrastructural Basis of the Mechanism of Rouleaux Formation." *Microvascular Research* 5:155–166.

repulsion becomes too strong, thereby requiring a minimal molecular size and thus molecular mass to induce RBC aggregation. Other reports in support of the bridging hypothesis have focused on quantification of macromolecular binding to the RBC surface (Brooks et al. 1980; Chien et al. 1977; Lominadze and Dean 2002). However, it should be noted that determination of polymer or protein adsorption to RBC is subject to numerous potential artifacts (e.g., trapped fluid between cells) and thus is quantitatively difficult to interpret (Janzen and Brooks 1989; Janzen and Brooks 1991).

3.1.2 DEPLETION HYPOTHESIS

The depletion model proposes quite the opposite. In this model, RBC aggregation occurs because of a lower localized protein or polymer concentration near the cell surface as compared to the suspending medium (i.e., relative depletion near the cell surface). The depletion concept was first introduced by Asakura and Oosawa (1954) more than half a century ago. These authors argued that if a surface is in contact with a polymer solution and the loss of configurational entropy of the polymer is not balanced by adsorption energy, a depletion layer develops near the surface as schematically shown in Figure 3.3. Within this layer, the polymer concentration is lower

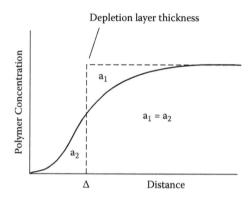

FIGURE 3.3 Schematic illustration of the polymer concentration of a depleted macromolecule as a function of the distance from the surface. The depletion layer thickness Δ is defined via the equality of the two areas a1 and a2.

than in the bulk phase. Thus, as two surfaces approach, the difference of solvent chemical potential (i.e., the osmotic pressure difference) between the intercellular polymer-poor depletion zone and the bulk phase results in solvent displacement into the bulk phase and hence depletion interaction. Due to this interaction, an attractive force develops that tends to minimize the polymer-reduced space between the two surfaces (Fleer et al. 1993; Jenkins and Snowden 1996).

Several previous reports have dealt with the experimental and theoretical aspects of depletion aggregation, often termed *depletion flocculation*, as applied to the general field of colloid chemistry (Feign and Napper 1980; Jenkins and Vincent 1996; Vincent 1990; Vincent et al. 1986). However, macromolecular depletion as a mechanism promoting red blood cell aggregation has received considerably less attention (Bäumler and Donath 1987; Bäumler et al. 1999; Neu and Meiselman 2001; Neu and Meiselman 2002). As outlined previously, this model argues that if the RBC surface is in contact with a solution containing macromolecules or proteins and the loss of configuration entropy of these molecules at the surface is not balanced by adsorption energy, a depletion layer will develop near the surface (stage I in Figure 3.4) (Joanny et al. 1979; Vincent 1990).

Thus, if two RBC with depletion layers approach one another, the difference of solvent chemical potential (i.e., the osmotic pressure difference) between the intercellular polymer-poor depletion zone and the bulk phase leads to the displacement of solvent into the bulk phase. In turn, this leads to depletion interaction and thus an attractive force develops that tends to minimize the polymer-poor space between the cells (stage II in Figure 3.4), thereby resulting in rouleau formation (stage III in Figure 3.4). As with the bridging model, disaggregation forces are electrostatic repulsion, membrane strain, and mechanical shearing. In contrast to the bridging model, rouleau formation via depletion does not require dynamic deformations (Figure 3.2) since the adjacent cells can slide along each other in order to maximize the contact area and thus the loss in free energy (black arrows in Figure 3.4).

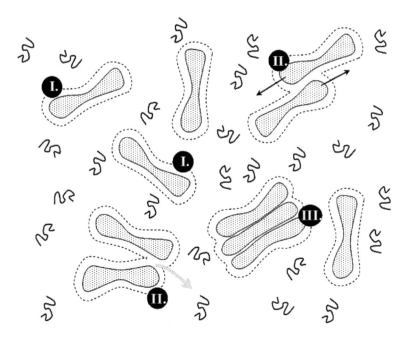

FIGURE 3.4 Schematic picture of how macromolecular depletion interaction brings adjacent cells into intimate contact: (I) macromolecules are depleted from RBC surfaces; (II) attraction develops when depletion layers overlap due to solvent displacement from the depletion zone into the bulk phase (grey arrow), and adjacent cells slide along each other to maximize the contact area and thus the loss in free energy (black arrows); (III) the resultant intimate cell–cell contact facilitates RBC aggregation.

3.1.3 BRIDGING VERSUS DEPLETION

As outlined previously, there are studies in support of the bridging model that either favor aggregation induced by a nonspecific binding of macromolecules (Brooks et al. 1980; Chien 1975) or by a specific binding site or mechanism (Lominadze and Dean 2002). However, as noted previously, determination of macromolecular adsorption of polymers and proteins to RBC is subject to numerous potential artifacts and thus the resulting data are difficult to interpret (Janzen and Brooks 1989; Janzen and Brooks 1991). Despite several attempts over the past few decades to quantify macromolecular binding, conclusive data are lacking. Conversely, several reports favoring the depletion model over the bridging model have been published (e.g., Armstrong et al. 2004; Bäumler et al. 1996; Neu and Meiselman 2002). The current understanding of the depletion hypothesis related to RBC aggregation is discussed in the following section.

3.2 THE DEPLETION THEORY FOR RED BLOOD CELL AGGREGATION

Before it is possible to calculate surface affinities between RBC in solutions containing nonadsorbing macromolecules, one has to define the nature of the cell–cell interaction. The exterior of the RBC, the glycocalyx, consists of a complex layer of proteins and glycoproteins and bears a net negative charge that is primarily due to ionized sialic acid groups (Seaman 1975). It can thus be expected that cells in close proximity will also interact via steric interactions due to the overlapping glycocalyces. However, in the theoretical model employed herein, only depletion and electrostatic interactions need to be considered. As shown in the following text, owing to the high electrostatic repulsion, cell–cell distances at which minimal interaction energy (i.e., maximal surface affinity) occurs are always greater than twice the thickness of the cell's glycocalyx. Thus, steric interactions between the glycocalyx on adjacent RBC can be neglected. In addition, it should be noted that the calculated total interaction energies are in the range of a few $\mu J/m^2$, whereas for cell separations greater than twice the glycocalyx thickness, van der Waals interactions are in the range of 10^{-2} $\mu J/m^2$ and thus small enough to be neglected (Neu and Meiselman 2002).

3.2.1 DEPLETION INTERACTION

To estimate this depletion attraction between two approaching surfaces, one has to multiply the change in the volume of the depletion zone by the osmotic pressure drop. If we assume a step profile for the free polymer, as shown schematically in Figure 3.5 (Fleer et al. 1984; Vincent et al. 1986), then the depletion interaction energy per area, w_D, can be calculated as a function of the cell–cell separation d:

$$w_D = -\Pi \cdot (2\Delta - d) \tag{3.1}$$

when $d/2 < \Delta$ and equals zero for $d/2 > \Delta$. In order to calculate the osmotic pressure Π, a virial equation (i.e., a power series in concentration) can be employed by neglecting coefficients higher than the second virial coefficient (B_2) since the concentrations relevant for RBC aggregation are comparably small (Neu and Meiselman 2002):

$$\Pi = \frac{RT}{M_2} c_2 + B_2 (c_2)^2 = -\frac{(\mu_1 - \mu_1^0)}{v_1} \tag{3.2}$$

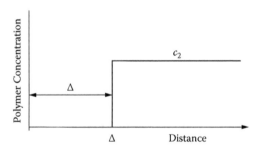

FIGURE 3.5 Stylized concentration-distance step profiles near a hard surface to calculate the depletion energy.

where R, T, v_1, and M_2 are the gas constant, absolute temperature, molecular volume of the solvent, and the molecular mass of the polymer. The chemical potential of the solvent in the polymer solution is μ_1 and is μ_1^0 in a polymer-free solution; c_2 represents the bulk polymer concentration.

However, examination of depletion layers also requires distinguishing between so-called *hard* and *soft* or *hairy* surfaces. Hard surfaces are considered smooth and do not allow polymer penetration into the surface as illustrated in Figure 3.5. On the other hand, soft surfaces, such as the RBC glycocalyx, are characterized by a layer of attached macromolecules that can be penetrated in part or entirely by the free polymer in solution (Jones and Vincent 1989; Neu and Meiselman 2002; Vincent et al. 1986). Figure 3.6 presents a stylized representation of the glycocalyx and polymer concentrations adjacent to the surface of a cell or particle with a soft surface. The segment densities of this attached layer (e.g., the cell glycocalyx, subscript a) and the free polymer (subscript 2) are indicated by ϕ, the glycocalyx thickness by δ, and the penetration depth of the free polymer into the attached layer by p. For a perfectly hard surface (see Figure 3.5), the penetration would equal zero and for a perfectly soft surface that provides no hindrance to the free polymer, it would be equal to the glycocalyx thickness δ.

The depletion interaction energy w_D can now be estimated by assuming a step profile for the free polymer as shown schematically in Figure 3.7 (Fleer, Scheutjens, and Vincent 1984; Vincent et al. 1986):

$$w_D = -2\Pi\left(-\frac{d}{2} + \delta - p\right) \tag{3.3}$$

when $(d/2-\delta+p)<\Delta$ and zero for $(d/2-\delta+p)>\Delta$. Thus, provided that the depletion layer thickness, the glycocalyx layer thickness, and the penetration of the free polymer into the glycocalyx are known, the depletion energy can be calculated via Equation 3.3.

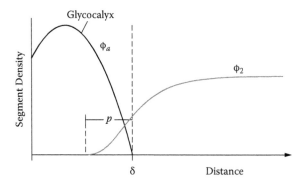

FIGURE 3.6 Schematic illustration of the concentration-distance profiles near the RBC surface in the presence of a depleted polymer. ϕ_a and ϕ_2 are the segment density of the glycocalyx and the depleted polymer, δ is the thickness of the glycocalyx, and p is the penetration depth of the free polymer into the glycocalyx.

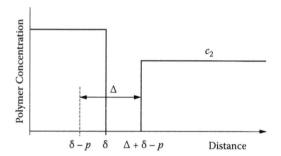

FIGURE 3.7 Concentration-distance step profiles near a soft surface (e.g., cell glycocalyx or attached polymer layer) as an approach to calculate depletion interaction energy. δ is the thickness of the glycocalyx, p is the penetration depth of the free polymer into the glycocalyx, and c_2 the bulk polymer concentration in the suspending medium.

3.2.2 DEPLETION LAYER THICKNESS

There have been a number of attempts to obtain expressions that allow calculation of the form of the concentration profile or the depletion layer thickness near surfaces (Jenkins and Snowden 1996). Here we will introduce an analytical expression that was derived by Vincent (1990). This approach is based upon calculating the equilibrium between the compressional or elastic free energy and the osmotic force experienced by polymer chains at a nonabsorbing surface. It yields the depletion layer thickness as a function of the bulk polymer concentration c_2 and the molecular mass of the polymer M_2:

$$= -\frac{1}{2}\frac{\Pi}{D} + \frac{1}{2}\sqrt{\left(\frac{\Pi}{D}\right)^2 + 4\,_0{}^2} \qquad (3.4)$$

where Π is the osmotic pressure of the bulk polymer solution. The parameter D is a function of the bulk polymer concentration c_2 and is given by:

$$D = \frac{2k_BT}{_0{}^2}\left(\frac{c_2 N_a}{M_2}\right)^{\frac{2}{3}} \qquad (3.5)$$

Here k_B and N_a are the Boltzmann constant and Avogadro number. Δ_0 is the depletion thickness for vanishing polymer concentration and is equal to $1.4 \cdot R_g$, where R_g is the polymer's radius of gyration (i.e., one-half the root mean square diameter of the hydrated polymer).

Figure 3.8 presents calculated depletion layer thicknesses for dextran and poly(ethylene glycol), PEG (Neu and Meiselman 2002). Dextran is a long chain of glucose units joined primarily by 1:6 alpha links with some 1:3 and 1:4 links, and PEG is a repeating linear chain of ethylene oxide. Both molecules are flexible, nonionic, water-soluble polymers that are known to induce RBC aggregation above a

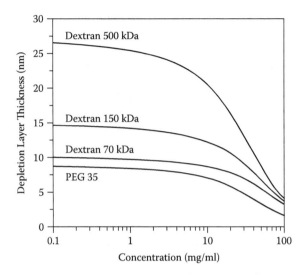

FIGURE 3.8 Theoretical dependence of the depletion layer thickness (Δ) on the bulk phase polymer concentration (c_2) for dextrans with molecular masses of 70 kDa, 150 kDa and 500 kDa and PEG with a molecular mass of 35 kDa. Modified with permission from Neu, B., and H. J. Meiselman. 2002. "Depletion-Mediated Red Blood Cell Aggregation in Polymer Solutions." *Biophysical Journal* 83:2482–2490.

certain threshold molecular mass (Boynard and Lelievre 1990; Brooks 1988; Chien and Lang 1987; Nash et al. 1987). Dextran and PEG are available in several molecular mass fractions and their physicochemical properties have been studied in detail, thus making them ideal model polymers to test the feasibility of any theoretical description of RBC aggregation. Table 3.1 presents osmotic virial coefficients (B_2), molecular mass, and radii of gyration (R_g) for 70-, 150-, and 500-kDa dextran and 35-kDa PEG.

As can be seen in Figure 3.8, the depletion layer thickness for the three dextrans and the PEG decreases only slightly with increasing concentration up to about

TABLE 3.1

Physicochemical Properties of Macromolecules[a]

	Dextran			PEG
M_2 [kDa]	70.00	150.00	500.00	35.00
B_2 [m⁵s⁻²kg⁻¹]	1.69	0.99	0.43	6.40
R_g [nm]	7.36	10.80	19.70	6.42

Source: Neu, B., and H. J. Meiselman. 2002. "Depletion-Mediated Red Blood Cell Aggregation in Polymer Solutions." *Biophysical Journal* 83:2482–2490.

[a] M_2 = weight average molecular mass; B_2 = second viral coefficient; R_g = radius of gyration

10 mg/ml, while at higher concentrations the depletion layer thickness decreases rapidly. As might be expected, the depletion layer thickness for a specific polymer type (e.g., dextran) increases with increasing molecular mass and hence with increasing polymer size or radius of gyration. However, in addition to molecular mass, the physicochemical properties of the polymers can also affect the depletion layer thickness. For example, PEG with a molecular mass of 35 kDa has a depletion layer thickness that is quite similar to the values calculated for dextran with 70 kDa. This near equality can be understood when comparing their sizes: 35-kDa PEG has a radius of gyration of 6.42 nm whereas that of dextran 70 kDa is only slightly larger at 7.36 nm (see Table 3.1).

3.2.3 MACROMOLECULAR PENETRATION INTO THE GLYCOCALYX

Lastly, to obtain a quantitative description of RBC depletion interaction energy, it is necessary to estimate the penetration of the polymer into the glycocalyx. Intuitively, the penetration depth p of the free polymer into the attached layer should depend on the polymer type, concentration, and molecular size, and one would expect that the penetration should be larger for smaller molecules and for higher concentration due to increasing osmotic pressure. In order to calculate p, one possible approach is to assume that the penetration of the polymer proceeds until the local osmotic pressure developed in the attached layer is balanced by the osmotic pressure of the bulk solution (Vincent et al. 1986).

However, in order to evaluate the change in free energy due to the mixing of the free polymer and the glycocalyx, it is necessary to identify the segment profile of the glycocalyx as well as that of the depleted polymer. Figure 3.9 indicates a

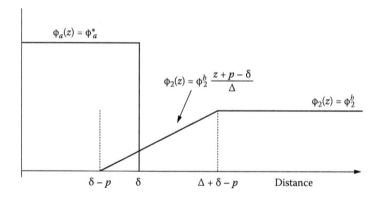

FIGURE 3.9 Schematic representation of concentration-distance profiles near a surface having a soft layer in which there is an uniform profile of attached macromolecules and a linear depletion layer profile for the polymer in solution. Subscripts a and 2 indicate the attached soft layer and the free polymer, respectively; z indicates the distance from the surface; other symbols are the same as in Figure 3.7. Modified with permission from Rad, S., and B. Neu. 2009. "Impact of Cellular Properties on Red Cell–Red Cell Affinity in Plasma-Like Suspensions." *European Physical Journal E* 30:135–140.

simple concentration-distance model in which there is a uniform profile of attached macromolecules for the glycocalyx (Vincent et al. 1986):

$$\phi_a(z) = \phi_a^*$$
(3.6)

and a linear profile of the depleted polymer in solution (Vincent et al. 1986):

$$\phi_2(z) = \frac{\phi_2^b}{\delta}(z + p - \delta).$$
(3.7)

As the glycocalyx is approached, the volume fraction of the polymer decreases from ϕ_2^b at $\Delta+\delta - p$ to zero at $\delta-p$. Using these profiles, the penetration p can be calculated (Vincent et al. 1986):

$$p = \frac{\Pi \cdot v_1 \cdot}{k_B T \phi_a^* \phi_2}$$
(3.8)

Here, v_1 is the molecular volume of the solvent and ϕ_a^* is given by

$$\phi_a^* = (1 - 2\chi)\phi_a$$
(3.9)

where ϕ_a represents the volume fraction of the attached macromolecules in the glycocalyx (Figure 3.9) and χ is the Flory interaction parameter for the polymer and solvent (Flory 1953). Thus for χ equal to zero, ϕ_a^* corresponds to the volume fraction of polymer in the glycocalyx.

The impact of glycocalyx volume fraction on polymer penetration is illustrated in Figure 3.10. In these calculations, the glycocalyx thickness was 5 nm and the bulk polymer concentration was 10 mg/ml. As can be seen, varying the glycocalyx volume fraction between 0 for a perfectly soft surface to 1 for a hard surface results in a distinct dependence of penetration on molecular mass. However, it should be noted that in this approach, δ is assumed to be independent of the bulk polymer concentration, whereas a more realistic approach would also consider that the attached polymers (i.e., the glycocalyx) collapse under the osmotic pressure of the bulk polymer (Jones and Vincent 1989). Unfortunately, it is difficult to accurately apply such a model to RBC in polymer or protein solution since too little is known about the physicochemical properties of the glycocalyx, and in particular, about the interaction between the glycocalyx and different polymers or proteins.

Given the previously mentioned lack of information about the glycocalyx, an alternative approach is to estimate the concentration dependence of the penetration depth via an exponential approach (Neu and Meiselman 2002):

$$p = \delta\left(1 - e^{-c_2/c_p}\right)$$
(3.10)

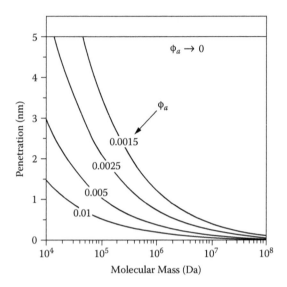

FIGURE 3.10 Effect of molecular mass on polymer penetration into the glycocalyx for different glycocalyx volume fractions ϕ_a assuming a constant bulk polymer concentration of 10 mg/ml and a glycocalyx layer thickness of 5 nm. Modified with permission from Rad, S., and B. Neu. 2009. "Impact of Cellular Properties on Red Cell–Red Cell Affinity in Plasma-Like Suspensions." *European Physical Journal E* 30:135–140.

where c_p is the penetration constant of the polymer in solution (i.e., when c_p equals c_2, p is 63% of δ). In this approach, δ is again assumed to be independent of bulk polymer concentration. Therefore, p is essentially a linear function of the polymer concentration at concentrations that are low relative to c_p and asymptotically approaches δ at high concentrations. As a first approximation, c_2 can be considered to be constant. However, for a more accurate description it would be necessary to express c_p as a function of the specific polymer and polymer concentration.

3.2.4 ELECTROSTATIC REPULSION BETWEEN RED BLOOD CELLS

Having established the depletion interaction energy between two soft surfaces, the last step is to estimate the electrostatic free energy between two cells. This can be calculated by simply considering an isothermal charging process:

$$E = \frac{1}{2} \int_0^d \int_0^\rho \Psi(\rho, z) d\rho dz \tag{3.11}$$

where ψ is the electrostatic potential between the cells, which is dependent on the charge density ρ. To calculate the electrostatic interaction energy between two cells, one first calculates the free energy of the two cells at a separation distance d, then deducts the free energy of two single cells (i.e., as $d \rightarrow \infty$).

To calculate the electrostatic potential ψ for red blood cells it is necessary to solve the Poisson-Boltzmann equation. The linear approximation can be employed for the problem under consideration since it is usually suitable for the moderate electrical potentials encountered for various cells (Bäumler et al. 1996). Assuming that both cells have the same constant charge and that this charge is distributed evenly within the glycocalyx (i.e., same profile as in Figure 3.7), ψ can be calculated for a single cell surface and for two cells at a separation distance d. However, it is possible to simplify this approach by approximating the electrostatic potential between two cells as a superposition of the potential of two single cells. This simplification is possible since the Debye length (κ^{-1}) is small compared to both the glycocalyx thickness δ and the calculated cell–cell distance d. Using this superposition the electrostatic interaction energy w_E is:

$$w_E = \frac{\sigma^2}{\delta^2 \varepsilon \varepsilon_0 \kappa^3} \begin{cases} \sinh(\kappa\delta)\left(e^{\kappa\delta-\kappa d} - e^{-\kappa d}\right) & d < 2\delta \\ (2\kappa\delta - \kappa d) - (e^{-\kappa\delta} + 1)\sinh(\kappa\delta - \kappa d) - \sinh(\kappa\delta)e^{-\kappa d} & d < 2\delta \end{cases}$$

(3.12)

where ε and ε_0 are the relative permittivity of the solvent and the permittivity of vacuum, and σ is the surface charge density.

3.2.5 RED BLOOD CELL AFFINITY

Having established the electrostatic interaction energy as well as the depletion energy between two soft surfaces in a solution containing nonadsorbing polymers, it is now possible to calculate the total interaction energy (w_T) (Neu and Meiselman 2002):

$$w_T = w_D + w_E$$

(3.13)

The effects of cell–cell separation distance (d) on total surface interaction energy (w_T) are shown in Figure 3.11 for dextran 70 kDa, dextran 500 kDa, and PEG 35 kDa. The RBC glycocalyx thickness δ was held constant at 5 nm, a value of 0.036 C/m^2 was assumed for the RBC surface charge density σ, and a value of 0.76 nm was used as the Debye length κ^{-1} (Bäumler et al. 1996; Donath et al. 1996; Donath and Voigt 1985; Levine et al. 1983; Neu and Meiselman 2002). In this figure, the bulk polymer concentration c_2 was held constant at 10 mg/ml and various values of the penetration constant c_p were employed.

The results presented in Figure 3.11 clearly demonstrate the impact of the penetration constant as well as of polymer type and size (i.e., molecular mass). For example, dextran 70 kDa has a more pronounced dependence on the penetration constant than dextran 500 kDa: for a change of c_p from 0 to 100 mg/ml, there is more than a 3-fold increase in interaction energy for dextran 70 kDa, whereas dextran 500 kDa shows only about a 30% increase. This unequal dependency on the penetration constant seems consistent with polymer size versus glycocalyx thickness. Since the

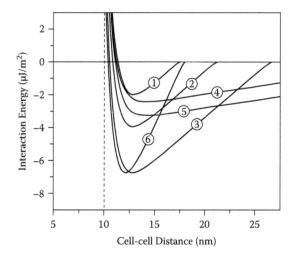

FIGURE 3.11 Total interactional energy as a function of RBC–RBC separation for cells in three polymers. Dextran 70 kDa (1) $c_p \rightarrow 0$; (2) $c_p = 10$ mg/ml; (3) $c_p = 100$ mg/ml; Dextran 500 kDa (4) $c_p \rightarrow 0$; (5) $c_p = 10$ mg/ml; PEG 35 kDa (6) $c_p = 10$ mg/ml. Bulk polymer concentration was held constant at 10 mg/ml. The dashed vertical line at 10 nm indicates the summed glyco-calyx thickness for both cells. Modified with permission from Neu, B., and H. J. Meiselman. 2002. "Depletion-Mediated Red Blood Cell Aggregation in Polymer Solutions." *Biophysical Journal* 83:2482–2490.

R_g of dextran 500 kDa is about four times greater than the 5-nm-thick RBC glycoca-lyx, the impact of the penetration constant on the interaction energy is rather small. Conversely, the R_g for dextran 70 kDa is only about 50% larger than the glycocalyx thickness, and thus glycocalyx penetration can markedly affect w_T.

Also shown in Figure 3.11 is the interaction energy (w_T) for PEG 35 kDa with a penetration constant c^p_2 of 10 mg/ml (curve 6). Clearly, PEG 35 kDa exhibits a higher maximal value of w_T than dextran 70 kDa with the same value for c^p_2. This difference of maximal w_T for PEG 35 kDa and dextran 70 kDa is due to their differ-ent physicochemical properties. Although these molecules have about the same R_g (Table 3.1) and thus about the same Δ for $c_2 = 10$ mg/ml (see Figure 3.8), the molar concentration of PEG 35 kDa is 2-fold greater than that of dextran 70 kDa. This dif-ference of molar concentration leads to about twice the osmotic pressure difference between the depletion layer and the bulk phase, and therefore twice the interactional energy can be expected for PEG 35 kDa.

3.2.6 RBC ADHESION ENERGY IN POLYMER SOLUTION

The effects of bulk phase polymer concentration on maximal interaction energy are shown in Figures 3.12 and 3.13. The maximal interaction energies represent the nadir of the w_T versus separation distance for each polymer. In Figure 3.12 the penetration was set equal to the glycocalyx thickness and thus penetration depth was indepen-dent of polymer concentration, whereas in Figure 3.13 the penetration was set equal

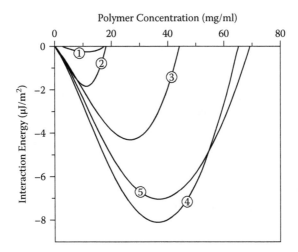

FIGURE 3.12   Effect of bulk phase polymer concentration on total interactional energy for RBC suspended in various molecular mass fractions of dextran and poly(ethylene glycol) when penetration p is equal to δ and thus all polymers fully penetrate the glycocalyx. (1) dextran 40 kDa, (2) dextran 70 kDa, (3) dextran 250 kDa, (4) dextran 500 kDa, (5) PEG 35 kDa. Modified with permission from Neu, B., and H. J. Meiselman. 2002. "Depletion-Mediated Red Blood Cell Aggregation in Polymer Solutions." *Biophysical Journal* 83:2482–2490.

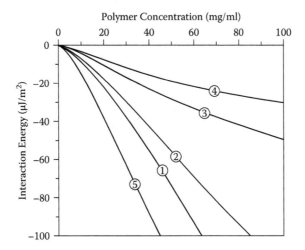

FIGURE 3.13   Effect of bulk phase polymer concentration on total interactional energy for RBC suspended in various molecular mass fractions of dextran and poly(ethylene glycol) when the penetration p is set equal to 0 and thus no polymers penetrate the glycocalyx. (1) dextran 40 kDa, (2) dextran 70 kDa, (3) dextran 250 kDa, (4) dextran 500 kDa, (5) PEG 35 kDa. From Neu, B., and H. J. Meiselman. 2002. "Depletion-Mediated Red Blood Cell Aggregation in Polymer Solutions." *Biophysical Journal* 83:2482–2490 with permission.

to zero, indicating no penetration of the free polymers into the glycocalyx regardless of the polymer concentration.

Within the concentration range shown in Figure 3.12, polymer penetration into the glycocalyx results in energy–polymer concentration relations that are bell-shaped and concave to the concentration axis. For a given polymer type, they also demonstrate increasing maximal values of the interaction energy with increasing molecular mass. In contrast, a lack of penetration into the glycocalyx (Figure 3.13) results in an essentially linear dependence of the interaction energy on polymer concentration as well as markedly higher levels of energy. In addition, smaller molecules result in higher values since the effect of the greater osmotic pressure difference outweighs the influence of the larger depletion layer for the larger molecules.

The effects of polymer molecular mass on interaction energy are shown in Figure 3.14a with the corresponding cell–cell separation distances shown in Figure 14b. The calculated interaction energies are again the nadir of the energy versus separation distance relation for each polymer and are shown for several polymer concentrations between 2.5 mg/ml and 40 mg/ml. The glycocalyx volume fraction was set equal to 0 representing an ideal soft glycocalyx (i.e., $p = \delta$).

As can be seen in Figure 3.14a for a completely penetrable soft glycocalyx, an attractive interaction energy can only be observed above a minimal molecular mass. Above this minimal molecular mass the cell affinity increases until it reaches a maximum followed by an asymptotic decline toward zero. For a polymer concentration of 2.5 mg/ml, the threshold for this attractive interaction energy is around 40 kDa and increases to 60 kDa at a bulk polymer concentration of 40 mg/ml. For the concentrations shown, the theoretical maximal interaction energy is observed at about a molecular mass of 200 kDa. This maximum depends to some extent on the bulk polymer concentration: increasing the concentration from 2.5 mg/dl to 40 mg/ml shifts the position from ~175 kDa to ~230 kDa. The total energy also demonstrates a strong dependence on the polymer concentration: for the concentration range shown (i.e., 2.5 to 40 mg/ml) the maximal interaction energy increases with increasing concentration from -0.6 to -7.9 $\mu J/m^2$.

The cell–cell separation distances shown in Figure 3.14b also demonstrate a strong dependence on polymer molecular mass and concentration: doubling the polymer concentration leads to an approximately one-nanometer decrease of the cell–cell distance. This decrease does not depend on the molecular mass of the depleted polymer. Further, it should be noted that for the molecular mass and concentrations used, the separation distance is always greater than twice the glycocalyx thickness of 5 nm. As mentioned previously, the latter observation justifies neglecting other forces in addition to electrostatic repulsion and depletion attraction.

The previous calculations clearly demonstrate that RBC affinity in dextran solutions shows a strong dependence on polymer concentration and molecular mass. To obtain a fuller understanding of the overall dependence of cell–cell affinity as a function of polymer size and concentration, the molecular mass and concentration dependence of red blood cell affinity are plotted in Figure 3.15. The dashed curves are contours of constant total interaction energy, in $\mu J/m^2$, for adjacent RBC with soft surfaces and a glycocalyx thickness of 5 nm. Within the concentration and molecular mass range shown, relations between energy and polymer concentration, as well as energy and

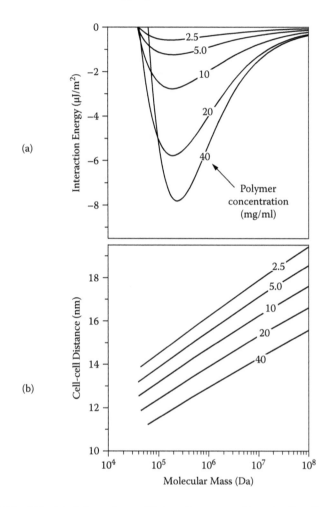

FIGURE 3.14 Calculated dependence of interaction energy (a) and cell–cell separation (b) on molecular mass at different suspending phase polymer concentrations between 2.5 mg/ml and 40 mg/ml; glycocalyx thickness $\delta = 5$ nm and an ideal soft surface was assumed. Modified with permission from Rad, S., and B. Neu. 2009. "Impact of Cellular Properties on Red Cell–Red Cell Affinity in Plasma-Like Suspensions." *European Physical Journal E* 30:135–140.

molecular mass, are again bell-shaped and concave to the concentration and molecular mass axis: peak cell affinity of -7.9 $\mu J/m^2$ occurs at 36 mg/ml and 220 kDa.

3.2.7 IMPACT OF CELL SURFACE PROPERTIES ON RED BLOOD CELL AFFINITY

It is of interest to consider the impact of cell surface properties on calculated RBC affinities. The effect of glycocalyx volume fraction on cell–cell affinities is shown in Figure 3.16 using the corresponding glycocalyx penetrations shown in Figure 3.10. For these calculations, the glycocalyx thickness Δ was set equal to 5 nm, the bulk polymer concentration c_2 to 10 mg/ml, and the glycocalyx volume fraction was varied

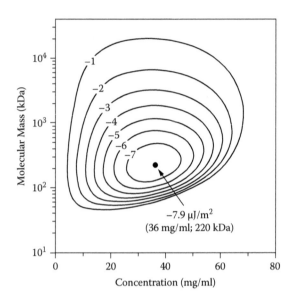

FIGURE 3.15 Lines of constant total interaction energy ($\mu J/m^2$) for adjacent RBC in dextran solutions as a function of polymer molecular mass and concentration for the limiting case of a soft surface with $p = \delta = 5$ nm. Curved lines are contours of constant interactional energy. Modified with permission from Rad, S., and B. Neu. 2009. "Impact of Cellular Properties on Red Cell–Red Cell Affinity in Plasma-Like Suspensions." *European Physical Journal E* 30:135–140.

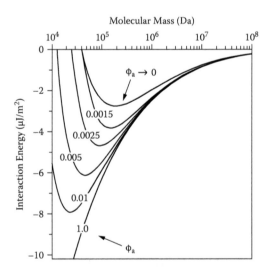

FIGURE 3.16 Dependence of interaction energy on molecular mass for different glycocalyx volume fractions, ϕ_a, assuming a constant bulk polymer concentration of 10 mg/ml and a glycocalyx layer thickness of 5 nm. Modified with permission from Rad, S., and B. Neu. 2009. "Impact of Cellular Properties on Red Cell–Red Cell Affinity in Plasma-Like Suspensions." *European Physical Journal E* 30:135–140.

between 0 and 1. As mentioned previously, attractive interaction energies can only be observed above a minimal molecular mass threshold; above this minimal molecular mass, cell affinity increases to a maximum followed by an asymptotic decline toward zero. The threshold for this minimal molecular mass decreases with increasing volume fraction and thus decreasing glycocalyx penetrability (Figure 3.10). Note that for a glycocalyx volume fraction of zero the threshold for attractive interaction energy is ~40 kDa, whereas increasing the volume fraction slightly to 0.005 decreases this threshold to 12 kDa.

The minimal interaction energy (i.e., the maximal affinity) and the molecular mass at which the maximal interaction energy can be observed, also demonstrate a strong dependence on glycocalyx volume fraction. Increasing the glycocalyx volume fraction leads to an increase in the maximal affinity as well as a decrease of the molecular mass for maximal interaction. For example, increasing the glycocalyx volume concentration from zero to 0.005 causes more than a 2-fold increase in the maximal interaction energy from $-2.8 \ \mu J/m^2$ to $-6.1 \ \mu J/m^2$, with the position of the maximum shifting from 170 kDa to 44 kDa. For the limiting case of a hard surface, the absence of a maximal affinity and a threshold molecular mass are predicted since the depletion interaction energy increases with decreasing molecular mass.

In Figure 3.17 the effects of glycocalyx thickness on interaction energy are plotted; in this figure the bulk polymer concentration was held constant and the limiting case of an ideal soft surface (i.e., $\phi_a \to 0$ or $p = \delta$) was assumed. Clearly, changes of glycocalyx thickness have a marked impact on cell–cell affinities: increasing thickness results in a steady decrease of affinity such that doubling the thickness from 5

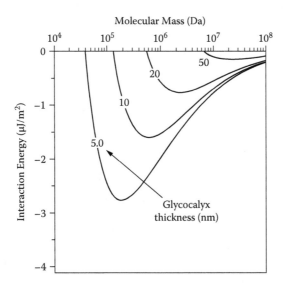

FIGURE 3.17 Dependence of interaction energy on molecular mass for different glycocalyx thicknesses for the limiting case of a soft surface ($\phi_a \to 0$) and bulk polymer concentration of 10 mg/ml. Modified with permission from Rad, S., and B. Neu. 2009. "Impact of Cellular Properties on Red Cell–Red Cell Affinity in Plasma-Like Suspensions." *European Physical Journal E* 30:135–140.

to 10 nm or from 10 to 20 nm leads to about a 50% decrease of peak energies. The threshold for the onset of attractive interaction also shifts to higher molecular mass because the depletion layer of smaller molecules will be within the glycocalyx.

3.3 EVIDENCE SUPPORTING DEPLETION HYPOTHESIS

The preferential exclusion of macromolecules leading to depletion flocculation has received considerable attention by colloid scientists during the past decades (Fleer et al. 1993). It is thus surprising that this phenomenon has been almost ignored when considering the stability of cell suspensions. In part, this lack of attention is due to previous and current problems associated with deciding if a depletion layer has occurred at cell surfaces. Determining the presence of a depletion layer is made difficult due to the very small scale of these layers, which are in the same size range as the hydrated size of the depleted macromolecule. This, in turn, means that the expected depletion layers for most plasma proteins known to induce aggregation are only a few nanometers thick and thus in the same size range as the RBC glycocalyx. Thus, it is not only impossible to detect such a layer with direct optical observations but, in addition, it is quite challenging to distinguish between weak absorption or a depletion effect since the latter might also involve some intermixing (i.e., penetration) of the macromolecule and the glycocalyx. In addition, even though past reports have described surface adsorption of dextran, and specific binding mechanisms between fibrinogen and RBC have been suggested (Lominadze and Dean 2002; Chien et al. 1977; Brooks 1988; Brooks 1973b), such results should be interpreted with great care. In fact, an extensive review of literature values by Janzen and Brooks (Janzen and Brooks 1991) has detailed likely technical artifacts (e.g., trapped fluid between RBC) and thus the extremely wide range of reported data for fibrinogen and dextran binding. Section 3.3 provides some recent experimental findings in support of the depletion theory.

3.3.1 Quantification of Depletion Layers via Particle Electrophoresis

Several studies have investigated the structure and extension of depletion layers at various surfaces using several techniques (Cowell et al. 1978; Donath et al. 1993; Krabi and Donath 1994; Vincent et al. 1986). One method that is applicable to macromolecular depletion at biological interfaces is the use of particle electrophoresis. In solutions of neutral soluble polymers that give rise to depletion layers, particles have an unexpectedly high *viscosity-corrected* mobility (Brooks and Seaman 1973; Brooks 1973a; Brooks 1973b; Neu and Meiselman 2001; Snabre and Mills 1985). This effect is due to the viscosity near the particle surface being lower than the bulk viscosity of the suspending medium due to the depletion effect (Bäumler and Donath 1987). Electro-osmotic flow decreases rapidly outside the electric double layer, so if the depletion layer thickness is comparable to or larger than that of the double layer, the influence of suspending phase viscosity on mobility is reduced.

Figure 3.18 presents results from a study that was directed toward validating the existence of the depletion layer by employing measurements of unit-gravity

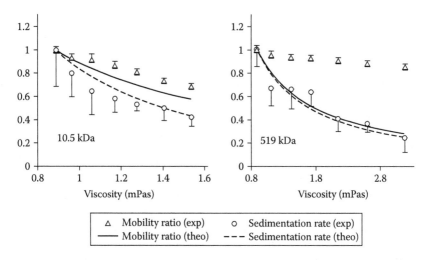

FIGURE 3.18 Experimental and theoretical values of RBC mobility and sedimentation rate versus suspending phase viscosity for cells in various solutions of (a) 10.5 kDa and (b) 519 kDa dextran. All mobility and sedimentation data are expressed relative to values obtained for cells in dextran-free buffer, and thus have a value of unity at a suspending medium viscosity of 0.89 mPas. Symbols are: Δ = experimental mobility ratio; o = experimental sedimentation ratio; solid line = predicted mobility; dashed line = predicted sedimentation; data are mean ± SD. Modified with permission from Neu, B., and H. J. Meiselman. 2001. "Sedimentation and Electrophoretic Mobility Behavior of Human Red Blood Cells in Various Dextran Solutions." *Langmuir* 17:7973–7975.

cell sedimentation and cell electrophoretic mobility for RBC in various polymer solutions (Neu and Meiselman 2001). These studies thus tested the hypothesis that regardless of polymer molecular mass, the effects of suspending medium viscosity on sedimentation could be predicted via the Stokes equation, whereas the electrophoretic mobility (EPM) would become less sensitive to suspending medium viscosity with increasing molecular mass (i.e., with increasing depletion layer thickness). As shown in Figure 3.18, the Stokes equation is applicable regardless of polymer mass, whereas the electrophoretic mobility follows the expected inverse relation only for small polymers (i.e., 10.5 kDa dextran). However, RBC electrophoretic mobility is essentially independent of medium viscosity for large polymers (i.e., 519 kDa dextran). The latter point is of particular interest because it demonstrates the existence of the depletion layer and its dependence on molecular size; dextran 10.5 kDa has a hydrated radius of about 3 nm compared to about 20 nm for the 519 kDa dextran (Neu and Meiselman 2001).

Consequently, particle electrophoresis has been used extensively to study macromolecules at biological interfaces. By changing the ionic strength of the suspending medium, it is even possible to probe the extent of the depletion layer. Variation of ionic strength varies the thickness of the double layer, and thus measurement of mobility at different ionic strengths allows estimation of depletion layer thickness (Donath et al. 1993). As long as the double layer is thinner than the depletion layer, the mobility will be higher than predicted based upon suspending medium viscosity.

With increasing double-layer thickness the influence of bulk viscosity increases whereas when the double-layer thickness becomes markedly thinner than the depletion layer, the measured mobilities are unaffected by bulk phase viscosity.

Figure 3.19 shows an example of the ionic strength approach in which the ratio of RBC mobility in polymer-free medium to that in solutions containing dextran are plotted versus Debye length (i.e., double-layer thickness) for three dextran fractions (Bäumler et al. 1996). According to electrophoretic mobility theory, it would be expected that the mobility ratios would scale inversely with suspending medium viscosity; the dashed horizontal lines in Figure 3.19 indicate the expected results. Instead, the ratio is well below the expected values for smaller Debye lengths (i.e., thinner double-layer thicknesses) thereby indicating a significantly lower viscosity close to the RBC surface. Note in particular the results for the 2,400 kDa dextran fraction: the predicted ratio is about 3.6 whereas the experimental values are 2 or less and thus the cells are moving much faster than predicted. Using this approach, it has been confirmed that the thickness of depletion layers is in the same range as the hydrated size of these polymers. This finding thus agrees with the concept of polymer depletion near the RBC surface and lends strong support to a *depletion model* mechanism for reversible RBC aggregation (Bäumler et al. 1996; Donath et al. 1997; Donath et al. 1989; Krabi and Donath 1994).

In addition to using neutral polymers, some studies have evaluated the depletion of proteins and polyelectrolytes at RBC surfaces by means of particle electrophoresis. However, the details of such experiments, as well as interpretation of the data, are not as straightforward as with neutral polymers. For charged

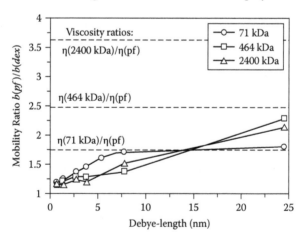

FIGURE 3.19 Ratio of the mobility for RBC in polymer-free (pf) media to that in solutions containing 2g/dl of three different dextran fractions as a function of Debye length (i.e., double-layer thickness). Also shown as horizontal dashed lines are predicted ratios based upon the viscosity of the suspending media. Modified with permission from Bäumler, H., E. Donath, A. Krabi, W. Knippel, A. Budde, and H. Kiesewetter. 1996. "Electrophoresis of Human Red Blood Cells and Platelets. Evidence for Depletion of Dextran." *Biorheology* 33:333–351.

polyelectrolytes and hence for proteins, it is also necessary to consider electro-static forces that can also affect depletion or adsorption (Bohmer et al. 1990; Fleer et al. 1993). Further, forces between the particle surface and the polyelectrolyte differ with ionic strength, and thus adsorption and depletion can vary with the extent of the double layer as well as the conformation of the charged macromol-ecule. For example, with polymers and surfaces of the same sign, only slight or no adsorption is often observed at low ionic strength, whereas with increasing ionic strength the adsorbed amount increases. This behavior is due to the electrostatic screening, which increases with increasing salt concentration; in the case of pure electroadsorption (i.e., when the surface charges and the polymer charges have opposite signs), the opposite effects of ionic strength are observed (Bohmer et al 1990; Fleer et al. 1993).

One consequence of the previously mentioned phenomena is that it is not possible to determine the thickness of the depletion layer for charged polymers by simply changing the ionic strength of the suspending media as is utilized for neutral poly-mers. Further, if only measured at a single ionic strength, it is sometimes not pos-sible to decide if apparent changes of the particle mobility should be attributed to depletion or adsorption. Thus, even though protein and polyelectrolyte depletion can be detected with particle electrophoresis, it is clear that for investigation of plasma proteins at biointerfaces it is still necessary to develop other techniques in order to obtain quantitative results.

3.3.2   COMPARISON OF EXPERIMENTAL FINDINGS WITH THE DEPLETION MODEL

3.3.2.1   Quantitative Comparison of the Adhesion Energies

Employing micropipette techniques to measure the extent of encapsulation of an RBC membrane sphere by an intact RBC, Buxbaum and coworkers (1982) deter-mined cell–cell surface affinities for normal human red blood cells in various dex-tran solutions. Their results, based on an analysis wherein the extent of encapsulation reflects surface affinity versus membrane shear elastic modulus, indicate biphasic affinity–concentration (c_2) relations with peak surface affinities of 4.9 $\mu J/m^2$ for 70-kDa dextran and 22 $\mu J/m^2$ for 150 kDa dextran (Figure 3.20).

In order to quantitatively compare these experimental findings with interactional energies calculated via the present model, the penetration constant c_p was varied until the calculated peak interactional energy for dextran 70 kDa or dextran 150 kDa equaled the value reported by Buxbaum et al. (1982) (i.e., c_p was the only parameter varied). This equality with their peak data occurred at $c_p = 0.70$ g/dl for dextran 70 and $c_p = 7.5$ g/dl for dextran 150; calculated $w_T - c_2$ relations based upon these c_p values are shown as solid curved lines in Figure 3.20. The shapes of the calcu-lated $w_T - c_2$ relations and the upper polymer concentrations at which w_T declines to zero are in excellent agreement with the experimental data reported by Buxbaum et al. (1982). In addition, the values of the polymer concentration required to achieve equality between the calculated and experimental peak w_T levels are consistent with the expected effect of polymer molecular mass on glycocalyx penetration (i.e., lower

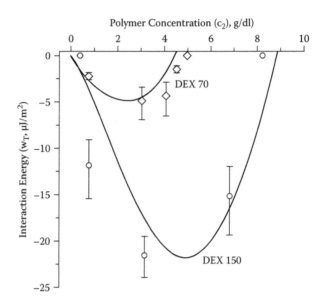

FIGURE 3.20 Comparisons between calculated (solid lines) and experimental values of interactional energy (w_T) for RBC suspended in various concentrations of dextran 70 kDa (DEX 70) or dextran 150 kDa (DEX 150). Redrawn with permission from Neu, B., and H. J. Meiselman. 2002. "Depletion-Mediated Red Blood Cell Aggregation in Polymer Solutions." *Biophysical Journal* 83:2482–2490.

value of c_p for dextran 70 kDa indicating greater penetration ability for smaller dextran molecules).

3.3.2.2 Dependence of Red Blood Cell Aggregation on Polymer Concentration

RBC aggregation has been studied using a variety of testing systems (see Chapter 4) and numerous investigators have described the effects of polymer concentration and molecular mass on RBC aggregation (Boynard and Lelievre 1990; Nash et al. 1987; Brooks 1988; Chien and Sung 1987). Representative aggregation data, obtained via light transmission and ultrasound backscattering methods for RBC in isotonic solutions of dextran 70 kDa and 500 kDa, are shown in Figure 3.21. These experimental data reflect two typical aspects of polymer-induced RBC aggregation: (1) biphasic, bell-shaped response to polymer concentration; (2) for a given polymer type (e.g., dextran), the extent or strength of aggregation increases with molecular mass. Owing to the empirical indices used to determine RBC aggregation, quantitative comparisons to calculated cell–cell affinities are precluded. However, the experimental findings shown in Figure 3.21 are in qualitative agreement with the shape and position of the calculated w_T results presented previously (Figures 3.12, 3.15, and 3.20).

The computed results for w_T, combined with experimental findings for RBC aggregation, clearly indicate that changes of interactional energy are mirrored by changes of RBC aggregation. Increased interactional energy increases RBC

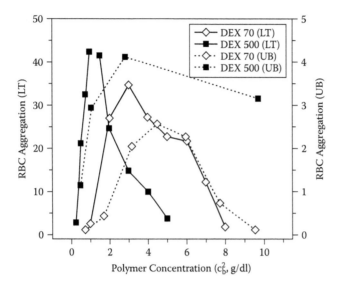

FIGURE 3.21 Polymer concentration-RBC aggregation results for cells suspended in solutions of dextran 70 kDa (DEX 70) or 500 kDa (DEX 500) recorded via light transmission (LT) (Nash, G. B., R. B. Wenby, S. O. Sowemimo-Coker, and H. J. Meiselman. 1987. "Influence of Cellular Properties on Red Cell Aggregation." *Clinical Hemorheology and Microcirculation* 7:93–108.) and ultrasound backscatter (UB) (Boynard and Lelievre 1990). Redrawn with permission from Neu, B., and H. J. Meiselman. 2002. "Depletion-Mediated Red Blood Cell Aggregation in Polymer Solutions." *Biophysical Journal* 83:2482–2490.

aggregation, whereas reduced interactional energy reduces aggregation. However, the relative importance of the factors that cause changes of w_T do differ between the ascending and descending regions of the aggregation–concentration relation. In the ascending region, w_T increases since the depletion layer thickness remains relatively constant, whereas the osmotic pressure difference increases. In the descending region, w_T decreases since the effects of the markedly reduced depletion layer thickness outweighs the effects of the increased osmotic pressure difference.

3.3.2.3 Dependence of RBC Aggregation on Polymer Molecular Mass

Even though most prior studies related to dextran-induced RBC aggregation have focused on molecular masses of less than 2 MDa (Armstrong et al. 2004), RBC aggregation as a function of molecular mass up to 28 MDa has been reported (Neu et al. 2008). In this study a series of measurements were conducted using five molecular mass dextran fractions between 70 kDa and 28 MDa at concentrations of up to 2 g/dl. The experimental data indicated that the aggregation–concentration relations for the 70-kDa to 2-MDa fractions had similar shapes, with a minimum concentration needed to initiate aggregation and a steady increase of aggregation with concentration. Increasing the molecular mass to 28 MDa also resulted in a minimum concentration to initiate aggregation, but in contrast to the other fractions, maximal aggregation induced by 28 MDa was markedly below the level found for smaller

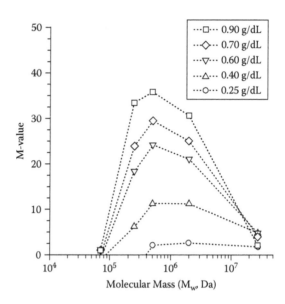

FIGURE 3.22 Dependence of RBC aggregation on dextran molecular mass at different suspending phase polymer concentrations. The M-value is a dimensionless index that increases with enhanced aggregation (see Chapter 4). Redrawn with permission from Neu, B., and H. J. Meiselman. 2008. "The Role of Macromolecules in Stabilization and De-Stabilization of Biofluids," in *Bioengineering in Cell and Tissue Research*, ed. by G. M. Artmann and S. Chien. New York: Springer; 387–408.

dextrans; very low aggregation was observed at all concentrations of the 28-MDa polymer.

The results of the previously mentioned study (Neu et al. 2008) were used to calculate the molecular mass dependence of aggregation at constant polymer concentration; Figure 3.22 presents calculated results at concentrations from 0.25 g/dl to 0.9 g/dl. Regardless of concentration, the curves are clearly bell-shaped, reaching maximal aggregation at a molecular mass of about 500 kDa. While in general the molecular mass for maximal aggregation does not show a significant dependence on polymer concentration, the relationship becomes broader and less defined at lower concentrations and suggests a shift toward higher molecular mass.

As illustrated in Figure 3.14, calculated depletion energies at equal bulk polymer concentrations and a glycocalyx thickness of 5 nm show a similar dependence on molecular mass. For a completely penetrable *soft* glycocalyx ($p = 5$ nm), depletion interaction only occurs above a minimum molecular mass (>16.5 kDa). The depletion energy then increases until it reaches a maximum at a molecular mass of about 100–200 kDa, following which it declines asymptotically toward zero. Thus, the previously mentioned theoretical result for a soft surface ($p = 5$ nm) is qualitatively similar to the experimental data presented in Figure 3.22: depletion interaction only occurs above a minimal molecular mass; the depletion energy then increases, reaches a maximum affinity, and then declines. This agreement between theoretically predicted behavior and experimental results lends credence to a mechanism governed by depletion interaction between soft surfaces. Note that the decrease of w_D and hence of aggregation tendency at the highest molecular mass does not agree with the nominally accepted steady increase of RBC aggregation with polymers of increasing molecular mass (Barshtein et al. 1998; Chien 1975; Pribush et al. 2007); this disagreement most likely arises due to the limited range of molecular mass utilized in these prior studies.

3.3.2.4 Comparison of Abnormal Aggregation with the Depletion Model

As outlined in Chapter 8, density-separated and hence age-separated RBC exhibit significant differences in aggregability in either plasma or polymer solutions. The aggregation of denser, older RBC is at least 2-fold greater than less dense, younger RBC, yet the differences are not related to altered cell volume, deformability, or surface immunoglobulin G (IgG) levels (Rampling et al. 2004). Based on the model presented herein, one possibility for such changes would be an age-dependent loss of surface charges and thus less repulsion between the cells; another possibility would be that the depletion layer increases at old cell surfaces due to surface alterations. With particle electrophoresis, it is possible to distinguish between these two alternatives (Bäumler et al. 1996; Neu and Meiselman 2001).

In Figure 3.23 the electrophoretic mobilities of young and old cells in polymer-free buffer and in buffer containing 3 g/dl of dextran 70 kDa are compared. The mobilities in polymer-free buffer show no significant difference, and since RBC mobilities are determined by their surface charge, there is no indication of an age-dependent alteration of electrostatic repulsion. In contrast, old RBC in 3 g/dl dextran 70 show a small but significantly higher EPM (+5%) than young cells. This difference is most likely due to a reduced local viscosity at the surface of these cells and therefore an increased depletion layer at the surface of the old RBC. Thus the

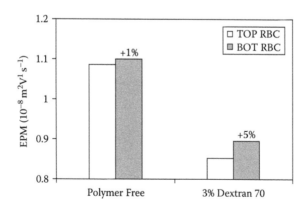

FIGURE 3.23 Electrophoretic mobility (EPM) of young (TOP) and old (BOT) RBC in polymer-free buffer and in buffer containing 3% dextran 70 kDa. Cells were separated by high-speed centrifugation and portions of the RBC column were isolated to yield young and old cells. Modified from Neu, B., S. Sowemimo-Coker, and H. Meiselman. 2003. "Cell–Cell Affinity of Senescent Human Erythrocytes." *Biophysical Journal* 85:75–84.

question: Are changes in the depletion layer leading to the 5% difference in mobility sufficient to explain the dramatic increase in aggregation for old RBC?

There are basically two possibilities (Figure 3.24) for an increasing depletion layer with cell age: a decreasing polymer penetration or a decreasing glycocalyx thickness, both leading to lower viscosities at the cell surface and thus higher mobilities in a polymer solution. Theoretical calculations showed that a decrease of 15% for the penetration (p) or the glycocalyx thickness (δ) would lead to a 5% increase in mobility and to a 70–80% increase of cell surface affinity, with the increased affinity sufficient to explain the differences in aggregation (Neu et al. 2003). In overview, the model for depletion-mediated aggregation provides a rational framework for explaining the observed equal EPM values in buffer, the unequal EPM values in polymer solutions, and the marked increases of polymer-induced aggregation for age-separated older RBC. That is, compared to less-dense TOP (young) cells, denser BOT (old) cells may have either a slightly thinner glycocalyx or slightly less polymer

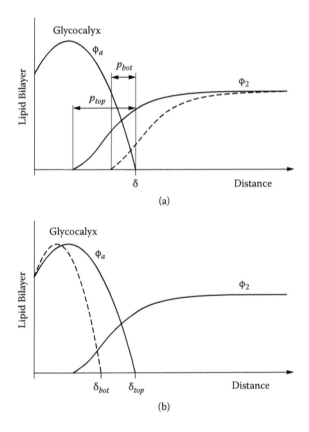

FIGURE 3.24 A schematic illustration of two possible explanations for the enhanced aggregability of old (i.e., BOT) RBC: (a) changes in the polymer penetration into the glycocalyx (i.e., $p_{bot} < p_{top}$); (b) changes in the thickness of the RBC glycocalyx during aging (i.e., $\delta_{bot} < \delta_{top}$). Redrawn with permission from Neu, B., S. Sowemimo-Coker, and H. Meiselman. 2003. "Cell–Cell Affinity of Senescent Human Erythrocytes." *Biophysical Journal* 85:75–84.

penetration into their glycocalyx, and thus greater polymer depletion and larger osmotic forces favoring aggregation.

3.4 CONCLUDING REMARKS

Although direct quantitative comparisons between experimentally measured extents of aggregation and theoretically derived attractive energies are very limited, the relations between these parameters seem consistent with experimental data. For example, the experimental findings presented in this chapter clearly indicate maximum aggregation at a molecular mass of about 500 kDa (Neu et al. 2008), which is in the same range as the molecular mass predicted theoretically (Figures 3.14 to 3.16). It is also known that maximal aggregation occurs in the range of 3–4 g/dl and that the threshold for aggregation is around 40 to 70 kDa (Chien 1975). All of these experimental observations are in qualitative agreement with the concept that RBC aggregation is induced via a macromolecular depletion interaction.

While the current understanding of depletion-mediated RBC interaction seems consistent with experimental observations, some aspects require further consideration. One drawback of the current model may arise from its simplicity. For example, in future studies it will be necessary to focus on more realistic treatments of the RBC glycocalyx (e.g., glycocalyx structure). Charge distribution within the glycocalyx also needs to be considered since it has a significant effect on the electrostatic interaction between adjacent surfaces (Lerche 1984). The effects of RBC deformability and membrane undulations have also not been considered. Whereas the latter is expected to have only a small effect on RBC aggregation, it is well known that RBC with decreased deformability exhibit decreased aggregation (Nash et al. 1987).

Further, the physicochemical parameters of the glycocalyx (i.e., glycocalyx volume concentration and thickness δ) should not be seen as absolute values and hence additional work is needed to identify how and to what extent these theoretical parameters reflect actual cellular properties. In addition, the current model only considers noncharged, neutral polymers (e.g., dextran) and has not yet been extended to charged polymers or proteins. Future efforts are required in order to resolve several unclear areas of human RBC aggregation, such as the reduced aggregation of neonatal red cells in both plasma and in polymer solutions (Linderkamp et al. 1984), or the wide donor-to-donor range of aggregation for RBC suspended in defined polymer or protein solutions (Neu and Meiselman 2008); details of these areas are considered in Chapter 8.

LITERATURE CITED

Armstrong, J. K., R. B. Wenby, H. J. Meiselman, and T. C. Fisher. 2004. "The Hydrodynamic Radii of Macromolecules and Their Effect on Red Blood Cell Aggregation." *Biophysical Journal* 87:4259–4270.

Asakura, S., and F. Oosawa. 1954. "On Interaction between Two Bodies Immersed in a Solution of Macromolecules." *Journal of Chemical Physics* 22:1255–1256.

Barshtein, G., I. Tamir, and S. Yedgar. 1998. "Red Blood Cell Rouleaux Formation in Dextran Solution: Dependence on Polymer Conformation." *Biophysical Reviews and Letters* 27:177–181.

Bäumler, H., and E. Donath. 1987. "Does Dextran Indeed Significantly Increase the Surface Potential of Human Red Blood Cells?" *Studia Biophysica* 120:113–122.

Bäumler, H., E. Donath, A. Krabi, W. Knippel, A. Budde, and H. Kiesewetter. 1996. "Electrophoresis of Human Red Blood Cells and Platelets. Evidence for Depletion of Dextran." *Biorheology* 33:333–351.

Bäumler, H., B. Neu, E. Donath, and H. Kiesewetter. 1999. "Basic Phenomena of Red Blood Cell Rouleaux Formation." *Biorheology* 36:439–442.

Bäumler, H., B. Neu, R. Mitlohner, R. Georgieva, H. J. Meiselman, and H. Kiesewetter. 2001. "Electrophoretic and Aggregation Behavior of Bovine, Horse and Human Red Blood Cells in Plasma and in Polymer Solutions." *Biorheology* 38:39–51.

Bohmer, M. R., O. A. Evers, and J. M. H. M. Scheutjens. 1990. "Weak Polyelectrolytes between 2 Surfaces—Adsorption and Stabilization." *Macromolecules* 23:2288–2301.

Boynard, M., and J. C. Lelievre. 1990. "Size Determination of Red Blood Cell Aggregates Induced by Dextran Using Ultrasound Backscattering Phenomenon." *Biorheology* 27:39–46.

Brooks, D. E. 1973a. "Effect of Neutral Polymers on Electrokinetic Potential of Cells and Other Charged-Particles. 2. Model for Effect of Adsorbed Polymer on Diffuse Double-Layer." *Journal of Colloid and Interface Science* 43:687–699.

Brooks, D. E. 1973b. "The Effect of Neutral Polymers on the Electrokinetic Potential of Cells and Other Charged Particles." *Journal of Colloid and Interface Science* 43:700–713.

Brooks, D. E. 1988. "Mechanism of Red Cell Aggregation." In *Blood Cells, Rheology and Aging*, ed. D. Platt. Berlin: Springer Verlag; 158–162.

Brooks, D. E., R. G. Greig, and J. Janzen. 1980. "Mechanism of Erythrocyte Aggregation." In *Erythrocyte Mechanics and Blood Flow*, ed. G. R. Cokelet, H. J. Meiselman, and D. E. Brooks. New York: A.R. Liss; 119–140.

Brooks, D. E., and G. V. F. Seaman. 1973. "Effect of Neutral Polymers on Electrokinetic Potential of Cells and Other Charged-Particles. 1. Models for Zeta Potential Increase." *Journal of Colloid and Interface Science* 43:670–686.

Buxbaum, K., E. Evans, and D. E. Brooks. 1982. "Quantitation of Surface Affinities of Red Blood Cells in Dextran Solutions and Plasma." *Biochemistry* 21:3235–3239.

Chien, S. 1975. "Biophysical Behavior of Red Cells in Suspensions." In *The Red Blood Cell*, ed. D. M. Surgenor. New York: Academic Press; 1033–1133.

Chien, S., R. J. Dellenback, S. Usami, D. A. Burton, P. F. Gustavson, and V. Magazinovic. 1973. "Blood Volume, Hemodynamic, and Metabolic Changes in Hemorrhagic Shock in Normal and Splenectomized Dogs." *American Journal of Physiology* 225:866–79.

Chien, S., and K. M. Jan. 1973. "Ultrastructural Basis of the Mechanism of Rouleaux Formation." *Microvascular Research* 5:155–166.

Chien, S., and L. A. Sung. 1987. "Physicochemical Basis and Clinical Implications of Red Cell Aggregation." *Clinical Hemorheology* 7:71–91.

Chien, S., S. Simchon, R. E. Abbott, and K. M. Jan. 1977. "Surface Adsorption of Dextrans on Human Red Cell Membrane." *Journal of Colloid and Interface Science* 62:461–470.

Cowell, C., R. Li-In-On, and B. Vincent. 1978. "Reversible Flocculation of Sterically-Stabilized Dispersions." *Journal of the Chemical Society-Faraday Transactions I* 74:337–347.

Donath, E., A. Budde, E. Knippel, and H. Bäumler. 1996. "'Hairy Surface Layer' Concept of Electrophoresis Combined with Local Fixed Surface Charge Density Isotherms: Application to Human Erythrocyte Electrophoretic Fingerprinting." *Langmuir* 12:4832–4839.

Donath, E., A. Krabi, M. Nirschl, V. M. Shilov, M. I. Zharkikh, and B. Vincent. 1997. "Stokes Friction Coefficient of Spherical Particles in the Presence of Polymer Depletion Layers: Analytical and Numerical Calculations, Comparison with Experimental Data." *Journal of the Chemical Society-Faraday Transactions* 93:115–119.

Donath, E., P. Kuzmin, A. Krabi, and A. Voigt. 1993. "Electrokinetics of Structured Interfaces with Polymer Depletion-A Theoretical-Study." *Colloid and Polymer Science* 271:930–939.

Donath, E., L. Pratsch, H. Bäumler, A. Voigt, and M. Taeger. 1989. "Macromolecule Depletion at Membranes." *Studia Biophysica* 130:117–122.

Donath, E., and A. Voigt. 1985. "Influence of Surface Structure on Cell Electrophoresis." In *Cell Electrophoresis*, ed. W. Schütt and H. Klinkmann. Berlin, NY: De Gruyter; 123–135.

Feign, R. I., and D. H. Napper. 1980. "Depletion Stabilization and Depletion Flocculation." *Journal of Colloid and Interface Science* 75:525–541.

Fleer, G. J., M. A. Cohen Stuart, J. H. M. H. Scheutjens, T. Cosgrove, and B. Vincent. 1993. *Polymers at Interfaces*. London: Chapman & Hall.

Fleer, G. J., J. H. M. H. Scheutjens, and B. Vincent. 1984. "The Stability of Dispersions of Hard Spherical Particles in the Presence of Nonadsorbing Polymer." In *Polymer Adsorption and Dispersion Stability*, ed. E. D. Goddard and B. Vincent. Washington, DC: ACS; 245–263.

Flory, P.J. 1953. *Principles of Polymer Chemistry*. Ithaca, NY: Cornell University Press.

Janzen, J., and D. E. Brooks. 1989. "Do Plasma Proteins Adsorb to Red Cells?" *Clinical Hemorheology.* 9:695–714.

Janzen, J., and D. E. Brooks. 1991. "A Critical Reevaluation of The Nonspecific Adsorption of Plasma Proteins and Dextrans to Erythrocytes and the Role of These in Rouleaux Formation." In *Interfacial Phenomena in Biological Systems*, ed. M. Bender. New York: Marcel Dekker; 193–250.

Jenkins, P., and M. Snowden. 1996. "Depletion Flocculation in Colloidal Dispersions." *Advances in Colloid and Interface Science* 68:57–96.

Jenkins, P., and B. Vincent. 1996. "Depletion Flocculation of Nonaqueous Dispersions Containing Binary Mixtures of Nonadsorbing Polymers. Evidence for Nonequilibrium Effects." *Langmuir* 12:3107–3113.

Joanny, J. F., L. Leibler, and P. G. Degennes. 1979. "Effects of Polymer-Solutions on Colloid Stability." *Journal of Polymer Science Part B-Polymer Physics* 17:1073–1084.

Jones, A., and B. Vincent. 1989. "Depletion Flocculation in Dispersions of Sterically-Stabilized Particles. 2. Modifications to Theory and Further Studies." *Colloids and Surfaces* 42:113–138.

Krabi, A., and E. Donath. 1994. "Polymer Depletion Layers as Measured by Electrophoresis." *Colloids and Surfaces a-Physicochemical and Engineering Aspects* 92:175–182.

Lerche, D. 1984. "Electrostatic Fixed Charge Distribution in the RBC-Glycocalyx and Their Influence upon the Total Free Interaction Energy." *Biorheology* 21:477–92.

Levine, S., M. Levine, K. A. Sharp, and D. E. Brooks. 1983. "Theory of the Electrokinetic Behavior of Human Erythrocytes." *Biophysical Journal* 42:127–135.

Linderkamp, O., P. Ozanne, P. Y. Wu, and H. J. Meiselman. 1984. "Red Blood Cell Aggregation in Preterm and Term Neonates and Adults." *Pediatric Research* 18:1356–1360.

Lominadze, D., and W. L. Dean. 2002. "Involvement of Fibrinogen Specific Binding in Erythrocyte Aggregation." *Febs Letters* 517:41–44.

Nash, G. B., R. B. Wenby, S. O. Sowemimo-Coker, and H. J. Meiselman. 1987. "Influence of Cellular Properties on Red Cell Aggregation." *Clinical Hemorheology and Microcirculation* 7:93–108.

Neu, B., and H. J. Meiselman. 2001. "Sedimentation and Electrophoretic Mobility Behavior of Human Red Blood Cells in Various Dextran Solutions." *Langmuir* 17:7973–7975.

Neu, B., and H. J. Meiselman. 2002. "Depletion-Mediated Red Blood Cell Aggregation in Polymer Solutions." *Biophysical Journal* 83:2482–2490.

Neu, B., and H. J. Meiselman. 2008. "The Role of Macromolecules in Stabilization and De-Stabilization of Biofluids." In *Bioengineering in Cell and Tissue Research*, ed. by G. M. Artmann and S. Chien. New York: Springer; 387–408.

Neu, B., S. Sowemimo-Coker, and H. Meiselman. 2003. "Cell-Cell Affinity of Senescent Human Erythrocytes." *Biophysical Journal* 85:75–84.

Neu, B., R. Wenby, and H. J. Meiselman. 2008. "Effects of Dextran Molecular Weight on Red Blood Cell Aggregation." *Biophysical Journal* 95:3059–3065.

Pribush, A., D. Zilberman-Kravits, and N. Meyerstein. 2007. "The Mechanism of the Dextran-Induced Red Blood Cell Aggregation." *European Biophysics Journal with Biophysics Letters* 36:85–94.

Rad, S., and B. Neu. 2009. "Impact of Cellular Properties on Red Cell–Red Cell Affinity in Plasma-Like Suspensions." *European Physical Journal E* 30:135–140.

Rampling, M. W., H. J. Meiselman, B. Neu, and O. K. Baskurt. 2004. "Influence of Cell-Specific Factors on Red Blood Cell Aggregation." *Biorheology* 41:91–112.

Seaman, G. V. F. 1975. "Electrokinetic Behavior of Red Cells." In *The Red Blood Cell*, ed. D. M. Surgenor. New York: Academic Press; 1135–1229.

Snabre, P., and P. Mills. 1985. "Effect of Dextran Polymer on Glycocalyx Structure and Cell Electrophoretic Mobility." *Colloid and Polymer Science* 263:494–500.

Vincent, B. 1990. "The Calculation of Depletion Layer Thickness as a Function of Bulk Polymer Concentration." *Colloids and Surfaces* 50:241–249.

Vincent, B., J. Edwards, S. Emmett, and A. Jones. 1986. "Depletion Flocculation in Dispersions of Sterically-Stabilised Particles ('soft spheres')." *Colloids and Surfaces* 18:261–281.

4 Measurement of Red Blood Cell Aggregation

Red blood cell (RBC) aggregation is a property of cells suspended in aggregating media, and is affected by both suspending phase and cellular properties (see Chapter 2). It should be noted that the suspending phase (i.e., plasma) and RBC properties may be altered under the influence of physiological and pathological processes as discussed in Chapter 8. Prominent changes in RBC aggregation can be detected in various diseases, including inflammatory states, vascular pathologies, and tissue perfusion problems inducing damage to RBC. Such alterations are reflected by various aspects of RBC aggregation. More specifically, these include: (1) the extent of aggregation (i.e., number of RBC in each aggregate), (2) the time course of aggregation (i.e., the rate of rouleaux formation), and (3) the magnitude of forces holding RBC together in aggregates.

Quantification of such aspects of RBC aggregation has been shown to be important from a clinical point of view as discussed in Chapter 8, and has resulted in the development of various methods and instruments to quantitate RBC aggregation. This chapter describes these methods in some detail and includes sections providing practical information to help laboratory practice related to the measurement of RBC aggregation and interpretation of the resulting data.

4.1 METHODS FOR QUANTIFICATION OF RED BLOOD CELL AGGREGATION

4.1.1 Erythrocyte Sedimentation Rate

Observing the *phase separation* of blood has been one of the oldest diagnostic methods and was even used in ancient Greece (Fåhraeus 1929). The measurement of erythrocyte sedimentation rate (ESR) in its modern sense was described by the Polish physician Edmund Faustyn Biernacki in 1897 (Biernacki 1897). He recognized that the rate of the separation of RBC from plasma was increased in blood of patients suffering from various diseases when compared to the rate for healthy people, with the magnitude of the increase determined by the severity of disease. He also pointed out that this increased sedimentation of RBC was due to increased concentrations of fibrinogen in plasma; the test is still known as Biernacki's Reaction in some countries. The sedimentation testing procedure was further developed by Robin Fåhraeus (Fåhraeus 1921) and by Alf Vilhelm Albertsson Westergren (Westergren 1921), resulting in the method currently being used in most clinical laboratories.

The ESR is generally accepted as a nonspecific indicator of inflammation. It is a very frequently requested laboratory test, mostly as a screening method. The results of the test are interpreted in terms of the presence and severity of inflammation or acute phase reaction, but are rarely associated with RBC aggregation by most physicians.

4.1.1.1 Measurement Procedure

ESR is measured using anticoagulated blood in standard glass tubes. Two separate methods have been utilized (i.e., Wintrobe and Westergren methods), with the differences between them related to the anticoagulation of the blood sample and the tubes used for the test (Raphael 1983).

4.1.1.1.1 Wintrobe Method

The blood sample is anticoagulated by ethylenediaminetetraacetic acid (EDTA, 1.5 mg/ml). A 10-cm long Wintrobe hematocrit tube is filled with 1 ml of anticoagulated blood, and the tube is placed vertically in a tube support. The height of the plasma column that is free of RBC is read after 1 h.

4.1.1.1.2 Westergren Method

The blood sample is anticoagulated by diluting it with 3% trisodium citrate at a ratio of 1/5 (1 volume of citrate, 4 volumes of blood). The Westergren tube, which is 300 mm long and has an internal diameter of 2.55 mm, is filled with blood and positioned vertically. The height of the plasma column that is free of RBC is read after 1 h of settling. This method requires a larger amount of blood since it uses a larger tube compared to the Wintrobe method. The test can also be performed using EDTA anticoagulated blood by dilution before loading into the Westergren tube.

The Westergren method has been widely used in many laboratories, and it has been accepted as a valid method by The International Committee for Standardization in Haematology (Bull et al. 1989; International Committee for Standardization in Haematology 1977). However, in 1993 this committee also recommended an alternative method of measuring ESR in undiluted blood samples, anticoagulated by EDTA, using the standard Westergren tubes (International Committee for Standardization in Haematology Expert Panel on Blood Rheology 1993). This method is known to be more sensitive to changes due to disease processes.

Measurement of ESR in clinical pathology laboratories has improved during the last several decades. An important step was the development of special disposable vacuum tubes (Seditainer®) that can be used for blood sampling into the correct amount of citrate (e.g., solution) and then placed vertically into specially designed tube racks (Figure 4.1). This product has been helpful in preventing possible contamination of personnel with blood during laboratory practice and significantly simplified the process. It has also improved the precision of the test owing to the accurate dilution of the blood samples.

Due to differences in tube geometry, the results of the ESR measurement using the special vacuum tubes are slightly different from the results obtained using Westergren tubes, (Bridgen and Page 1993; Patton et al. 1989). However, the values obtained by the two methods are strongly correlated (Figure 4.2) and a regression equation can be used to calculate the corresponding Westergren value using the sedimentation rate measured by the Seditainer method (Seditainer Product Sheet 1996).

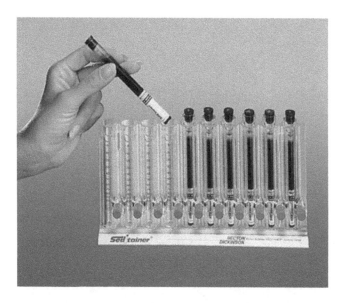

FIGURE 4.1 Vacuum tubes of 5 ml capacity containing 1.25 ml of 0.105 M buffered sodium citrate which can be inserted into the specially designed rack to measure erythrocyte sedimentation in 1 h after blood sampling into the tube. Reproduced with permission from Becton Dickinson Co. 1996. *Seditainer® Product Sheet*. Franklin Lakes, NJ: Becton Dickinson Co.

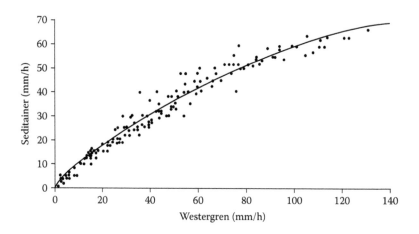

FIGURE 4.2 The relationship between erythrocyte sedimentation rate values measured by Westergren and Seditainer methods. This relationship is $y = 0.87x - 0.0027x^2$, where $x =$ Westergren value (mm/hr) and $y =$ Seditainer value for the same sample. The correlation coefficient was reported to be 0.998 in a study using 300 blood samples. Reproduced with permission from Becton Dickinson Co. 1996. *Seditainer® Product Sheet*. Franklin Lakes, NJ: Becton Dickinson Co.

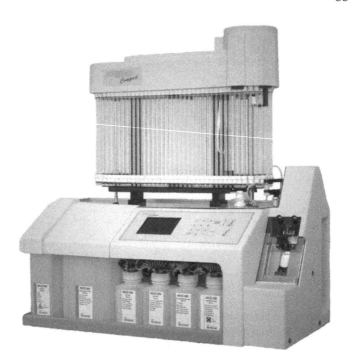

FIGURE 4.3 An automated instrument for the measurement of erythrocyte sedimentation rate (Starrsed®, Courtesy of Mechatronics, Hoorn, The Netherlands).

Automated systems for the measurement of ESR using the Westergren method are also available from various companies (Figure 4.3). These systems use undiluted, anticoagulated samples and make the citrate dilution prior to automatically loading the sample to the glass tube. The plasma–RBC column interface is detected by measuring optical density along the tube.

4.1.1.2 Mechanism of Red Blood Cell Sedimentation in Plasma

Sedimentation of particles in a suspension is explained by Stoke's Law (Lamb 1994), and these principles also apply to RBC in plasma or other suspending media. The mechanism of RBC sedimentation has been theoretically analyzed by various investigators (Mayer et al. 1992; Oka 1985). RBC tend to sediment in plasma under the influence of gravity because their density is greater than that of plasma (Phear 1957; Trudnowski and Rico 1974). The sedimentation rate of a particle (e.g., an RBC) is determined by the balance between its effective weight, which is related to its mass and buoyancy produced by the suspending medium (e.g., plasma), and the viscous fluid drag on the particle. This opposing force due to viscous drag is proportional to the surface area of the particle and thus the square of its radius. Since the difference between the densities of plasma and RBC is not large, sedimentation is very slow if the cells remain separated from each other. However, RBC in plasma aggregate and the surface area of the resulting rouleaux are smaller than the sum of the areas of the individual RBC in the aggregate (Raphael 1983). Therefore, aggregation of RBC enhances their sedimentation rate. Hence, the length of the RBC-free plasma column in 1 h, which reflects the rate of RBC sedimentation, indicates the extent of RBC aggregation (i.e., the degree of the change in particle surface area due to rouleaux formation). Dilution of blood with 3% sodium citrate in the Westergren method has a dual effect on the factors discussed here: (a) plasma density and viscosity are reduced by dilution thus favoring sedimentation, and (b) RBC aggregation is reduced due to the lower level of plasma factors promoting aggregation.

RBC aggregation is a fast process and is mostly completed in several minutes following filling of the sedimentation tube (phase 1). This process is not visible by the naked eye and does not result in an obvious separation of plasma during this very early period. However, the extent of RBC aggregation during the first phase (i.e., ~2 min after filling the tube) determines the rate of sedimentation in the second phase of the test, which may continue for 2 h after the initial phase (phase 2). This period is followed by a slower phase of sedimentation (phase 3) due to the compaction of RBC aggregates (Raphael 1983). The second phase of fast sedimentation has been accepted as the most important phase, but it should be remembered that the result obtained in this second phase is determined by the extent of RBC aggregation in phase 1.

4.1.1.3   Normal Ranges of Erythrocyte Sedimentation Rate

The normal range of ESR depends on gender, and is higher in females (Piva et al. 2001; Rampling 1988). The upper limit of ESR measured by the standard Westergren method is accepted to be 10 mm per h for men and 12 mm per h for women between 17 and 50 years of age (Rampling 1988). These values should be accepted as corresponding to the physiological rouleaux formation; ESR increases with aging in control populations (Miller et al. 1983). However, various factors may significantly affect ESR values and these should be considered to obtain an accurate estimation of RBC aggregation based on ESR.

4.1.1.4   Factors Affecting Erythrocyte Sedimentation Rate

ESR has been related to RBC indices such as mean cell volume (MCV) and mean corpuscular hemoglobin concentration (MCHC), mainly based on the expected effect of the cell size on buoyancy and cell density (see Section 4.1.1.2). Phear (1957) reported a significant correlation of Westergren ESR with MCV ($r = 0.61, p < 0.001$) but not with MCHC.

One of the most important factors affecting ESR values is the hematocrit of the sample. This effect not only reflects the influence of hematocrit on the aggregation process (Chapter 2, Section 2.1.2), but also includes the changes in the process of sedimentation during phase 2, which are not directly related to the rate or extent of aggregation. In brief, there is less interaction between aggregates (i.e., less hindered settling) if there are less RBC relative to plasma. This results in an inverse relationship between ESR and hematocrit, leading to an increase of ESR in samples with low hematocrit (Figure 4.4) (Poole and Summers 1952).

Various investigators have attempted to develop a mathematical approach to "correct" ESR values measured in samples with different hematocrit values (Hynes and Whitby 1938; Wintrobe and Landsberg 1935). Such correction methods work

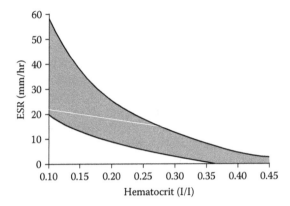

FIGURE 4.4 Dependence of erythrocyte sedimentation rate (ESR) on hematocrit. The two curves represent the expected range for normal RBC in autologous plasma. Redrawn using the data presented by Poole, J. C. F., and G. A. C. Summers. 1952. "Correction of ESR in Anaemia. Experimental Study Based on Interchange of Cells and Plasma Between Normal Anaemic Subjects." *British Medical Journal* 1: 353–356.

well with measurements using various dilutions of a given sample with autologous plasma. However, the equation constants characterizing this relationship change from sample to sample, especially in pathological blood samples (Bull and Brecher 1974), and therefore are not very useful in developing a mathematical method for hematocrit correction (Rampling 1988). More sophisticated methods for correcting ESR according to the hematocrit of the sample were proposed based on the continuous monitoring of the sedimentation process during the 1-h period (Rourke and Ernstene 1930); this approach is not used widely, most probably due to the complexity of data collection. There are newer mathematical approaches proposed for ESR corrections using the hematocrit and other RBC-related properties (e.g., hemoglobin concentration) (Borawski and Mysliwiec 2001; Pawlotsky et al. 2004), but none are widely accepted. Therefore, adjustment of sample hematocrit to a standard value (e.g., 0.4 l/l) is strongly recommended if the ESR measurement is being used as a method to estimate RBC aggregation.

The above discussion of the effect of hematocrit on ESR may also indicate that the sensitivity of the test is increased if lower hematocrit samples are used (Stuart 1991). This increased sensitivity might be an advantage, especially when using ESR to estimate RBC aggregation under certain special, mostly experimental, conditions. For example, ESR measurements in samples diluted with autologous plasma to obtain 0.1 l/l hematocrit suspensions exhibit highly increased and accelerated ESR. This approach can be used to estimate and compare RBC aggregation in blood samples from experimental animals (e.g., rat, guinea pig), which have very low aggregation, and therefore have nonmeasurable ESR at native hematocrit values (Yalcin et al. 2004).

While ESR measurements are usually performed at ambient temperature, a strong dependence of ESR on temperature was reported (Manley 1957). Manley (1957) also proposed a nomogram for the correction of ESR for the ambient temperature. The measurement temperature should be considered as an important factor significantly influencing RBC aggregation (Baskurt and Mat 2000). It is therefore important to report the ambient temperature at which the measurements were performed, especially if it is not within the usually accepted limits (e.g., 20 ± 2°C).

4.1.1.5 Other Approaches to the Measurement of Erythrocyte Sedimentation Rate

It has been suggested that monitoring the time course of the sedimentation process may provide more information compared to the single measurement after a given period of time (e.g., 1 h) (Ismailov et al. 2005; Pawlotsky et al. 2004). Pawlotsky et al. introduced the term Σ ESR for the sum of the plasma column lengths measured at 20, 30, 40, 50, and 60 min that are corrected for hematocrit and hemoglobin measured in the same sample (Pawlotsky et al. 2004). They claimed that Σ ESR is a more reliable indicator of inflammation compared to the standard Westergren method.

The length of the measurement period (i.e., 1 h for the Westergren method) is one of the important drawbacks of the ESR measurement. Various attempts have been developed to estimate ESR in shorter periods. ESR is known to be enhanced in an inclined tube and this enhancement, known as the Boycott effect (Boycott 1920), may reduce the time required for a "readable" sedimentation. Detailed analysis of the influence of tube inclination on ESR can be found elsewhere (Dobashi et al. 1994).

Alexy et al. (2009) proposed the calculation of a *sedimentation index* based on the difference between RBC column heights in one vertical tube of a U-shaped blood viscometer at 3 min following loading of the sample. Optoelectronic level sensors positioned adjacent to the tube of ~4 mm internal diameter were used to monitor the plasma–RBC interface. They reported a highly significant second-order polynomial regression between Westergren ESR values and the calculated index (Alexy et al. 2009).

An instrument has recently been introduced that is stated to measure ESR within a few minutes. This new method, called Test-1, is termed "a fully automated analyzer for the measurement of ESR" (Plebani et al. 1998, p. 334) using "photometrical capillary stopped flow kinetic analysis." However, the description of the measurement method indicates that this instrument is actually a photometric RBC aggregometer (see Section 4.1.6.9.5). There are numerous publications in the literature proving the very significant correlation between the ESR values generated by using regression models based on measured RBC aggregation indices and those directly measured by the Westergren method (Cha et al. 2009; Giavarina et al. 2006; Hardeman et al. 2010a; Ozdem et al. 2006; Romero et al. 2003). Hardeman et al. (2010a) reported a correlation coefficient of 0.9 when comparing the results of Test-1 versus the Westergren method done by an automated instrument for samples from 680 patients with various rheumatoid disorders. However, they mentioned that Test-1 results deviate from Westergren results in samples with high sedimentation rates (Hardeman et al. 2010a). It does not seem to be appropriate to say that this instrument is "measuring ESR," as it is not a direct measurement of sedimentation.

4.1.1.6   Erythrocyte Sedimentation Rate as a
Method to Estimate Aggregation

It follows from the previous discussion that ESR measured in samples with hematocrit adjusted to a given value can be used to estimate RBC aggregation (Potron
et al. 1994). Significant correlations have been reported between ESR and RBC
aggregation quantitated using various instruments built to measure RBC aggregation (Baskurt et al. 2009c; Rampling and Martin 1989). This approach has been
proven to be especially useful with pathological or experimentally modified blood
samples in which other methods fail to work, mainly due to highly enhanced aggregate strength and insufficient disaggregation mechanisms (see Section 4.1.6.8.2)
(Yalcin et al. 2004).

However, in most cases, ESR is not a method of choice to assess RBC aggregation
as a clinical laboratory practice or in research laboratories. The major reasons for
this statement are listed below:

a. ESR measurement is time consuming, usually requiring at least 1 h.
b. The sensitivity of this method may not be satisfactory simply due to the
   technical difficulties in detecting the differences between sedimentation
   levels in nonpathological samples. It may not be equally sensitive to the
   alterations of various factors determining the rate and extent of aggregation, at least with standard ESR procedures. Rampling reported that ESR
   measured in 1 h failed to correlate with plasma fibrinogen, while an 18-h
   ESR did coordinate for fibrinogen concentrations between 100 and 1000
   mg/dl (Rampling et al. 1984).
c. ESR has a strong hematocrit dependence. A prominent increase of ESR
   might be the result of anemia (i.e., decreased hematocrit) or an acute phase
   reaction, and these conditions cannot be differentiated by evaluating only
   ESR.
d. ESR may provide an estimate of the extent of RBC aggregation but does
   not yield data relevant to the time course of this process. RBC aggregation
   has a time course that is mostly completed within several minutes (Section
   4.1.6.2.2). This phase of the ESR measurement is not detectable but the
   extent of aggregation (i.e., the size of aggregates in terms of the number of
   RBC per aggregate) determines the rate of settling in the second phase (as
   discussed in Section 4.1.1.2). Recording the time course of ESR over 1 h
   does not provide any information about the time course of RBC aggregation
   within the early phase (i.e., phase 1).

4.1.2   Zeta Sedimentation Ratio

The *zeta sedimentation ratio* (ZSR) is a parameter determined by a special procedure of centrifuging blood samples and was developed to deal with the drawbacks
of ESR, especially the length of the time required for the measurement of ESR and
hematocrit dependence of the results. Bull and Brailsford introduced this method in
early 1970s (Bull and Brailsford 1972).

The method was suggested to be sensitive to the zeta potential of RBC in suspension, which explains its name. It was argued that fibrinogen and gamma globulins in plasma decrease the zeta potential (i.e., the electrical potential resulting from the negative charges on the surface of RBC) and increase the rouleaux formation accordingly (Bull and Brailsford 1972). Although this explanation is not totally satisfactory according to our current understanding of the RBC aggregation mechanisms (see Chapter 3), the endpoints of the increased asymmetrical macromolecules in the suspending phase of RBC (i.e., enhanced aggregation with increased macromolecular concentration) are in agreement.

4.1.2.1 Measurement Procedure

Blood is introduced into a special glass tube (internal diameter = 2 mm, length = 75 mm and thus a larger diameter than a standard microhematocrit tube) and centrifuged in a near-vertical position at very low g (7–8 g) using a specially designed centrifugal instrument called a Zetafuge® (Figure 4.5). The first phase of centrifugation continues for 45 s, during which RBC move toward the outer edge of the tube and rouleaux formation starts. The Zetafuge head stops after 45 s, the direction of rotation is reversed, the tube is rotated 180 degrees about its long axis, and the next 45 s cycle starts automatically. There are 4 cycles in a complete run, with direction of rotation changing every 45 s. This procedure is thus an accelerated sedimentation process related to the low g force acting on the RBC rouleaux; the RBC follow a

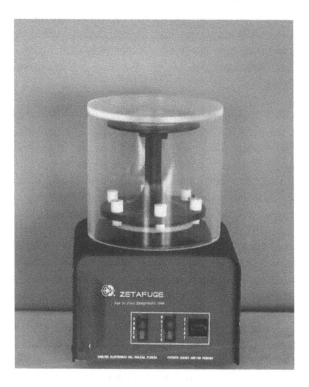

FIGURE 4.5 Zetafuge® (Coulter Electronics).

zigzag path during the four 45-s periods. After centrifugation is completed, the ratio of the RBC column length to the initial length of the blood column is determined and is termed the *zetacrit*. The same tube is then centrifuged at high *g* using a standard hematocrit centrifuge, and the true hematocrit is determined. Finally, the ZSR is calculated as the ratio of hematocrit to the zetacrit, with the results expressed as a fraction or percentage.

The Zetafuge is not a temperature-controlled instrument, and therefore the measurements are performed at ambient temperature as in case of the ESR.

4.1.2.2   Normal Ranges of Zeta Sedimentation Ratio

The ZSR was found to be 41–54% in blood samples from normal humans (Morris et al. 1975) with the normal range similar for male and female subjects (Bull and Brailsford 1972; Saleem et al. 1977); ratios from 55–65% are accepted to represent enhanced RBC aggregation (Bull and Brailsford 1972). Bull and Brailsford (1972) also demonstrated that the ZSR is linearly related to the suspending phase components inducing RBC aggregation (e.g., fibrinogen). Furthermore, ZSR values were not influenced by the hematocrit of the sample unlike the very significant dependence of ESR measured by the Westergren method (Figure 4.6).

4.1.2.3   Correlation of Zeta Sedimentation Ratio with Erythrocyte Sedimentation Rate

ZSR values have been shown to significantly correlate with ESR measured in the same samples using the Wintrobe method; the correlation coefficients were 0.88 and 0.92 for female and male human subjects, respectively (Raich and Temperly 1976). A good correlation between ZSR and the Westergren method was also reported (Saleem et al. 1977).

4.1.2.4   Zeta Sedimentation Ratio as a Method to Estimate Aggregation

The method to measure ZSR was developed as an alternative to the measurement of ESR, and has the advantage of being faster and insensitive to hematocrit differences. The ZSR never became a widely used method in clinical pathology laboratories as a replacement for the Westergren method for monitoring acute phase reactions. However, it has been used in a limited number of clinical studies as an alternative method to monitor the acute phase reaction (Bennish et al. 1984; Morris et al. 1977). ZSR has also been used for more specific approaches to RBC aggregation (Baskurt et al. 1997b; Baskurt et al. 1997a; Dadgostar et al. 2006; Ozanne et al. 1983; Stoeff et al. 2008). These studies included samples from various patients and from various species covering a wide range of aggregation properties (Baskurt et al. 1997a). The ZSR was also useful in detecting the differences in RBC aggregation in a standard suspending medium containing high-molecular-mass dextrans as the aggregating agent (Baskurt et al. 1997b; Baskurt et al. 1997a), thereby indicating its usefulness in studying RBC aggregability (see Section 4.3). The ZSR was successfully used to demonstrate the alterations in RBC aggregation by covalent binding of polyethylene glycol to the membrane surface (Jovtchev et al. 2008).

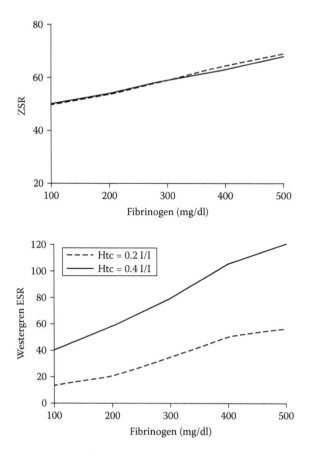

FIGURE 4.6   Effect of fibrinogen concentration in the suspending phase and hematocrit on the zeta sedimentation ratio (ZSR) and Westergren erythrocyte sedimentation rate (ESR). In both panels the dashed line is for 0.2 l/l hematocrit and the solid line for 0.4 l/l. Redrawn using the data by Bull, B. S., and J. D. Brailsford. 1972. "The Zeta Sedimentation Ratio." *Blood* 40:550–559.

In overview, the ZSR seems as useful as ESR for estimating RBC aggregation. The limitations of both methods include determining only an overall extent of RBC aggregation but not the time course of the process as discussed for ESR (Section 4.1.1.6).

4.1.3   LOW-SHEAR VISCOMETRY

RBC aggregation is the main determinant of blood viscosity at low shear rates (see Chapter 5) and hence viscometry at appropriately low shear rates has been used as a method to assess the aggregation properties of RBC suspensions. This approach has been widely used, especially during the early days of hemorheology, before the specific methods for measuring RBC aggregation were developed.

Various approaches have been proposed to calculate aggregation indexes (AIs) for RBC aggregation based on blood viscometry:

a. The difference between low-shear viscosity (η_L) and high-shear viscosity η_H) scaled by high-shear viscosity (Bull et al. 1986; Bull et al. 1989) mathematically expressed as

$$AI = \left(\eta_L - \eta_H\right)/\eta_H \qquad (4.1)$$

b. The ratio of relative apparent viscosities in aggregating (η_{rA}) and nonaggregating (η_{rNA}) suspending media at the same hematocrit measured at low-shear rate (Brooks et al. 1974)

$$AI = \eta_{rA}/\eta_{rNA} \qquad (4.2)$$

4.1.3.1 Measurement Procedure

Viscosity measurements should be performed using an instrument capable of measurements at various shear rates covering the *low* range suitable for detecting the influence of RBC aggregation. This range refers to shear rates below 1 s^{-1} (Rampling 1988). Some widely used viscometers (e.g., Wells-Brookfield cone-plate viscometer) cannot accurately measure blood viscosity in this range and therefore cannot be used to quantitate RBC aggregation as described previously. In addition to rotational viscometers capable of measurements in the low shear rate range (e.g., Contraves LS series), newly developed scanning capillary viscometer is also suitable for low shear rate measurements with sufficient sensitivity and precision (Alexy et al. 2005).

The technique of viscometry is described in detail in Chapter 5. These procedures are also critically discussed in the guidelines prepared by the Expert Panel on Blood Rheology of the International Committee for Standardization in Haematology in 1986 (Bull et al. 1986) and recently revised along with the current technical developments (Baskurt et al. 2009a). The reader is referred to these documents for the details and precautions related to the blood viscosity measurement. Measurements at a temperature of 37°C are strongly recommended.

The method of Brooks et al. (1974) requires the measurement of low shear rate viscosity in both aggregating medium (e.g., plasma) and nonaggregating medium. The latter can be isotonic phosphate-buffered saline (PBS) at appropriate pH (i.e., 7.4). RBC should be separated from autologous plasma by gentle centrifugation (e.g., 400 *g*, for 5 min), washed once with PBS, and resuspended in the same. The hematocrit of both aggregating and nonaggregating RBC suspensions (i.e., suspensions in autologous plasma and PBS) must be adjusted to a selected hematocrit (e.g., 0.4 l/l) by removing or adding autologous plasma after centrifugation (Baskurt et al. 2009a) (also see Section 4.2.2). This method also requires the measurement of the viscosity of the suspending media (e.g., plasma and PBS) to calculate the relative viscosities (i.e., the ratio of suspension viscosity to the suspending phase viscosity). This can be done using the same viscometer used for measuring suspension viscosities at the same temperature. As detailed in Chapter 5, the possibility of artifacts due to surface films should be considered for plasma viscosity measurements using rotational viscometers (Baskurt et al. 2009a). Alternatively, suitable capillary viscometers can be used for measuring the viscosity of these Newtonian fluids.

The exact shear rates used as low and high for measuring the viscosity values for the previously mentioned calculations should be reported together with the aggregation indexes.

4.1.3.2 Alternative Methods

Kawakami et al. developed a method of estimating RBC aggregation using a *damped-oscillation rheometer*. The device consists of a glass tube of 10 mm diameter filled with the sample. The damping oscillations following a rotational displacement were recorded and analyzed to calculate the *logarithmic damping factor*, which was reported to correlate strongly with ESR values (Kawakami et al. 1994).

Lee et al. (2007) estimated *yield shear stress* using data obtained by a low-shear viscometer (Contraves LS30) at shear rates between 0.51 and 27.7 s^{-1}. The calculation was done by fitting the viscosity data to the Casson Equation:

$$\tau^{1/2} = K\gamma^{1/2} + \tau_y^{1/2} \tag{4.3}$$

where τ and γ are corresponding shear stress and shear rate, and τ_y is yield shear stress. Yield shear stress was reported to significantly correlate with ESR and the aggregation parameters measured using a Myrenne aggregometer (Lee et al. 2007) (See Section 4.1.6.9.1).

4.1.3.3 Limitations of Low-Shear Viscometry to Quantitate Red Blood Cell Aggregation

Previous studies have clearly shown that aggregation indexes based on low-shear viscometry calculations correlate well with RBC aggregation quantitated using other methods (Brooks et al. 1974; Rampling and Martin 1989). This correlation is especially valid in suspensions containing RBC that are normal in terms of morphology and deformability. Alterations in RBC aggregation due to changes in the suspending phase composition and/or cellular factors not affecting RBC morphology and deformability (e.g., surface properties) can then be accurately detected by this approach. However, low shear rate viscometry and aggregation indexes based on this approach may, under some conditions, fail to correctly indicate alterations in RBC aggregation (Baskurt and Meiselman 1997; Lacombe and Lelievre 1987).

It is well known that RBC deformability is a determinant of the degree of RBC aggregation (see Chapter 2, Section 2.3.5.2). Therefore, reduced RBC deformability is expected to result in decreased aggregation tendency. Glutaraldehyde (GA) treatment at very low concentrations was used as an experimental method to modify RBC deformability, resulting in a concentration-dependent decrease of RBC deformability (Figure 4.7a). RBC aggregation was also influenced by this treatment, with

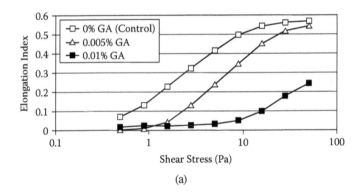

(a)

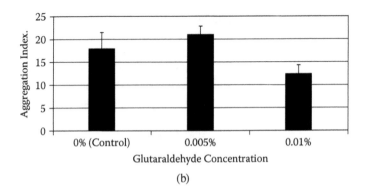

(b)

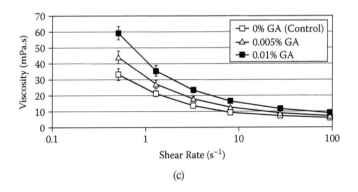

(c)

FIGURE 4.7 Effect of glutaraldehyde (GA) treatment at 0.005% and 0.01% concentrations on RBC elongation indexes (a) aggregation indexes (b) and viscosity values measured between 0.512 and 94.5 s⁻¹ using a Contraves LS-30 viscometer (c). See Section 4.1.3.3 for discussion. Redrawn using data from Baskurt, O. K., and H. J. Meiselman. 1997. "Cellular Determinants of Low-Shear Blood Viscosity." *Biorheology* 34:235–247.

aggregation indexes measured by a photometric method (see Section 4.1.6) being significantly decreased for RBC treated with 0.01% GA and resuspended in autologous plasma (Figure 4.7b). Given the decrease of RBC aggregation for these GA-treated RBC, low shear rate viscosity should also be expected to decrease due to decreased RBC aggregation. However, Figure 4.7c indicates that viscosities measured at 0.51 s^{-1} shear rate were increased in both GA-treated suspensions compared to control while viscosities at high shear are very similar. Baskurt and Meiselman (1997) also studied the influence of other agents modifying RBC mechanical properties and morphology, such as treatment with low-dose hydrogen peroxide, and demonstrated the failure of low-shear viscometry in detecting the resulting alterations in RBC aggregation.

The previous interpretation problem is related to the general influence of impaired RBC deformability on the viscosity of RBC suspensions regardless of the shear rate used for measurement. However, for only slight decreases of deformability, the cells do deform at high shear and thus suspension viscosities are only slightly modified. Accordingly, the merit of aggregation indexes calculated using viscosity values measured at low and high shear rates can be affected (Baskurt and Meiselman 1997), and thus low-shear viscosity values and aggregation indexes based on such values should be used with this weakness in mind.

4.1.4 MICROSCOPIC AGGREGATION INDEX

Direct observation of anticoagulated blood samples, either microscopically or by the naked eye, have served as a method to detect and estimate the degree of RBC aggregation since the nineteenth century (Fåhraeus 1929). A simple test recommended by a Dr. Jones in 1843 was to observe the "dotted appearance" of a drop of patient's blood taken from a fingertip and pressed between two pieces of glass plates, as summarized by Robin Fåhraeus (1929). Apparently, the "distinctness" of the dotted appearance was accepted as an indicator of the inflammatory status, which could also be monitored by the degree of buffy coat development. Similar approaches have also been described and used more recently, such as the grading system ranging from 0 to 4 based on the shape and size of RBC rouleaux observed under the microscope (Engeset et al. 1966). The method for quantification of RBC aggregation based on microscopic observations was refined by Chien and Jan (1973) and is done by estimating the average number of RBC per aggregate.

4.1.4.1 Measurement Procedure

Dilute RBC suspensions of 0.01 l/l hematocrit are prepared in the aggregating (e.g., autologous plasma) and nonaggregating (e.g., Ringer-albumin solution, isotonic phosphate-buffered saline) suspending media. The number of cellular units (including single cells and RBC rouleaux of any size) (Figure 4.8) are counted in each suspension using a hemocytometer in a humidified chamber at 37°C after the settling of RBC is complete. Standardization of the waiting time after filling the hemocytometer (e.g., 15 min) is recommended. The cellular units within a given hemocytometer area can be counted in both suspensions by direct observation or by using microphotographs or video images. The cell count in the nonaggregating suspending media

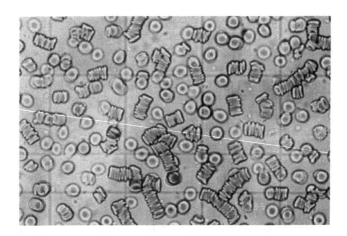

FIGURE 4.8 Cellular units in autologous plasma at 0.01 1/l hematocrit showing single RBC and rouleaux of various sizes. Reproduced with permission from Rampling, M. W. 1988. "Red Cell Aggregation and Yield Stress." In *Clinical Blood Rheology*, ed. G. D. O. Lowe, 45–64. Boca Raton, FL: CRC Press.

provides an estimate of the total number of RBC in the volume corresponding to the microscopic area used for counting. Alternatively, the RBC count in the nonaggregating suspension can be determined using an electronic hematology analyzer. In that case the number of cellular units counted in the hemocytometer area should also be converted to the corresponding units (e.g., RBC/μl) used by the hematology analyzer. The Microscopic Aggregation Index (MAI) is then calculated by dividing the RBC count by the number of cellular units in the aggregating suspension (Chien and Jan 1973). Thus, in the absence of aggregation, the MAI has a value of 1.0 and higher values as aggregation is increased.

4.1.4.2 Usage of Microscopic Aggregation Index in Modern Hemorheology

Given the development of more accurate, sensitive, and automated methods for the quantification of RBC aggregation, the MAI is not a very widely used method in modern hemorheology laboratories. The disadvantages of the method include:

a. The MAI is an estimate of the overall extent of RBC aggregation, as also noted for ESR and ZSR, but does not provide any information about its time course.
b. The manual counting of cellular units can be time-consuming.

However, the method has several features that may be considered advantages:

a. It is a simple procedure and requires no special laboratory equipment, and therefore can be performed in any clinical or research laboratory with basic laboratory skills.
b. The MAI is a direct measure of aggregate size in terms of the number of RBC per aggregate and therefore provides a more realistic estimate of the

extent of rouleaux formation compared to the indirect methods discussed previously (e.g., ESR, ZSR, low-shear viscometry). The MAI correlates well with other indexes of RBC aggregation, and Rampling has reported a good correlation of MAI with fibrinogen concentration (Rampling 1988).

c. The MAI might be useful in estimating the degree of aggregation in nonhuman blood samples. Most laboratory animals (e.g., mouse, rat, guinea pig) have very low levels of RBC aggregation in autologous plasma that cannot be detected with the RBC aggregometers used in hemorheology laboratories (see Section 4.1.6.9). However, RBC aggregation can be quantified by this method in blood samples from such species.

d. The MAI can be used to estimate the degree of aggregation in pathological or experimentally modified RBC suspensions (Baskurt et al. 2004). The accurate measurement of RBC aggregation in photometric aggregometers strongly depends on complete dispersion of aggregates at the start of the measurement (Section 4.1.6.1). Pathological or experimentally modified RBC suspensions are frequently characterized by highly increased forces required for disaggregation, thereby making RBC aggregometers unusable for these samples. The MAI was successfully used to quantitate aggregation in suspensions with enhanced aggregation caused by copolymer coating of the cells (See Chapter 2, Section 2.3.5.4) (Yalcin et al. 2004).

4.1.5 Image Analysis Techniques

Techniques for quantification of aggregation based on the average number of RBC per aggregate using microscopic images were further developed by applying image analysis techniques. This approach has been applied to estimate aggregate size in RBC suspensions following the full-development of RBC aggregation (i.e., at stasis), during flow at various shear rates or during transition from flow to stasis.

4.1.5.1 Image Analysis of Red Blood Cell Aggregates at Stasis

Computerized image analysis techniques have been used to analyze microscopic images of RBC aggregates in blood films. This method can be used as an equivalent of the MAI as discussed in Section 4.1.4 simply by digitally estimating the volume of each RBC aggregate and dividing this by the volume of each RBC, thus providing aggregate size; aggregate morphology can also be investigated. Foresto et al. (2000) used a dilute RBC suspension (i.e., 0.02 l/l hematocrit) and recorded digital images of aggregates within 5 minutes after preparing a film on a slide. An *aggregate shape parameter* (ASP) is then calculated using the projected area (A) and perimeter (P) of each aggregate measured by digital processing:

$$ASP = 4 \cdot \pi \cdot (A / P^2) \tag{4.4}$$

This parameter is maximum for a spherical aggregate (i.e., 1.0 for a perfect sphere) and smaller for a cylindrically shaped aggregate similar to the rouleaux observed in blood samples from healthy individuals. Foresto et al. (2000) reported ASP to be 0.28 ± 0.15 and 0.65 ± 0.18 for normal and diabetic blood samples, respectively.

Mchedlishvili et al. (1993) used a semiautomated method to calculate an index of RBC aggregation as the ratio of the area occupied by aggregated RBC to the total area covered by RBC, whether aggregated or not. They used 1/200 diluted blood and 200-μm-thick blood films to visualize the cells, then employed texture analysis software to selected areas in the images (Mchedlishvili et al. 1993). This method is termed the *Georgian technique* in the literature.

Rotstein et al. (2001) developed an alternative technique for quantitating RBC aggregation using digital image analysis. Their *slide test* is performed using blood samples after 1 to 4 dilution with 3.8% sodium citrate solution. One drop of diluted blood is applied on a microscope slide and held at an angle of 30 degrees to allow blood to flow slowly under the influence of gravity and to form a thin film. This slide is then observed under 200X magnification and digital images recorded and analyzed with specially developed software (INFLAMET™). The parameter proposed for the quantification of aggregation is the *vacuum radius* (Vr), calculated based on the relative areas occupied by RBC aggregates and plasma in the images (Figure 4.9). A more effective compaction due to enhanced aggregation leaves larger areas occupied by plasma only, corresponding to higher Vr values (Rotstein et al. 2001). Vr values measured in blood samples obtained from a large group of normal subjects as well as patients with various diseases including inflammatory conditions correlated significantly with ESR and plasma fibrinogen concentration (Maharshak et al. 2002). A significant correlation was also demonstrated to exist between Vr values and aggregate size estimates obtained under shear in a flow chamber (see Section 4.1.5.2) (Berliner et al. 2004). It has been suggested that this method be used as a rapid screening test for inflammatory reactions of various degrees (Berliner et al. 2005; Urbach et al. 2005) and to serve as an alternative to ESR. The method has

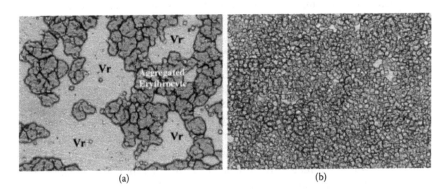

(a)   (b)

FIGURE 4.9 Appearance of RBC in the slide test demonstrating the vacuum radius (Vr). Vr value was 12 for the sample with high ESR (50 mm/h) (a) and 1.2 for the sample with normal ESR (6 mm/h) (b). Reproduced with permission from Rotstein, R., R. Fusman, S. Berliner, D. Levartovsky, O. Rogowsky, S. Cohen, E. Shabtai, et al. 2001. "The Feasibility of Estimating the Erythrocyte Sedimentation Rate within a Few Minutes by Using a Simple Slide Test." *Clinical and Laboratory Haematology* 23:21–25.

been extensively used to evaluate acute phase reactions in a wide variety of clinical disorders (Almog et al. 2005; Fusman et al. 2001; Kesler et al. 2006; Zilberman et al. 2005).

Developments in medical image processing have also been applied to the analysis of blood smears in order to quantitate RBC aggregation. Examples include the application of wavelet transforms (Kavitha and Ramakrishnan 2007) and fractal dimensions (Oancea 2007) for detailed analysis of the patterns resulting from aggregate formation. Jayavanth and Singh (2004) used a digital image analysis technique to investigate the size distribution and shape of RBC aggregates during sedimentation in a vertically oriented 100-μm-thick sample chamber. Analysis of the sequentially recorded microscopic images provides information regarding the time course of aggregate size and shape changes during the sedimentation process (Jayavanth and Singh 2004).

4.1.5.2 Image Analysis of Red Blood Cell Aggregation under Dynamic Conditions

Implementation of modern digital image analysis equipment and software has allowed the analysis of RBC aggregation under dynamic conditions. This approach provides information about the status of RBC aggregation under various shearing conditions and is thus more relevant in terms of the *in vivo* flow environment. A variety of flow systems combined with appropriate microscopic techniques have been used to model the desired shearing environment. Schmid-Schönbein et al. (1973) used a transparent cone and plate combination mounted on an inverted microscope to visualize RBC aggregates during the application of selected shear stresses. Counter-rotation of the cone and plate enabled the observation and quantification of selected RBC aggregates during shear. Shiga et al. (1983) applied TV image analysis to study the number of RBC aggregates and the area covered by aggregates as a function of time during aggregation. Their system included a cone-plate shearing unit attached to an inverted microscope and a "particle analyzer" that allowed the assessment of aggregate properties under various shearing conditions.

Parallel-plate flow chambers combined with photomicrography techniques have also been used for this purpose (Chen et al. 1995; Chen et al. 1994). The system includes a flow chamber of 1×20 cm dimensions with a gap of 10–50 μm, a variable-speed pump to generate controlled flow in the chamber, and a video-microscopy system. Desired levels of shear stress in the flow chamber can be generated by varying the flow of dilute RBC suspensions (i.e., 0.06–0.1 l/l hematocrit) while monitoring the pressure difference across the chamber (ΔP). The shear stress (τ) in a flow chamber with a given gap (D) and length (L) can be calculated using the Stokes equation (Chen et al. 1994):

$$\tau = P\frac{D/4}{L} \tag{4.5}$$

The images obtained under various shearing conditions are transferred to a digital computer and analyzed to calculate several parameters reflecting aggregate size

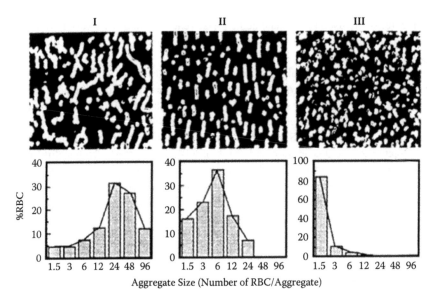

FIGURE 4.10   Image of red blood cells subjected to 0.0125 (I), 0.05 (II), and 0.2 (III) Pa shear stress. The histograms below demonstrate the size distribution of RBC aggregates for each level of shear. Reproduced with permission from Chen, S., G. Barshtein, B. Gavish, Y. Mahler, and S. Yedgar. 1994. "Monitoring of Red Blood Cell Aggregability in a Flow-Chamber by Computerized Image Analysis." *Clinical Hemorheology* 14:497–508.

(Chen et al. 1994), shape (Chen et al. 1995), and their dependence on the applied shear stress. Aggregate size is estimated in terms of the number of RBC per aggregate, using a specially developed image analysis software (Ben Ami et al. 2001; Berliner et al. 2004). The calculated parameters include the average aggregate size at a low level of shear stress, aggregate size distribution (Figure 4.10), the dependence of aggregate size on shear stress (Ben Ami et al. 2001; Berliner et al. 2004), and the shear stress required for disaggregation (Maharshak et al. 2009). Additionally, aggregation kinetics can be studied by analyzing the video images sequentially recorded following cessation of flow (Figure 4.11) (Barshtein et al. 2000; Chen et al. 1994). Chen et al. (1995) used the ASP as described in Section 4.1.5.1 to assess aggregate morphology under the influence of shear stress at various levels. They reported a close association between aggregate size and ASP. However, this dependence was dissimilar for RBC aggregation induced by high molecular mass dextrans, which are characterized by different morphology. Furthermore, RBC aggregates in the samples from β-thalassemia major patients exhibit a similar characteristic, suggesting that ASP might be a valuable parameter in discriminating pathological aggregation (Chen et al. 1996).

4.1.5.3   Advantages of Methods Based on Computerized Image Analysis

Application of image analysis techniques is actually an extension of calculating the Microscopic Aggregation Index, but with significant improvements. These improvements include the automation of the procedure and the associated speed of laboratory

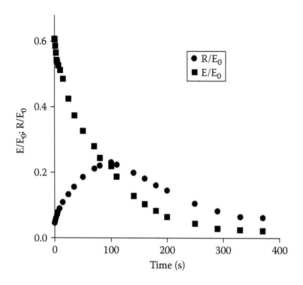

FIGURE 4.11 The ratio of the number of rouleaux (R) or single erythrocytes (E) to the total number of erythrocytes (E_0) as a function of time after the cessation of flow in a flow chamber. The number of aggregates reaches a maximum at around 100 s after the cessation of flow, while the number of single erythrocytes continues to decrease as the aggregation process continues. The decreased number of rouleaux after peak level reached at 100 s is due to a secondary aggregation of rouleaux resulting in larger aggregates. Reproduced with permission from Barshtein, G., D. Wajnblum, and S. Yedgar. 2000. "Kinetics of Linear Rouleaux Formation Studied by Visual Monitoring of Red Cell Dynamic Organization." *Biophysical Journal* 78:2470–2474.

practice and the ability to study RBC aggregation under dynamic conditions (e.g., under various shear stresses, during transition between different shearing conditions). It is therefore possible to study the factors that are important determinants of the effects of RBC aggregation on *in vivo* blood flow dynamics.

4.1.6 Photometric Methods

It was recognized about sixty years ago that the optical behavior of blood is mainly determined by RBC, not only by their relative volume (i.e., hematocrit), but also their status in the suspension (Zijlstra 1958). In principle, light beams are either absorbed or reflected (i.e., backscattered) if they hit RBC in a suspension. Alternatively, light beams can be transmitted through the blood film if they traverse the gaps between RBC that are occupied by suspending media. RBC deform under high shear stresses in flowing blood. They are oriented along flow stream lines in an elongated form and blood films of several hundred microns become more transparent (i.e., light transmittance through the blood increases and reflection of light decreases) because of the increased size of plasma gaps in between the cellular components of blood. RBC elongation and orientation decrease under low shear stresses or stasis (i.e., shape recovery), which decreases light transmittance or increases reflection. If the

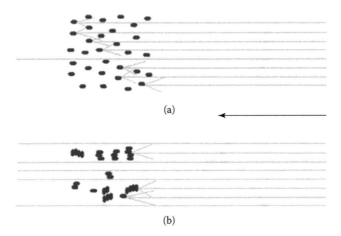

FIGURE 4.12 Light transmittance and reflection through suspensions with dispersed (a) and aggregated (b) red blood cells. Arrow indicates the direction of incident light beams.

shear forces are sufficiently low, RBC start to form rouleaux and this aggregation leaves fluid gaps between aggregates (Kaliviotis and Yianneskis 2008). Light transmittance or backscattering is a function of the effective average size of these gaps and therefore is indicative of the degree of aggregation (Figure 4.12). Therefore, the intensity of transmitted or reflected light through or from a blood sample can be used to monitor aggregation under a given shearing condition. This can be done using specially designed photometric instruments suitable for measuring transmitted or reflected light intensity; these instruments are called *photometric rheoscopes.*

RBC aggregation is a dynamic process that has a typical time course, and the optical properties of suspensions closely reflect this time course. Figure 4.13 demonstrates changes in reflected light intensity from a ~300 micrometer-thick blood film between two concentric glass cylinders. Blood was sheared at a shear rate of 500 s^{-1} during the first 10 s by rotating one of the cylinders, and then the rotating cylinder was abruptly stopped. The small and short-lasting increase was followed by a prominent decrease in light reflection. The first phase (i.e., the short increment in light reflectance) after the cessation of shearing represents the shape recovery of RBC while the later, decreasing phase of reflected light represents the RBC aggregation process (Dobbe et al. 2003). It can be seen that the aggregation phase also has two components: (a) a fast change in light reflection during the first ~10 s of aggregation and (b) a slower change in reflected light during the later phase of the time course.

The time course of the intensity of reflected light during RBC aggregation as shown in Figure 4.13 was termed a *syllectogram* by Zijlstra et al. (1958). This term was derived from the Greek word meaning "to gather"; light transmittance time courses were also called syllectograms. Light transmittance syllectograms also closely reflect the aggregation status of RBC in suspension, although the changes are in the reverse direction: RBC shape recovery is indicated by a brief decrement in light transmittance while RBC aggregation is reflected by increased light transmittance (see Figure 4.15). However, although both reflection and transmittance

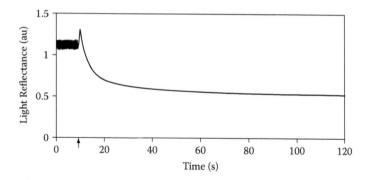

FIGURE 4.13 Time course of the intensity of reflected light from a ~300-micrometer-thick blood film between two glass cylinders. Light intensity was measured in arbitrary units (au). The blood film was first sheared at 500 s^{-1} by rotating one of the cylinders then stopping the shearing at 10 s (indicated by arrow). This curve is known as a syllectogram.

syllectograms indicate the time course of RBC aggregation, these two curves are not mirror images of each other (see Section 4.1.6.4).

Syllectograms can be mathematically analyzed and a number of parameters reflecting both time course and magnitude of RBC aggregation can be calculated as detailed in Section 4.1.6.2. However, for an accurate analysis of aggregation by this approach, the procedure needs to be started with effective disaggregation to disperse all existing RBC aggregates. This is achieved by applying sufficiently high shear stresses to the blood sample at the initial stage of the measurement procedure using various shearing geometries.

4.1.6.1 Disaggregation Mechanisms Used in Aggregation Measurements

RBC suspensions are disaggregated in shearing systems made of transparent surfaces to allow the measurement of light transmittance or backscattering through the blood film under investigation. The main objective of such designs is to generate a sufficiently high and controllable shear rate or stress in the sample. Figure 4.14 demonstrates various shearing systems designed for this purpose.

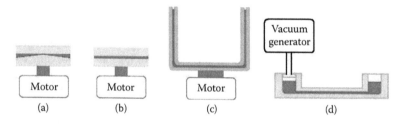

FIGURE 4.14 Disaggregation mechanisms used for recording syllectograms: (a) cone-plate, (b) plate-plate, (c) couette, (d) parallel plate flow channel.

The blood film can be sheared between a stationary plate and a rotating circular cone (Kiesewetter et al. 1982) with a gap of <100 micrometers (Figure 4.14a). The cone angle is selected to maintain a constant shear rate across the gap between the cone and plate; shear rates up to 500 s^{-1} can be achieved by rotating the cone by a motor at the desired rotational speed. Both the plate and cone are transparent and light transmittance through the sample between them can be measured by placing a light source and photo sensor on opposite sides of the system.

The cone can be replaced by a circular plate that is rotated by a motor (Figure 4.14b) (Baskurt et al. 1998). The gap between the two plates is usually about several hundred micrometers. Obviously, the shear rate is not constant in the fluid filling the gap but rather is at minimum at the central region of the rotating plate and increases toward the edge of the plate. However, light transmittance can be measured within a small limited area at a fixed radial position; shear rate can be calculated based on rotational speed, the distance of the measurement from the center of the plate, and the gap between the two plates.

Couette systems consisting of two transparent concentric cylinders are suitable for light transmission or backscattering measurements during aggregation (Groner et al. 1980; Hardeman et al. 2001). One of the cylinders, either outer or inner, is rotated by a motor at a speed calculated to generate the desired shear rate, which can be varied over a wide range (Figure 4.14c).

Parallel-plate flow chambers are also used for aggregation measurements (Shin et al. 2006). Disaggregation in such systems can be achieved by the rapid flow of an RBC suspension through a channel several hundred micrometers high, either by vacuum application or by pumping (Figure 4.14d). Shin et al. suggested that externally vibrating the flow channel might be an effective mechanism for disaggregation (Shin et al. 2006); this method has not been proven effective under all circumstances. Alternatively, Shin et al. also used a mechanical stirring mechanism in a microfluidic system (Shin et al. 2009a), which is further discussed in Section 4.1.6.9.3.

The measurement can also be done using a transparent tube through which blood can be pumped for disaggregation (Tomita et al. 1986). Tomita et al. (1986) developed a sensor assembly to measure light transmittance through a transparent tube that can be used on a disposable plastic tube or even on an in situ blood vessel. It has been recently demonstrated that RBC aggregation can be measured in a glass tube of 1-mm diameter (i.e., a standard microhematocrit tube) following disaggregation by briefly vibrating the blood column in the tube (Baskurt et al. 2011). In this system the vibration was achieved by generating vacuum pulses using a miniature solenoid.

All of the shearing geometries described above are suitable for measurements using both light transmittance and backscattering approaches. The shearing can be abruptly halted by stopping the motor or disconnecting from the vacuum generator or the pumping system. Electrical or mechanical breaking can also be applied if the stoppage of the shearing mechanisms is not fast enough. Alternatively, shear rate can be changed to a lower level from the high level used for disaggregation.

4.1.6.2 Analysis of Syllectogram—Calculation of Aggregation Parameters

A syllectogram carries two types of information about the aggregation process which starts with the change of shearing from high to either low shear rate or stasis:

(1) the extent of aggregation, corresponding to the aggregate size in terms of the number of individual RBC per aggregate, at selected time points (e.g., 5 s, 10 s, 120 s) after shape recovery and the start of the aggregation process (Figures 4.13 and 4.15); (2) the time course of the aggregation process. A number of parameters can be calculated using syllectograms to express the aggregation behavior of blood samples. Analysis of syllectograms for aggregation parameters is performed using the aggregation phase of the curves beginning at the end of the shape recovery process, where shape recovery is the rapid decrement of light transmittance or sudden increment in light reflection following the disaggregation period, reaching minimum or maximum, respectively, within a second. This minimum or maximum (for light transmittance or reflectance, respectively) represents the starting point of aggregation. Various combinations of the parameters described below are implemented by instruments available for the measurement of RBC aggregation (Baskurt et al. 1998; Bauersachs et al. 1989; Hardeman et al. 2001; Kiesewetter et al. 1982; Shin et al. 2009a).

4.1.6.2.1 Parameters Reflecting the Overall Extent of Aggregation

The overall extent of RBC aggregation during a selected period can be estimated by calculating the difference between light intensities at that time point and at the start of the aggregation process. This parameter, termed *amplitude* (AMP) (Figure 4.15), does not provide any information about the time course of the aggregation process and is expressed in arbitrary units (au).

The area under the aggregation phase of the light transmittance time curve also reflects the extent of RBC aggregation (area B in Figure 4.15a) and is also sensitive to the rate of change of transmitted light intensity. A blood sample with a faster aggregation process is represented by a greater *area under curve* compared to a sample with slower aggregation even if the AMP values for the two are identical. *Area under curve* or *surface area* (SA) is used as an aggregation index by some instruments (e.g., the M and M1 parameters of the Myrenne aggregometer; see Section 4.1.6.9.1). This area is usually calculated during the first 5 or 10 s of the aggregation process and is again expressed in arbitrary units. It should be noted that the same concept could be used for analyzing the light reflectance time course (Figure 4.15b). However, in that case the *area above curve* (area A in Figure 4.15b) represents the rate-extent combination of the aggregation process.

4.1.6.2.2 Parameters Reflecting the Time Course of Aggregation

An index that reflects the rate of the aggregation process, independent of the extent of aggregation, can be calculated by dividing the area B for light transmittance (Figure 4.15a) or the area A for light reflectance (Figure 4.15b) by the sum of areas A plus B. This index is generally known as the Aggregation Index (AI) and is expressed as the percentage of the total area.

$$AI = \frac{B}{A+B} \times 100 \ \ \text{(For light transmittance)} \qquad (4.5)$$

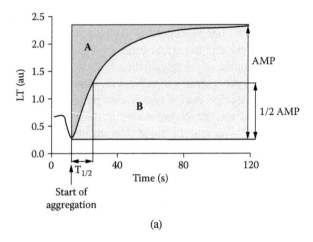

(a)

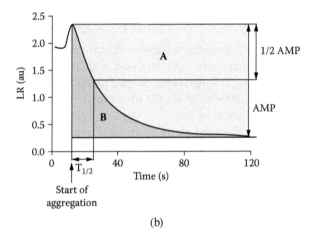

(b)

FIGURE 4.15   Analysis of the time course of light transmittance (LT) (a) and of light reflectance (LR) (b). A and B denote the areas above and below the syllectogram curve and are shaded in different gray tones. LT and LR are expressed in arbitrary units (au). Please see the text for explanations.

$$AI = \frac{A}{A+B} \times 100 \quad \text{(For light reflectance)} \tag{4.6}$$

The *aggregation half time* ($T_{1/2}$), in seconds, is another measure of the rate of aggregation. This parameter corresponds to the time required to reach the transmitted or reflected light intensity that corresponds to one-half of the total change at the end of the measurement period (i.e., one-half the AMP parameter) (Figure 4.15). This calculation should be done using syllectograms recorded until the comple-

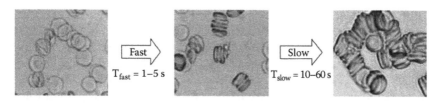

FIGURE 4.16 Formation of two-dimensional rouleaux corresponds to a fast change in light transmittance or reflectance, while three-dimensional aggregate formation is related to the slower change in light intensity. The time constant for the fast phase is usually a few seconds, while the slow phase has a time constant in the range of 10–60 seconds.

tion of the aggregation process, which is usually achieved in 120 s, for meaningful representation of the time course.

A careful inspection of the syllectogram reveals that the time course of aggregation is characterized by two phases with different rates of change of either transmitted or reflected light intensity. These curves can be accurately represented by double exponential equations in the form:

$$I_t = a + b \cdot e^{-t/T_{fast}} + c \cdot e^{-t/T_{slow}} \tag{4.7}$$

where I_t is the light intensity at time t, a corresponds to the light intensity at the end of the measurement period, b and c are the magnitude of changes in light intensity at fast and slow rates, respectively, and T_{fast} and T_{slow} are the time constants for the earlier (fast) and later (slow) phases of change in light intensity (Hardeman et al. 2001). These parameters can easily be calculated for digitized syllectograms using a nonlinear curve-fitting algorithm. The length of the syllectogram should be sufficient to include the whole aggregation time course, which takes ~120 s in most samples, in order to obtain meaningful representation of the process, especially for the slow component. T_{fast} has been reported in the range of 1–5 s while T_{slow} was in the range of 10–60 s for normal blood samples, depending on the measurement system geometry and the recorded parameter (i.e., light reflectance or light transmittance) (Baskurt et al. 2009d).

The fast phase represented in the syllectogram reflects the primary rouleaux formation (i.e., formation of two-dimensional RBC aggregates), while the slow phase reflects the secondary aggregation of the two-dimensional aggregates into three-dimensional structures (Figure 4.16) (Hardeman et al. 2001).

It has been reported that various disease processes affect T_{fast} and T_{slow} differently: Both time constants were found to be increased in pulmonary hypertension, while T_{fast} was decreased and T_{slow} was increased in chronic glomerulonephritis, nephrotic syndrome, and systemic lupus erythematosus (Priezzhev et al. 1999).

4.1.6.3 Other Parameters Measured Based on the Optical Properties of Red Blood Cell Suspensions

In addition to the parameters obtained by syllectogram analysis, the shear rate just sufficient to prevent red blood cell aggregation can also be measured using the photometric instruments equipped with suitable shearing systems. This parameter corresponds to the magnitude of forces holding RBC in aggregates and is termed the *disaggregation shear rate* (abbreviated as γT_{min} or γ_{thr}) (Bauersachs et al. 1989; Deng et al. 1994; Hardeman et al. 2001). Note that the phrase *just sufficient* is incorporated into the definition and thus this shear rate is sufficient to disperse aggregates without causing RBC deformation. The measurement principle is based on the dependence of transmitted or reflected light intensity on both the extent of aggregation and the deformation-orientation of RBC to flow stream lines. Under fully disaggregated flow conditions (i.e., with shear stresses in the range sufficiently high to disperse all RBC aggregates), light transmittance increases with shear rate due to more effective deformation-orientation of RBC. There is a level of shear rate at which light transmission is minimized due to minimal orientation-deformation of RBC and prevention of aggregation by this critical level of shear forces. Below this critical level of shear rate, the shear forces affecting RBC are not sufficient to prevent RBC aggregation; light transmission is increased at shear rates below this critical level. This critical shear rate corresponds to the disaggregation shear rate (Figure 4.17). The principle applies to light reflectance, except light reflection decreases below or above the disaggregation shear rate due to increased aggregation or orientation, respectively.

In order to accurately detect this critical shear rate, the photometric shearing apparatus should be capable of altering the applied shear rate between <10 s^{-1} to >400 s^{-1}. The software programs developed for measuring disaggregation shear rate first record transmitted or reflected light intensity at a number of predefined shear levels (e.g., 7 shear rates between 8 s^{-1} and 300 s^{-1}; Bauersachs et al. 1989), with shearing sufficient for full disaggregation and orientation (e.g., 500 s^{-1}) following each level. The program then calculates the difference between the light intensity measured at that level and the following high-shear period, identifies the shear rate resulting in the maximum difference in light intensity, and determines four additional shear rates around this, based on a simple algorithm and repeats the procedure for these

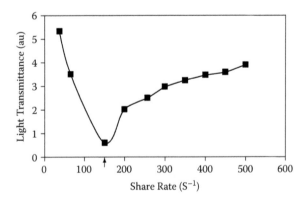

FIGURE 4.17 Plot of light transmittance versus shear rate demonstrating the critical shear rate for disaggregation termed the disaggregation shear rate, γT_{min}. The arrow indicates disaggregation shear rate.

shear rates. The final calculation of critical shear rate is governed by the software: (1) The procedure developed by Bauersachs et al. (1989) fits the shear rate–light intensity difference data to a third-order polynomial, differentiates the equation, and *calculates* the disaggregation shear rate. (2) The procedure used by Hardeman et al. (2001) selects one of the 11 shear levels having the highest shear rate–light intensity difference. The normal range of disaggregation shear rate reported is in the range of $70-150$ s^{-1}.

Shin et al. (2009b) measured the threshold shear stress based on light transmittance in a parallel-plate flow channel. Various levels of shear stress were generated by vacuum application on one side of the flow channel while monitoring the pressure gradient (see Figure 4.14d). The critical shear stress value for disaggregation was reported to be ~200 mPa and hematocrit independent (Shin et al. 2009b).

Another parameter related to the dependence of reflected or transmitted light intensity from or through blood samples is the *flow-to-stasis aggregation ratio* (FSAR). This is calculated by dividing the light intensity measured during the application of a ~ 3 s^{-1} shear rate by the light intensity at stasis. Both periods of measurements are preceded by a 10 s period of high shear rate (e.g., 500 s^{-1}) (Bauersachs et al. 1989).

4.1.6.4 Light Reflectance versus Transmittance

It is obvious from the previous discussion that the same principles of analysis are applicable to both light reflectance and transmittance time curves. However, it has been known for several decades that light transmittance and reflectance time curves are not just mirror images of each other but rather have significantly different characteristics (Gaspar-Rosas and Thurston 1988). Figure 4.18 demonstrates light transmittance and reflectance time courses recorded simultaneously during aggregation in a Couette type measurement system following disaggregation at 500 s^{-1}. The magnitude of the overall changes in light intensity cannot be expected to be comparable between the two recordings and simply reflects the influence of the combination of several factors including the characteristics of photo sensors, their alignment angles, and electronic settings of the circuits recording each signal. However, time courses are also significantly different from each other, with the light reflection time course having a higher rate of change during the initial phase of aggregation. The double-exponential equations (Section 4.1.6.2.2; Equation [4.7]) representing both curves are presented here and demonstrate approximately a four-fold difference in T_{slow}.

$$\text{Light transmittance: } I_t = 6.03 - 3.05 \cdot e^{-t/2.41} - 2.96 \cdot e^{-t/60.17}$$

$$\text{Light reflectance : } I_t = 1.40 + 3.19 \cdot e^{-t/1.63} + 0.61 \cdot e^{-t/16.22}$$

Both reflected and transmitted light intensities recorded simultaneously were representing the time course of the same aggregation process.

Gaspar-Rosas and Thurston (1988) reported similar differences between reflected and transmitted light intensity time courses from or through 127 to 508 µm-thick

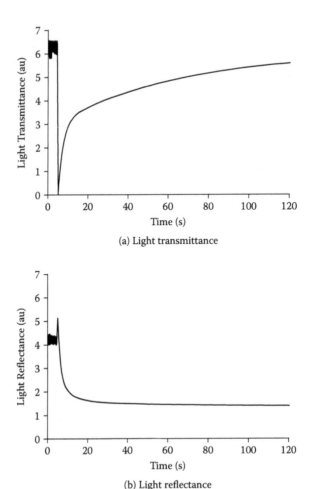

FIGURE 4.18 Time course of light transmittance (a) and reflectance (b) recorded simulta-neously during aggregation of normal blood in a Couette-type shearing system.

blood films during aggregation. They stated that light reflection is predominantly determined by the behavior of RBC close to the surface exposed to light, whereas light transmittance is affected by the entire thickness of the RBC film. This concept was supported by the significant dependence of the time constants of light transmis-sion time course on the thickness of the blood film, whereas much lower degrees of dependence were reported for light reflection (Gaspar-Rosas and Thurston 1988). It is very likely that different levels of constraints affect rouleaux formation at different depths from the surface, being more effective for the surface layers. This results in the development of larger RBC aggregates in the deep layers of the blood film, longer time periods with continuing changes of optical properties, and therefore longer time constants for light transmittance.

Nam et al. (2010) used polarized light for photometry during RBC aggregation and observed a similar difference between transmitted and reflected light time courses (Figure 4.19). However, the reflected light intensity time course became similar to that of light transmittance if they placed an orthogonal polarizer in the optical path of the photo sensors recording light reflectance. Orthogonal polarizers eliminate the polarized light that is directly reflected from the surface layers of the blood film, while only multiscattered light, reflected from deeper layers of the blood film and which is no longer polarized, can pass through the orthogonal polarizer. This observation also supports the suggestion that the difference between the time courses of transmitted and reflected light intensities is due to the dominance of RBC populations at different depths of the blood film under investigation (Nam et al. 2010).

The differences detailed in this section suggest that the parameters derived from light reflectance and light transmittance time courses (see Section 4.1.6.2), may not be comparable with each other. Furthermore, Baskurt et al. (2009d) demonstrated that the power of transmittance- and reflectance-based parameters for detecting alterations in RBC aggregation may also differ. In general, parameters reflecting the overall extent of RBC aggregation (AMP, SA) based on light transmittance were characterized by higher standardized differences for samples with normal and experimentally altered aggregation (i.e., higher power for detecting a difference between the two samples (Stuart et al. 1989) while parameters reflecting the time course of aggregation (AI, $T_{1/2}$, T_{fast}, T_{slow}) have a higher power if calculated using light reflectance time courses (Baskurt et al. 2009d).

The previously mentioned results also indicate that the differential power of various calculated parameters depends on the manner by which aggregation is modified. That is, calculation methods for aggregation parameters should be modified to achieve greater sensitivity for detection of a particular type of aggregation (e.g., highly increased rate of aggregation due to inclusion of high-molecular-mass dextran in the suspending medium); one approach would be to calculate the AMP parameter over shorter lengths of time (Baskurt et al. 2009d). Thus the characteristics of transmittance- and reflectance-based parameters should be considered when selecting the measurement method and the aggregation parameters for a given study.

4.1.6.5   Choice of Light Wavelength for Recording Syllectograms

Most photometric systems designed to quantitate RBC aggregation utilize light at a single wavelength, usually in the region corresponding to red or above (>600 nm). Either laser diodes or simple light-emitting diodes (LEDs) are successfully used for this purpose in modern instruments. The rationale behind this selection is related to the specific absorbance spectrum of hemoglobin, which is around 500–550 nm. If a light beam with a wavelength close to this region is used for the measurement, hemoglobin absorbance may strongly influence the change in optical properties of the blood (Roggan et al. 1999).

Figure 4.20 presents light transmittance syllectograms recorded simultaneously at four different wavelengths between 500 and 800 nm during RBC aggregation in a 230 micrometer thick blood film using a parallel-plate flow

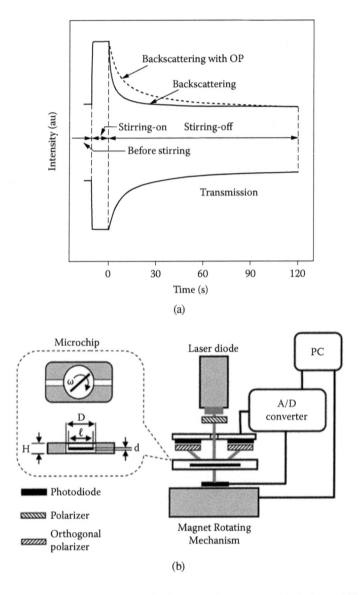

(a)

(b)

FIGURE 4.19 Light transmittance and reflectance time courses (a) during red blood cell aggregation recorded in a microfluidic chamber equipped with a magnetic stirrer for disaggregation (b). The laser beam was polarized before being directed to the blood sample. The difference between transmitted and reflected light time courses was abolished by the orthogonal polarizers that eliminated polarized light directly reflected from the surface layers of the blood film. Reproduced with permission from Nam, J-H., Y. Yang, S. Chung, and S. Shin. 2010. "Comparison of Light Transmission and Backscattering Methods in the Measurement of RBC Aggregation." *Journal of Biomedical Optics* 15: 027003.

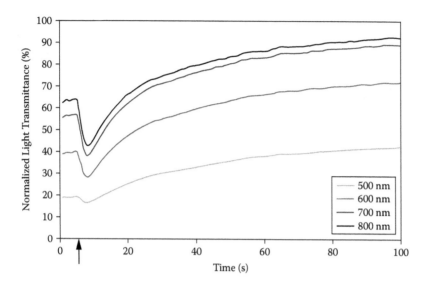

FIGURE 4.20 Light transmittance syllectograms recorded simultaneously for four different wavelengths between 500 to 800 nm during red blood cell aggregation in a 230 micrometer-thick blood film. Arrow indicates the stoppage of flow corresponding to a shear rate of ~500 s⁻¹. Light transmittance was normalized according to the light intensity for distilled water at each wavelength and expressed as a percentage. Time constants (T_{fast} and T_{slow}) representing each curve are presented below.

chamber. Venous blood at 0.45 l/l hematocrit with a pO_2 ~48 mmHg corresponding to ~83% oxygen saturation of hemoglobin was used for these studies. Disaggregation was achieved by flow corresponding to a shear rate of 500 s⁻¹ and aggregation was started after an abrupt stoppage of flow. Obviously, light transmittance was significantly affected at the 500-nm wavelength due to hemoglobin absorbance, resulting in a smaller amplitude compared to those recorded at longer wavelengths. This attenuation is expected to influence the sensitivity of the calculated parameters. However, more importantly, the time course of transmitted light intensity also differed at different wavelengths. T_{fast} and T_{slow} values (see Section 4.1.6.2.2) for the syllectograms recorded at wavelengths of 500 to 800 nm were as follows:

800 nm	T_{fast}: 8.30 s	T_{slow}: 40.09 s
700 nm	T_{fast}: 8.30 s	T_{slow}: 40.80 s
600 nm	T_{fast}: 8.60 s	T_{slow}: 44.50 s
500 nm	T_{fast}: 9.29 s	T_{slow}: 52.40 s

The most significant effects were observed at the 500-nm wavelength: a 12% change in T_{fast} and a 30% difference for T_{slow} compared to the time constants for 800 nm. Therefore, usage of light sources with wavelengths close to this range should be avoided in the measurement of RBC aggregation.

As expected from Figure 4.20, the influence of oxygenation level on light transmission depends on the wavelength of the light beam. This influence of oxygenation
is prominent when comparing measurements using red light sources, whereas the
dependence of light transmission and related aggregation indexes on oxygenation
status becomes less pronounced if infrared light (wavelength >800 nm) is used rather
than red at ~700-nm wavelength (Figure 4.21). The difference in absorbance between
hemoglobin and oxyhemoglobin is highest at ~660 nm, while the isosbestic point
(i.e., wavelength at which hemoglobin and oxyhemoglobin have the same molar
absorptivity) is 800 nm with smaller differences between the two above this wavelength (Roggan et al. 1999). Thus photometric measurements would not be affected

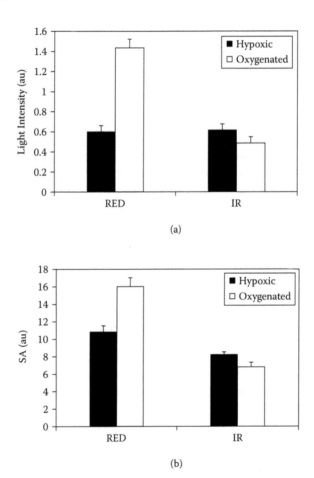

(a)

(b)

FIGURE 4.21 Light transmittance during disaggregation (a) and surface area under the
syllectogram (SA) as an aggregation index (b) for hypoxic (pO$_2$: ~40 mmHg, ~75 saturation)
and oxygenated (pO$_2$: ~80 mmHg, ~95% saturation) samples. Light transmittance through
the sample in a 1-mm-diameter glass tube was recorded simultaneously using two separate
photo sensors and LEDs at two different wavelengths of 700 nm (RED) and 918 nm (IR).
Disaggregation was achieved by flow through the tube at a rate corresponding to ~500 s^{-1}.

by oxygenation status if obtained at the isosbestic point (~800-nm wavelength) and the influence would be minimized in the infrared region. Tomita et al. (1986) also suggested that using light at wavelengths corresponding to the infrared region may minimize the influence of hemoglobin oxygenation on measured aggregation parameters. The influence of oxygenation status of the blood sample on aggregation parameters is further discussed in Section 4.1.6.8.3.

4.1.6.6   Influence of Measurement Chamber Geometry on Aggregation Parameters

Syllectograms are influenced by the thickness of blood films under investigation, as discussed in Section 4.1.6.4. This is especially prominent for light transmission time curves (Gaspar-Rosas and Thurston 1988), but was also observed for light reflectance time curves (Dobbe et al. 2003). Obviously, the measurement chamber should provide sufficient space for the full development of three-dimensional RBC aggregates in order to quantitate the process accurately. It has been shown that aggregate dimensions can reach ~500 µm, especially under pathological conditions (Lademann et al. 1999). Furthermore, wall effects during aggregation should be eliminated by using sufficiently large measurement chambers (Pribush et al. 2004a). This can be achieved if the thickness of the sample film exceeds 30 times the mean particle dimensions (Barnes 1995), which corresponds to ~250 µm for human RBC suspensions. The measurement chamber should thus provide a blood film thickness exceeding this limit at the site of light transmittance or reflectance measurement.

However, it has been demonstrated that the aggregation process is affected by the changes of blood film thickness even above this limit. Dobbe et al. (2003) studied the dependence of aggregation parameters on the gap width of a Couette system and reported that aggregation half time ($T_{1/2}$) increased and aggregation index (AI) decreased with increasing gap width from 0.37 to 2 mm. This dependence of aggregation parameters on measurement chamber geometry is one of the reasons for the difficulty in relating aggregation parameters obtained using different instruments.

4.1.6.7   Effect of Hematocrit on Measured Parameters

The RBC aggregation process is influenced by hematocrit (Agosti et al. 1988). This is a physiologically important relationship, since the hematocrit of blood in various segments of the vasculature differs significantly (Baskurt et al. 2006; Desjardins and Duling 1987) and thus the kinetics of RBC aggregation can vary locally within the circulation (Kim et al. 2005). This concept is further discussed in Chapter 7. Therefore, the hematocrit of the sample should be considered as an important determinant of RBC aggregation and be taken into account during the measurement of aggregation parameters.

Additionally, hematocrit also affects the optical properties of RBC suspensions and interferes with measured parameters based on light transmittance or light reflectance. Light transmittance through fully disaggregated blood samples (i.e., sheared at ~500 s^{-1} for 10 s) is inversely related to its hematocrit as demonstrated in Figure 4.22. Light reflectance also depends on hematocrit and is enhanced with increasing hematocrit. Obviously, transmitted or reflected light intensity under standardized

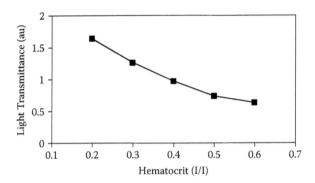

FIGURE 4.22 Light transmittance during complete disaggregation through blood samples with hematocrits of 0.2 to 0.6 l/l.

measurement conditions can be used to estimate the hematocrit of the sample by employing an appropriate regression equation (Jung et al. 1987). The influence of hematocrit is clearly shown by the parameters calculated based on the time course of recorded light intensity following disaggregation (Figure 4.23). Syllectogram AMP decreased with increasing hematocrit (Figure 4.23a), while the effect of hematocrit on the surface area under the syllectogram curve (SA) is biphasic for many measurement systems, being highest between 0.3-0.4 l/l (Figure 4.23b) (Baskurt et al. 1998; Deng et al. 1994; Shin et al. 2009a).

The parameters reflecting the time course of aggregation are also sensitive to hematocrit (Figures 4.23c and 4.23d). AI increases with hematocrit while $T_{1/2}$ decreases (Baskurt et al. 1998; Hardeman et al. 2001; Shin et al. 2009a). While not shown in Figure 4.23, the faster, initial phase of aggregation is accelerated with increased hematocrit as indicated by a decreased T_{fast}, while the slower phase time constant (T_{slow}) is slightly increased with hematocrit above 0.4 l/l. (Baskurt et al. 1998; Deng et al. 1994).

The disaggregation shear rate (Section 4.1.6.3) is also affected by the hematocrit of the sample: the critical shear rate required to prevent RBC aggregation is significantly higher in blood samples with low hematocrit and decreases with increased hematocrit (Deng et al. 1994; Shin et al. 2009b). This dependence of disaggregation shear rate on hematocrit is due, at least in part, to the increase of blood viscosity with hematocrit, thereby achieving the critical shear stress at lower shear rates with samples having higher hematocrit. Shin et al. (2009b) demonstrated that the critical shear stress, calculated as the product of shear rate and blood viscosity, was not influenced by hematocrit.

The significant dependence of most aggregation parameters on the hematocrit of the sample underlines the need for attention to this property when quantitating RBC aggregation. Adjustment of hematocrit to a standard value (e.g., 0.4 l/l) prior to the measurement is the recommended approach (see Section 4.2.2) A mathematical correction can also be employed, if the hematocrit of the sample is known, using regression equations developed specifically for the parameter and device being used. It should be noted that hematocrit could also be estimated using the transmitted or

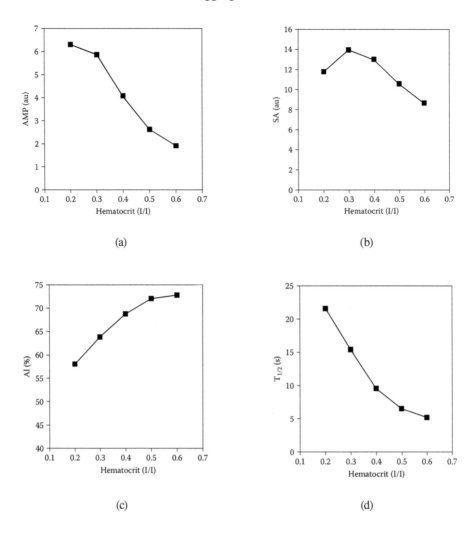

FIGURE 4.23 Syllectogram amplitude (AMP) at 120 s (a), surface area under syllectogram (SA) (b), aggregation index (AI) (c), and aggregation half time ($T_{1/2}$) (d) as a function of the sample hematocrit. See Section 4.1.6.2 for the explanation of the aggregation parameters.

reflected light intensity through or from suspensions and this hematocrit value used for automatic corrections.

4.1.6.8 Optimal Measurement Conditions and Precautions

A number of factors significantly influence the measured aggregation parameters and thus these should be carefully controlled during the quantification of RBC aggregation. The importance of sample hematocrit was discussed previously in Section 4.1.6.7. Other factors are detailed in the following text.

4.1.6.8.1   Measurement Temperature

RBC aggregation is a temperature-dependent process (Dintenfass and Forbes 1973; Maeda et al. 1987; Neumann et al. 1987; Singh and Stoltz 2002). Aggregation indexes (e.g., SA or M index measured by Myrenne aggregometer) were found to be increased while aggregation time constants decreased as temperature was raised from 5 to 37°C (Baskurt and Mat 2000; Singh and Stoltz 2002). This is also true for indexes measured by other instruments (e.g., the Laser-Assisted Optical Rotational Cell Analyzer [LORCA]), and Lim et al. (2009) reported that the threshold shear stress in a parallel-plate flow channel exhibits a strong temperature dependence. The influence of temperature on aggregation kinetics is partly related to the increase of plasma viscosity with decreased temperature (Singh and Stoltz 2002). Plasma viscosity is an opposing factor for RBC movement in a suspension, with such movement of cells toward each other an important factor in the aggregation process (Chien and Sung 1987; Skalak and Zhu 1990).

Baskurt and Mat (2000) demonstrated that significant differences of aggregation in blood samples from septic and control subjects could only be demonstrated if the measurements were conducted over a certain temperature range. This temperature dependence is especially relevant to AI and disaggregation shear rate measurements based on light reflectance: differences in these parameters in septic and healthy rats could only be detected if the measurements were conducted at temperatures >25°C (Figure 4.24).

RBC aggregation measurements should be conducted at constant temperature and the recommended temperature is 37°C (Baskurt et al. 2009a). The measurement chambers of the photometric instruments might be temperature controlled. Alternatively, the entire instrument can be operated in a temperature-controlled environment if the instrument cannot maintain a constant temperature in the measurement chamber. The sample can be preheated to 37°C, especially if the sample volume is large (~1 ml for some instruments, see Section 4.1.6.9), to shorten the equilibration time after loading the sample into the instrument.

4.1.6.8.2   Disaggregation Process

RBC should be fully disaggregated at the start of aggregation measurements for both light transmission and light reflectance instruments. Disaggregation mechanisms used in various instruments are discussed in Section 4.1.6.1. The shear rate and thus shear stress should be high enough for complete disaggregation of the sample since incomplete initial disaggregation can cause incorrect results: aggregation indices reflecting the overall extent of aggregation as well as those indicating its time course are affected.

In some instruments, the shear rate can be controlled by changing the speed of the motor driving the rotating part of the shearing section. In such instruments, efficacy of the disaggregation phase can be tested by repeating aggregation tests with the shear rate increasing stepwise for each measurement. Results from such a test should indicate that AMP or SA parameters increase with increasing disaggregation shear rate until reaching a level beyond which aggregation parameters remain constant. The existence of a plateau of aggregation parameters beyond a certain level of shear rate assures that operating the instrument at or above this level induces complete

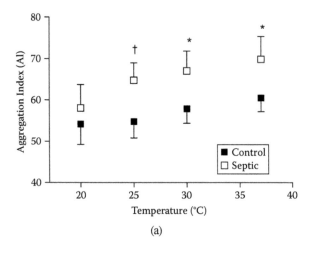

(a)

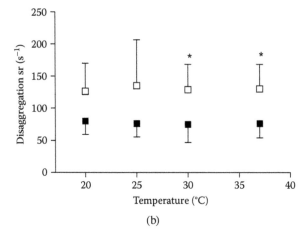

(b)

FIGURE 4.24 Aggregation index (AI) (a) and disaggregation shear rate (b) measured in blood samples from healthy and septic rats measured in a Couette shearing system by monitoring light reflectance (LORCA) at temperatures between 20 and 37°C. Difference from Control; *: p<0.05, †: p<0.01. Reproduced with permission from Baskurt, O. K., and F. Mat. 2000. "Importance of Measurement Temperature in Detecting the Alterations of Red Blood Cell Aggregation and Deformability Studied by Ektacytometry: A Study on Experimental Sepsis in Rats." *Clinical Hemorheology and Microcirculation* 23:43–49.

disaggregation during the initial phase of the measurement. Shearing at > 400 s⁻¹ is sufficient for total disaggregation for most samples.

4.1.6.8.3   Oxygenation Status of the Samples
As discussed in Section 4.1.6.5, light transmittance or reflectance as well as aggregation parameters are affected by hemoglobin absorption of light. This absorption behavior

of hemoglobin underlies the influence of the wavelength of the light used for syllecto-gram recording. It is also obvious that this dependence can be affected by changes in the oxygen saturation of hemoglobin (Figure 4.21) due to the shift of the absorbance spectrum of hemoglobin with oxygenation-deoxygenation status. Figure 4.25 presents the surface area above the syllectogram (SA) recorded in an aggregometer using light reflectance intensity. The SA parameter was significantly higher after oxygenation of the sample (pO_2: 142.0 ± 3.1) compared to that measured after deoxygenation (pO_2: 28.6 ± 1.7) (Uyuklu et al. 2009b) when measured in an instrument (i.e., LORCA) that uses a red diode laser at 670 nm (red) as the light source.

Conversely, Uyuklu et al. (2009b) reported that measurements done with the same oxygenated and deoxygenated samples as in Figure 4.25 but using an instrument based on light transmittance (Myrenne aggregometer) showed no significant change of the M parameter, which is equivalent to the SA. However, it should be noted that the Myrenne aggregometer uses an infrared light source, with this wavelength of light likely explaining the insensitivity of the aggregation parameter to the oxygen-ation status of the sample (see Section 4.1.6.5 and Figure 4.21).

Given the results shown in Figure 4.25, standardization of the oxygenation sta-tus of hemoglobin is not relevant for devices using infrared light (e.g., Myrenne aggregometer) but is an important aspect of aggregation measurement for devices using visible light (e.g., He-Ne laser) for reflectance or transmittance. The easiest way to achieve a standard pO_2 level is to fully oxygenate blood samples prior to the measurement. This can be done by gently mixing a small volume of the sample in a large test tube containing ambient air: ten minutes on a tube roller for one ml of blood in a 20-ml polypropylene tube is usually sufficient for ~99% oxyhemoglobin saturation. An advantage of essentially complete oxygenation is that this level would

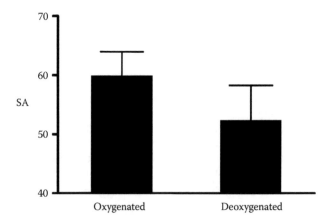

FIGURE 4.25 Surface area above syllectogram (SA) measured using a light reflectance aggregometer (LORCA) for oxygenated and deoxygenated aliquots of the same blood sample. The aggregometer uses a 670-nm diode laser. Redrawn from Uyuklu, M., H. J. Meiselman, and O. K. Baskurt. 2009b. "Effect of Hemoglobin Oxygenation Level on Red Blood Cell Deformability and Aggregation Parameters." *Clinical Hemorheology and Microcirculation* 41:179–188.

not change by exposure to ambient air during the measurement. In general, deoxygenation of samples is not recommended since it is difficult to maintain the low oxygen saturation during a measurement because most instruments are not capable of testing blood in a closed, gas-tight chamber.

4.1.6.9 Comparison of Instruments

There are several photometric instruments currently available that have been developed for quantitating RBC aggregation. Additionally, studies done using custom-built and experimental instruments have been published (Baskurt et al. 1998; Shin et al. 2006; Baskurt et al. 2011), and instruments that are not commercially available have been patented (Tomita 1989). Light transmittance or reflectance monitoring were utilized in these different instruments, and the shearing systems used for disaggregating the sample differed between devices. Commercially available photometric aggregometers are briefly described in the following text followed by a comparison between them.

4.1.6.9.1 Myrenne Red Blood Cell Aggregometer

The Myrenne aggregometer (Figure 4.26) is probably the most widely used hemorheological instrument. It was originally developed by H. Schmid-Schönbein (Schmid-Schönbein 1979) and is now manufactured by Myrenne GmbH (Roetgen, Germany, http://www.myrenne.com/de/produkte/medizintechnik.html). The shearing section of the instrument consists of a 2-degree cone that is rotated by a DC motor and a plate (i.e., a glass microscope slide) positioned on top of it; a small amount of blood (30–50 μl) is placed in the gap between the cone and plate. An infrared LED is used as the light source and the light beam passes through the sample at a distance 1.5 mm from the apex of the cone at which the gap is ~50 μm (Kiesewetter et al. 1982).

The measurement procedure starts with shearing of the sample at a fixed, high shear rate (600 s⁻¹) by rotating the cone for 10 s. The light transmission signal is then integrated, either following the complete stoppage of the cone (i.e., zero shear

FIGURE 4.26 Myrenne erythrocyte aggregometer. (Courtesy of Myrenne GmbH, Roetgen, Germany.)

rate) or after reducing the rotational speed to a very low level corresponding to a shear rate of 3 s^{-1}; the shear rate depends on the push-button (i.e., M or M1) used to start the measurement (see Figure 4.26) and the integration time is user selectable to be 5 or 10 s. The numerical value resulting from the integration of the transmitted light signal, which corresponds to the surface area under syllectogram, SA, as discussed in Section 4.1.6.2.1, is reported as the aggregation index. The aggregation index measured during stasis is termed the *M index*, while the one measured at low shear rate is the *M1 index*. These indexes correspond to the integrated light transmission intensity measured in arbitrary units, but are always reported as dimensionless indexes. Usually, the M and M1 indexes are measured sequentially without changing the sample and both M and M1 indexes can be determined within a minute. The cone and plate need to be manually cleaned following a measurement.

The Myrenne aggregation indexes (M and M1) reported in the literature have a wide range, even for blood from healthy controls (Vaya et al. 2003). This wide distribution is, at least in part, related to the difference in hematocrit of the whole blood samples. The user-selectable integration time (i.e., 5 or 10 s) also contributes to the confusion in the literature, especially if the authors do not report the time selected. M values for control specimens are usually reported to be between 3 and 6 if measured within 5 s and between 8 and 18 if measured within 10 s. The range of M1 values for 5- and 10-s integration times corresponds to 6–10 and 20–28, respectively.

Vaya et al. (2003) compared correlations of M and M1 indexes measured using two different integration times with plasma fibrinogen and lipid concentrations; hematocrit was adjusted to 0.45 l/l prior to the measurements. They reported that the M1 parameter measured in 5 s had the highest correlation coefficients compared to M1 measured at 10 s and M index measured at 5 and 10 s and recommended its use for clinical studies (Vaya et al. 2003).

The Myrenne aggregometer is a fast, easy-to-operate instrument and suitable for clinical usage. It is not possible to obtain information about the time course of aggregation with the basic instrument, which could be a problem in basic science studies. However, this instrument can be modified to be controlled by an external computer that also records and analyzes the syllectogram (Bauersachs et al. 1989; Deng et al. 1994). The modified instrument can be used to measure all the parameters described in Section 4.1.6.2 including the disaggregation shear rate. The Myrenne aggregometer is not temperature controlled and therefore as sold is suitable only for operation at ambient temperature. However, a temperature sensor and a heating element can be installed in the measurement chamber of the device and used for maintaining constant temperature, preferably 37°C, during the measurement procedure (Baskurt et al. 2009c).

4.1.6.9.2 LORCA as an Erythrocyte Aggregometer

The Laser-Assisted Optical Rotational Cell Analyzer (LORCA) was developed by M. Hardeman at the Academic Medical Center (Amsterdam) (Hardeman et al. 1994) and is being manufactured by R&R Mechatronics (Hoorn, The Netherlands; http://www.mechatronics.nl/products/lorrca/index.htm). The LORCA is a combination of an ektacytometer used for measuring RBC deformability and a photometric system for measuring aggregation (Figure 4.27). The device employs a shearing apparatus consisting of two coaxial glass cylinders (i.e., a Couette system) with a ~300-µm gap

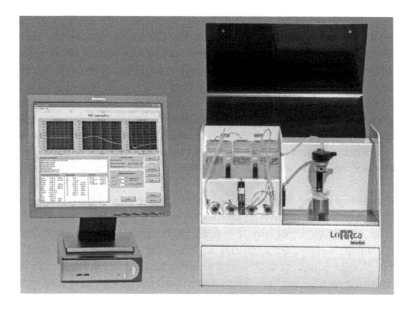

FIGURE 4.27 Laser-assisted optical rotational cell analyzer. (LORCA, courtesy of R&R Mechatronics, Hoorn, The Netherlands.)

between them. This device requires about 1 ml of sample, with shearing obtained by rotating the outer cylinder at the desired speed via a stepper motor that is controlled by an external computer. A 670-nm laser beam generated by a laser diode is directed to the sample in the gap between the cylinders and backscattered, and reflected light is recorded by two photodiodes built into the inner stationary cylinder (Figure 4.28). The measurement system of the LORCA is temperature controlled and can be operated between ambient temperature and 37°C. Although the instrument can maintain higher temperatures, usage of temperatures above 37°C is not recommended for routine studies.

The LORCA is operated with the aid of an external computer that controls the rotation of the outer cup and records and analyzes the light reflection signal to calculate aggregation parameters. The typical measurement procedure starts with a

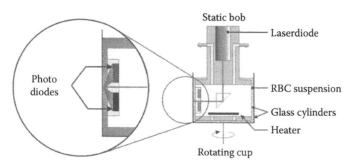

FIGURE 4.28 Shearing and photometric apparatus of LORCA system. (Courtesy of R&R Mechatronics, Hoorn, the Netherlands.)

disaggregation period of 10 s, then an abrupt stoppage of the rotating cup occurs, followed by recording of the syllectogram for 120 s. The parameters calculated by LORCA software using this syllectogram include the surface areas above (SA, corresponding to Myrenne M index) and below the syllectogram, the aggregation index (AI), syllectogram amplitude (AMP), aggregation half-time ($T_{1/2}$), and aggregation time constants T_{fast} and T_{slow} (see Section 4.1.6.2 for detailed description). Disaggregation shear rate and the flow-to-stasis ratio are also measured if selected to be done by the operator. A full analysis of a sample takes ~5 minutes. Older versions of the instrument require that the inner and outer cylinders be washed and dried manually whereas the newer version of the device has a built-in, automatic washing system.

The shear rate during disaggregation is user-selectable from the software and can be applied over a wide range (i.e., up to 3500 s^{-1}). This ability to select high shear rates is an advantage, especially if samples with strong aggregation tendency are being used (e.g., samples from patients with multiple myeloma) (Zhao et al. 1999); complete disaggregation for such samples cannot be achieved at shear rates of ~500 s^{-1} as used for routine analysis. The selected disaggregation shear rate should be reported with the results.

4.1.6.9.3 RheoScan-A

The RheoScan-A is a newer instrument developed by S. Shin (2009a) and manufactured by Rheomeditech, (Seoul, Korea; http://www.rheoscan.com). The instrument operates with a disposable plastic unit (Figure 4.29) having a disc-shaped sample chamber that is 4 mm in diameter and 0.3 mm high containing a metal bar. The instrument is computer controlled and has a motor that rotates a magnet beneath the sample chamber, thereby rotating the tiny metal bar in the sample chamber to achieve disaggregation as the initial step of the measurement procedure (see Figure 4.29). A diode laser at 635 nm is used as the light source and the optical path is perpendicular to the disc-shaped measurement chamber. A sensor positioned opposite to the laser diode measures light transmission through the sample.

After loading the small amount of sample (<10 μl), the measurement starts with the rotation of the metal bar in the measurement chamber at 900 rpm for 10 s. Following a sudden stop of the rotating mechanism, light transmittance is recorded for 120 s and then the recorded syllectogram is analyzed as described in Section 4.1.6.2. The RheoScan-A cannot measure disaggregation threshold, but the RheoScan-D, a slit-flow ektacytometer manufactured by the same company, can measure the critical shear stress for disaggregation; a combination model (Rheoscan-AnD) that can use slit flow and circular chambers is also available.

The disaggregation mechanism (i.e., rapidly spinning metal bar) was found to be effective for total disaggregation even in hyperaggregating blood samples (Shin et al. 2007). It has also been reported that this mechanism does not cause any RBC hemolysis (Shin et al. 2009a), at least when rotating the bar at 900 RPM. The disaggregation conditions cannot be expressed as shear rate or stress and therefore are not comparable with other instruments. The instrument in its current configuration is not temperature controlled.

(a)

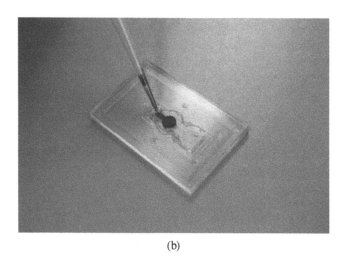

(b)

FIGURE 4.29 RheoScan-A erythrocyte aggregometer (a) and the disposable microfluidic unit with the metal stirring bar (b). (Courtesy of Rheomeditech, Seoul, South Korea.)

4.1.6.9.4 Sefam Erythroaggregometer

Another instrument that has been used for the measurement of RBC aggregation is the Sefam erythroaggregometer, which was developed by J-F. Stoltz (Donner et al. 1988; Stoltz et al. 1987) and produced by Regulest, France but is no longer manufactured. It has a Couette type shearing system and it monitors light reflectance during the aggregation process following disaggregation. The instrument is temperature controlled.

Some parameters calculated by this instrument differ from other instruments (Freyburger et al. 1993; Zhao et al. 1999). The primary aggregation time is calculated as the reciprocal of the slope of the syllectogram between 0.5–2 s after the sudden stop of shearing. Aggregation indexes S10 and S60 are calculated by integrating the area above the syllectogram during 10 and 60 s after the disaggregation period. The instrument also measures the partial disaggregation threshold as the shear rate for maximum light reflectance, corresponding to the disaggregation shear rate as described in Section 4.1.6.3.

4.1.6.9.5 Test-1

Test-1 is an instrument marketed as "a fully automated analyzer for the measurement of ESR" (Plebani et al. 1998, p. 334) and is therefore discussed briefly in Section 4.1.1.5. However, it is really an RBC aggregometer as indicated by the description of the working principle as "photometrical capillary stopped flow kinetic analysis" (Plebani and Diva 2000, p. 622). The instrument measures infrared light transmittance for 20 s through a small amount of whole blood sample in a flow cell. The instrument reports results converted to ESR values calculated using a regression equation.

4.1.6.9.6 Comparison of the Photometric Red Blood Cell Aggregometers

Table 4.1 summarizes important features of the three instruments currently available. These properties apply to the basic instrument that can be purchased and do not include possible custom modifications.

TABLE 4.1

Comparison of the Three Currently Available Instruments for RBC Aggregation Measurement

	Myrenne Aggregometer	LORCA	RheoScan-A
Shearing system	Cone-plate	Coaxial cylinders	Magnetic stirring
Disaggregation shear rate	Fixed, 600 s^{-1}	User-selectable, <3500 s^{-1}	NA
Measurement mode	Light transmittance	Light reflectance	Light transmittance
Light source wavelength	Infrared	670 nm	635 nm
Sample size	30 µl	1000 µl	10 µl
Parameters (see Section 4.1.6.1)	M, M1	AMP, SA, AI, $T_{1/2}$, T_{fast}, T_{slow}, γT_{min}, F/S ratio	SA, AI, $T_{1/2}$, T_{fast}, T_{slow}
Temperature control	No	Yes	No

TABLE 4.2

Instrument Precision Expressed as Coefficient of Variation (%) for Various Parameters Measured Using the Three Currently Available Instruments for RBC Aggregation Measurement.

	Myrenne Aggregometer	LORCA	RheoScan-A
M (SA)	2.76%	2.75%	7.41%
M1	3.08%	—	—
AI	—	2.44%	2.43%
AMP	—	4.20%	6.79%
$T_{1/2}$	—	4.11%	4.51%

Source: Data taken from Baskurt, O. K., M. Uyuklu, P. Ulker, M. Cengiz, N. Nemeth, T. Alexy, S. Shin, M. R. Hardeman, and H. J. Meiselman. 2009c. "Comparison of Three Instruments for Measuring Red Blood Cell Aggregation." *Clinical Hemorheology and Microcirculation* 43:283–298.

Table 4.2 presents intra-assay variation (i.e., instrument precision) for the parameters provided by these devices as the coefficient of variation (CV) for 10 repeated measurements on the same sample; a CV less than 5% is necessary for a parameter to be acceptable (Baskurt et al. 2009a; Stuart et al. 1989). CV values were found to be lower than 5% for all parameters measured by the Myrenne aggregometer and LORCA, while SA and AMP parameters obtained using the RheoScan-A had a higher CV that exceeded the 5% limit (Baskurt et al. 2009c). The instrument precision values for the Myrenne and LORCA were in agreement with previously reported figures by other investigators (Hardeman et al. 2001; Hardeman et al. 1994; Rampling and Martin 1989). Shin et al. (2009a) reported CV levels between 0.46 to 3.2% for the aggregation parameters presented in Table 4.2, using RheoScan-A in a separate study.

Baskurt et al. (2009c) also compared the power of aggregation parameters in detecting the difference between two groups with experimentally altered intensity of aggregation. Standardized differences (i.e., difference between the mean values of the two groups divided by the common standard deviation of the two groups) can be used as a measure of the power of a parameter (Stuart et al. 1989). This approach is similar to the power analysis of a method (Cohen 1988). It was reported that parameters reflecting the time course of aggregation (AI and $T_{1/2}$) were characterized by a somewhat higher power to detect the differences between normal RBC suspensions and those prepared in diluted plasma (i.e., decreased aggregation); results obtained for the three instruments (Myrenne, LORCA, and RheoScan-A) did not significantly differ. Experimentally enhanced RBC aggregation achieved by including 0.5% dextran 500 (500 kDa) in the suspending medium significantly influenced the LORCA SA and AMP parameters that reflect the extent of aggregation. The Myrenne aggregometer M and M1 indexes had a power similar to the LORCA, but the RheoScan-A failed to detect this enhanced aggregation. It was suggested that this insensitivity of RheoScan-A might be due to incomplete disaggregation of the RBC suspensions in the dextran 500 suspending media (Baskurt et al. 2009c).

Zhao et al. (1999) compared the Myrenne aggregometer, LORCA, and Sefam erythroaggregometer and reported similar performance for detecting experimentally induced changes in aggregation (e.g., suspending RBC in various hyperaggregating media). However, they reported biological variation, but not instrument precision in terms of CV of repeated measurements on the same sample, and therefore the results are not comparable with the study mentioned previously. An important observation was the failure of the Myrenne aggregometer in detecting the very prominent enhancement in RBC aggregation in multiple myeloma patients, while the other two instruments reported significantly enhanced aggregation parameters. Falco et al. (2005) also reported a similar failure of the Myrenne aggregometer in detecting the highly enhanced aggregation in samples with plasma fibrinogen over 400 mg/dl, while the Sefam erythroaggregometer results were correlated with fibrinogen concentration over this level. This failure seems related to incomplete disaggregation at the initial phase of the measurement, leading to lower aggregation indexes in strongly aggregating samples (Pieragalli et al. 1985; Zhao et al. 1999). Hardeman et al. (2010b) compared RBC aggregation measured by the LORCA aggregometer and the Test-1. They reported highly significant correlations among various aggregation parameters reported by the LORCA (e.g., AI, $T_{1/2}$) and Test-1 results using blood samples from 75 patients (Hardeman et al. 2010b).

4.1.7 ULTRASOUND BACK-SCATTERING

An ultrasound wave directed to a suspension is scattered by the suspended particles (Cloutier 1999). The theoretical background for this phenomenon is provided by Rayleigh scattering theory, including the dependence of scattering on the frequency of the ultrasound signal (Shung and Thieme 1993). The power of the backscattered signal is related to the fourth power of the ultrasound frequency and the square of the volume of scattering particles (Yu et al. 2009). However, this relationship is only strictly correct if the scattering particles are smaller than the wavelength of the signal (Cloutier 1999).

Ultrasound backscattering from RBC suspensions was demonstrated to be shear dependent (Cloutier et al. 1996; Haider et al. 2004). Furthermore, it was also demonstrated to be affected by the fibrinogen or high molecular mass dextran concentrations in the suspending medium (Kitamura et al. 1995; Rouffiac et al. 2003; Shung and Reid 1979; Yu and Cloutier 2007), macromolecules that are known to correlate with the intensity of RBC aggregation. This dependence can be understood considering the increase in effective particle volume with RBC aggregation leading to increased power of ultrasound backscattering (Cloutier and Qin 1997). Ultrasound backscattering was proposed to be sensitive even to the formation of small aggregates of 2–3 RBC (Cloutier and Shung 1993). Monitoring ultrasound backscattering has therefore been used as a method to estimate aggregate size (i.e., average number of RBC per aggregate) and to investigate the dynamics of aggregation (Boynard and Lelievre 1990; Cloutier et al. 1996; Rouffiac et al. 2003; Yu et al. 2009).

Ultrasound backscattering has been used to quantitate *in vitro* RBC aggregation. These investigations employed various shearing mechanisms (e.g., Couette, tube flow) (Rouffiac et al. 2003; Yu et al. 2009) and analyzed backscattered power either under various shear rates (Haider et al. 2004; Rouffiac et al. 2003; Shehada et al. 1994; Yu et al. 2009) or during the time course of aggregation following a shearing period for disaggregation (Rouffiac et al. 2003), as discussed for photometric methods in Section 4.1.6.

Various parameters calculated based on ultrasound backscattering power have been proposed for the evaluation of RBC aggregation. A widely used parameter is the backscattering coefficient (BSC), defined as the power backscattered by a unit volume of scatterers per unit incident intensity, per unit solid angle (Shung et al. 1976). Therefore, it is independent of factors other than RBC aggregation that affect backscattered ultrasound power. BSC was demonstrated to follow a typical time course during aggregation resembling the syllectograms discussed for photometric methods (Section 4.1.6). Rouffiac et al. (2003) used a mathematical approach very similar to that described for analyzing syllectograms (Section 4.1.6.2) for analyzing BSC-time records following a disaggregation period in a Couette-type shearing apparatus. Their comparison of aggregation indexes calculated using BSC data with those obtained using a laser backscattering aggregometer (Sefam erythroaggregometer, Section 4.1.6.9.4) revealed significant correlations.

Yu and Cloutier (2007) developed *the structure factor size estimator model,* based on the dependence of BSC on ultrasound frequency, which can be used to calculate two parameters related to the microstructure of RBC aggregates. These parameters are the *packing factor (W)* and *the ensemble averaged aggregate isotropic diameter (D)*, which can be evaluated under various shearing conditions (Yu et al. 2009). The parameter D corresponds to an index of aggregate size in terms of the number of RBC per aggregate (Yu and Cloutier 2007). Figure 4.30 demonstrates the shear dependence of the W and D parameters in various porcine RBC suspensions. Neither parameter was influenced by shear rate for a nonaggregating RBC suspension prepared in phosphate-buffered saline. Conversely, a highly aggregating suspension of poloxamer-coated RBC (see Chapter 2, Section 2.3.5.4) exhibited a strong dependence on shear rate compared to normal blood.

Other approaches, including pattern recognition techniques, have also been used to estimate aggregate size by measuring backscattered ultrasound power (Aggelopoulos et al. 1997; Boynard and Lelievre 1990; Karabetsos et al. 1999). B-mode echo images were used to visualize the distribution in backscattering power during flow in order to estimate RBC aggregation properties (Shehada et al. 1994). Ultrasound interferometry is another approach to investigate RBC aggregation *in vitro* (Razavian et al. 1991). This approach is based on A-mode echography in a specially designed *echo cell* during the sedimentation of RBC on a solid surface positioned on top of an ultrasound transducer. It has been suggested that by monitoring the echo signal from the sedimenting suspension, the initial aggregation process, the sedimentation rate of aggregates, and the packing of sedimented RBC can be studied.

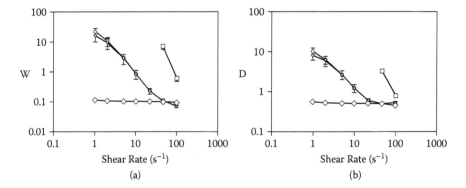

FIGURE 4.30 Packing factor (*W*) and aggregate isotropic diameter (*D*) calculated by structural factor size estimator model based on ultrasound backscattering coefficient measurements from porcine RBC suspensions over a range of shear rates. Both the *W* and *D* parameters did not exhibit dependence on shear rate in nonaggregating RBC suspensions (◇), while suspension of poloxamer-coated RBC with highly enhanced aggregation (□) had higher *W* and *D* values with a strong dependence on shear rate compared to RBC suspensions with normal aggregation (▽○). Reproduced with permission from Yu, F. T. H., E. Franceschini, B. Chayer, J. K. Armstrong, H. J. Meiselman, and G. Cloutier. 2009. "Ultrasonic Parametric Imaging of Erythrocyte Aggregation Using the Structure Factor Size Estimator." *Biorheology* 46:343–363.

Ultrasound-based methods are especially promising for studying RBC aggregation *in vivo*. Increased echogenicity in certain regions of the vasculature has been well known by clinicians for several decades, with these observations linked with RBC aggregation, especially under low flow conditions (Cloutier et al. 1996). Although there were promising efforts for determining RBC aggregation *in vivo* using ultrasound backscattering (Cloutier et al. 1997; Kitamura and Kawasaki 1997), well-established methods do not yet exist. B-mode echography is the preferred method for *in vivo* studies, but this approach provides only qualitative data (Yu et al. 2009). An important limitation of *in vivo* studies of RBC aggregation is the control of shearing conditions (Rouffiac et al. 2003), thus making it extremely difficult to obtain comparable parameters based on ultrasound backscattering. However, flow in a blood vessel can be temporarily stopped by external manipulations, and this technique may provide the opportunity to study the time course of RBC aggregation in a manner similar to the methods used *ex vivo*. The aggregation parameters developed by Rouffiac et al. (2003) using BSC recorded after flow cessation were designed to be used under such conditions.

4.1.8 ELECTRICAL PROPERTIES OF RBC SUSPENSIONS

The electrical impedance of RBC suspensions is determined by plasma resistivity, hematocrit, and the geometry and orientation of RBC and can be evaluated in terms of the Maxwell–Fricke equation (Fricke 1924). This approach considers the significant contribution of RBC, mostly due to the electrical behavior of their membranes, which have very high resistivity and some capacitance. The electrical properties (e.g., impedance, capacitance) of blood are well known to depend on whether the sample is undergoing flow or at stasis (Gaw et al. 2008; Hoetink et al. 2004). This behavior has been explained by changes in the geometrical factors due to RBC deformation/orientation and aggregation (Beving et al. 1994; Pribush et al. 2004b; Pribush et al. 1999; Zhao and Jacobson 1997). Interestingly, the electrical impedance of an RBC suspension can exhibit a syllectogram-like time course during the aggregation process (Baskurt et al. 2009b; Pribush et al. 2007).

The electrical properties of RBC suspensions can be recorded under conditions similar to that used for syllectometry (Section 4.1.6). That is, the RBC suspension under investigation is sheared at a level sufficient for complete disaggregation and electrical impedance is recorded following a sudden cessation of shearing. Both impedance and capacitance derived from impedance data increase immediately following the sudden stop of shearing, reaching a maximum within a second, then decline during the rest of the period of several hundred seconds (Figure 4.31). Both impedance and capacitance time courses have been suggested to carry information about the time course of RBC aggregation (Baskurt et al. 2009b; Pribush et al. 2000; Pribush et al. 2004b). Various shearing geometries have been used for studying the electrical properties of RBC suspensions, including a rectangular flow chamber (Pribush et al. 1999; Pribush et al. 2004b), a Couette type viscometer (Antonova et al. 2008), and straight tubes (Baskurt et al. 2009b). Different levels of shearing can be obtained by pumping the blood at calculated flow rates through the flow systems (Baskurt et al. 2009b; Baskurt et al. 2010; Pribush et al. 2000) or by rotating the bob of the Couette system (Antonova and Riha 2006).

An important factor in recording electrical parameters is the electrode configuration in order to minimize effects related to factors such as electrode polarization and the influence of high current density at the electrode interface (Baskurt et al. 2010; Hoetink et al. 2004). Measuring electrical impedance across a glass tube of 1 mm ID and 75 mm length using metal electrodes with a large surface area in contact with blood at both ends of the tube minimizes electrode effects due to the relatively high impedance and low current density over a useful frequency range (Baskurt et al. 2010). The test signal frequency is another important factor in monitoring RBC aggregation by the electrical properties of cell suspensions: it should be in the β-dispersion range (Holder 2005; Pribush et al. 1999). Test frequencies in the range of 2–200 kHz have been used in studies of RBC aggregation (Antonova and Riha 2006; Baskurt et al. 2010; Pribush et al. 1999). The electrical properties of RBC suspensions are dependent on hematocrit (Baskurt et al. 2010), and therefore the hematocrit of blood samples should be adjusted to a standard value prior to electrical measurement.

Baskurt et al. (2010) employed the mathematical approach applied to syllectograms (see Section 4.1.6.2) to derive aggregation indexes from electrical impedance and capacitance time courses. The parameters calculated using electrical properties (i.e., impedance, capacitance) were compared with those calculated using simultaneously recorded light transmittance. Interestingly, the aggregation parameters (SA, AI, $T_{1/2}$) calculated using the time course of capacitance during RBC aggregation were significantly correlated with the same indexes calculated using light transmittance data and those reported by commercial photometric

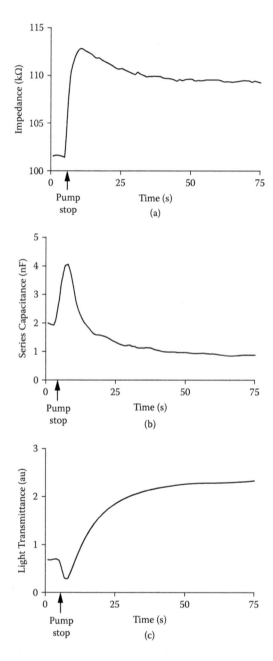

FIGURE 4.31 Electrical impedance (a), series capacitance (b), and light transmission (c) recorded simultaneously during red blood cell aggregation in 0.4 l/l hematocrit blood in a 1 mm ID by 67 mm-long glass tube. Impedance was monitored using a test signal of 100 kHz, while a 700 nm wavelength LED was used as the light source. The arrow indicates the sudden stop of flow at a shear rate of ~500 s^{-1}.

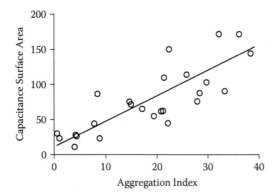

FIGURE 4.32 Correlation between RBC aggregation parameters obtained by electrical and photometric devices. The surface area above the capacitance-time curve was measured using a 1 mm ID glass tube and the aggregation index measured by the photometric Myrenne aggregometer (r: 0.84, $p < 0.0001$). Drawn using data from Baskurt, O. K., M. Uyuklu, and H. J. Meiselman. 2010. "Time Course of Electrical Impedance during Red Blood Cell Aggregation in a Glass Tube: Comparison with Light Transmittance." *IEEE Transactions on Biomedical Engineering* 57:969–978.

aggregometers (Figure 4.32). However, indexes calculated using the time course of impedance failed to exhibit a significant correlation (Baskurt et al. 2010) and thus monitoring the capacitance of RBC suspensions was recommended for the assessment of RBC aggregation.

Kaliviotis et al. (2010) compared RBC aggregation parameters based on image analysis with data obtained based on electrical conductometry measured at 2 kHz in a Couette-type shearing system. They reported good agreement of the data obtained by the two approaches, indicating that conductometry at the low frequency they used might also be a promising method for assessing RBC aggregation (Kaliviotis et al. 2010).

4.1.9 MEASUREMENT OF RED BLOOD CELL AGGREGATION *IN VIVO*

It has been suggested that *in vivo* RBC aggregation can be monitored and even quantified. Ultrasound backscattering is one of the methods proposed for *in vivo* measurements (see Section 4.1.7) since this approach can be developed as a noninvasive method applicable to large blood vessels and also to microvessels as small as 50 μm in diameter (Cloutier 1999).

Photometric methods have also been suggested to be applicable to assessment of *in vivo* RBC aggregation. Tomita et al. (1986) claimed that the *densitometer head* for measuring light transmittance across cylindrical tubes could also be applied to *in vivo* blood vessels, although this method requires dissection of the vessel and is therefore an invasive approach. Light transmittance of blood *in vivo* has a pulsatile nature, suggesting cyclic changes in RBC aggregation within the cardiac cycle (Shvartsman and Fine 2003). More pronounced optical changes can be recorded following temporary occlusion of blood vessels *in vivo*; such signals can be used to assess RBC aggregation noninvasively using instrumentation similar to that used for pulse oximetry (Shvartsman and Fine 2003).

Lipowsky estimated the degree of RBC aggregation based on light backscattering during *in vivo* viscometry in order to study the relationship between shear rate and aggregation in postcapillary venules (Pearson and Lipowsky 2004). It was demonstrated that the power spectrum of light intensity transmitted through blood vessels with diameter <10 micrometer shifted to lower frequencies, thus indicating increased RBC aggregation (Soutani et al. 1995).

Pop et al. proposed that the electrical properties of blood monitored by electrodes mounted on catheters could be used to estimate RBC aggregation *in vivo* (Pop et al. 2004b; Pop et al. 2004a; Pop et al. 2003), and reported a significant correlation between blood viscosity and electrical impedance. Furthermore, they used a specially designed central venous catheter equipped with electrodes for monitoring electrical impedance as an approach for the detection of early signs of infection or inflammation by continuous online monitoring of hemorheological alterations (Iliev et al. 2008).

4.2 SAMPLE PREPARATION FOR RED BLOOD CELL AGGREGATION MEASUREMENT

4.2.1 BLOOD SAMPLING AND STORAGE FOR MEASURING RED BLOOD CELL AGGREGATION

Blood sampling for investigation of hemorheological parameters, including RBC aggregation, should be done with special care as summarized in the guidelines for the measurement of hemorheological parameters (Baskurt et al. 2009a). Sampling time and conditions should be standardized and reported, inasmuch as RBC aggregation is significantly affected by plasma composition, which is influenced by diurnal variations, hydration of the subject, recent food/fluid intake, physical activities, and environmental conditions. A usual sampling time for human subjects is between 8.00 a.m. and 10.00 a.m. after an overnight fast, with sampling for human subjects usually done from an antecubital vein. A tourniquet may be applied for locating the vein; hemorheological parameters are less affected if the sampling is completed within 90 s of application without removing the tourniquet (Connes et al. 2009). Narrow-bore needles should be avoided since RBC may be damaged by the high shear stresses generated in the needle during sampling; 21g needles are recommended. Commercially available vacuum tubes for blood sampling are also recommended since they generate a standard vacuum for blood flow; use of these tubes may also result in more accurate and reproducible anticoagulant concentrations in the samples.

Blood sampling should be done using a nondiluent anticoagulant, except the samples for a classical Westergren sedimentation test in which 1/5 diluted samples in 3% trisodium citrate are used (Section 4.1.1.1.2). The recommended anticoagulant for the measurement of RBC aggregation is ethylenediaminetetraacetic acid (EDTA) at a concentration of 1.5 mg/ml blood. The most widely used alternative to EDTA is heparin at a concentration of 15 IU/ml. However, heparin was shown to slightly enhance RBC aggregation at this recommended concentration (Jan 1986; Kameneva et al. 1994) but not at lower concentrations (Martinez M. et al. 2000). It should be noted that anticoagulants might have differential effects on RBC rheology of blood of various species (Nemeth et al. 2009).

The recommended maximum storage time after the sampling of blood for RBC aggregation measurement is 4 h (Uyuklu et al. 2009a). If the delay between sampling and the measurement exceeds 1 hr, the sample should be refrigerated at 4°C. The sample should be warmed to the measurement temperature prior to the measurement.

4.2.2 HANDLING OF THE SAMPLES AND HEMATOCRIT ADJUSTMENT

Blood samples should be handled with special care to avoid RBC damage and changes in the suspending phase (i.e., plasma) composition. However, hematocrit adjustment to a standard value in the normal range (e.g., 0.4 l/l) prior to the measurement is strongly recommended if a lower hematocrit value is not required by the specific method (e.g., methods based on microscopic analysis) (Baskurt et al. 2009a).

Hematocrit adjustments should be done by adding or removing calculated amounts of autologous plasma to or from whole blood samples. Complete separation of the RBC pellet and plasma and then recombining to achieve the desired hematocrit, as well as washing RBC with isotonic buffers, should be avoided unless there is a special reason for such manipulations (e.g., need to remove white blood cells and/ or platelets). Instead, plasma should be removed or added to the tube containing the original blood sample followed by resuspending the RBC in the same tube.

For example, to adjust the hematocrit of a 1 ml sample to 0.4 l/l starting with an original value of 0.35 l/l, the sample volume should be reduced to 0.875 ml (0.35/0.4) by removing 0.125 ml of plasma. This volume of plasma can be removed using a pipette after partial separation of plasma by gentle centrifugation (e.g., 1500 g, 5 min at room temperature), then resuspending the RBC in the same tube. Adjustment of hematocrit to 0.4 l/l in a 1 ml sample with a 0.48 l/l original hematocrit requires the addition of autologous plasma to bring the total volume to 1.2 ml (0.48/0.4) by adding 0.2 ml of plasma obtained by centrifugation of a *separate aliquot* of the same blood sample. It is strongly recommended that the hematocrit of samples before and after adjustments be checked using a microhematocrit centrifuge rather than an electronic hematology analyzer, especially for samples with significant changes in RBC mechanical properties; when using automated analyzers, impaired RBC deformability can result in a falsely elevated hematocrit value.

If RBC washing is necessary during sample preparation, this should be done after removing the plasma and buffy coat following gentle centrifugation (e.g., 1500 g, 5 min). The RBC pellet, minus the buffy coat, is resuspended in the wash solution, centrifugation is repeated, and the wash solution aspirated and replaced with a new aliquot. Isotonic phosphate-buffered saline (PBS, 290 ± 10 mOsmKl, 10 mM phosphate, pH: 7.4) is recommended for wash procedures for most samples (Keidan et al. 1987); two wash steps are usually sufficient. Other buffers such as HEPES can be used for wash procedures if PBS is not appropriate for the study (e.g., measurements using calcium-containing media). However, possible adverse effects on cell morphology and/or rigidity should be evaluated with a preliminary study if the buffer is being used for the first time. If the final suspending medium is not the wash buffer, the RBC should be washed once in this medium (e.g., isotonic polymer solutions) then resuspended at the desired hematocrit.

The morphology of RBC should be evaluated in the final RBC suspensions to determine if they are damaged or influenced by the wash and resuspension procedures. This can be done using a wet-mount preparation of a dilute suspension of RBC in the medium. Since RBC morphology can be markedly affected by direct contact with glass surfaces, an appropriate volume of this dilute suspension should be transferred to a thin well of silicone grease on a glass slide, and then covered by a cover slip; RBC can be viewed in the mid-plane of such freshly prepared slides. Modification of the blood handling procedures should be carried out if there are major changes of RBC morphology (e.g., extensive echinocyte or stomatocyte formation). Addition of albumin to the suspending medium at a concentration of 0.1% can reduce such shape alterations. Details of the wash procedure and any changes in RBC morphology should be reported.

Aggregation measurements using some instruments can be affected by the oxygenation status of the samples and therefore standardization of oxyhemoglobin saturation when using these instruments is highly recommended as discussed in Section 4.1.6.8.3.

4.3 ASSESSMENT OF RED BLOOD CELL AGGREGABILITY

As detailed in Chapter 1, the term *aggregability* refers to the intrinsic tendency of RBC to aggregate in various suspending media (Baskurt et al. 2009a; Baskurt and Meiselman 2009; Neu and Meiselman 2007; Rampling et al. 2004). This property of the RBC reflects cellular factors contributing to the aggregation process (see Chapter 2, Section 2.3).

It is important to note that the determination of RBC aggregability requires a *comparison* between two cell populations in the same suspending medium, and thus an isolated aggregation measurement, by itself, is not sufficient for this purpose. The measurement can be made using any appropriate suspending medium that promotes aggregation, including compatible plasma and isotonic solutions containing high molecular mass proteins or polymers. Comparison of aggregation parameters measured in the same suspending medium eliminates the influence of suspending medium factors and provides information regarding differences in cellular properties contributing to the aggregation process (Baskurt and Meiselman 2009).

The standard suspending media should be selected based on two important properties:

a. RBC properties should not be affected by the suspending medium. It should be isotonic and RBC shape in the suspending medium confirmed to be normal.

b. Aggregate morphology should be similar to that seen with normal blood (i.e., regular rouleaux formation). The medium should not induce extremely strong aggregation that results in aggregates that cannot be dispersed by the disaggregation shear rate used by a specific instrument. Further, if the suspending medium causes very intense aggregation, all samples may exhibit the same high level of aggregation regardless of differences in cellular properties.

The suspending media for determining aggregability can be prepared using high-molecular mass polymers such as dextrans, with the intensity of aggregation a function of both polymer molecular size and concentration (Figure 4.33) (Baskurt et al. 2000). The two widely used suspending media are 3% dextran 70 (70 kDa) and 0.5% dextran 500 (500 kDa) dissolved in PBS. Ben Ami et al. (2001) reported that RBC aggregate sizes in 0.5% dextran 500 were similar to those in normal, autologous plasma and therefore this medium is suitable for the study of aggregability; other macromolecules can also be used as long as the two properties mentioned just previously are considered. It is recommended that, if available, the polydispersity of the polymer (i.e., ratio of weight-average to number-average molecular mass) should be stated and the measured viscosity of the polymer solution reported (Baskurt et al. 2009a). Suspensions in these media should be prepared using RBC washed twice in PBS, at the hematocrit used for measurements in autologous plasma.

It should be noted that shear forces required for complete disaggregation during the measurement procedure might be higher in polymer-containing suspending media (Baskurt et al. 2009c). Disaggregation shear rates might be selected at a higher level, if the employed instrument allows such selections.

Ben Ami et al. (2001) calculated the contribution of plasma factors (PF) to RBC aggregation in pathological and control samples using aggregation parameters measured in autologous plasma and 0.5% dextran 500:

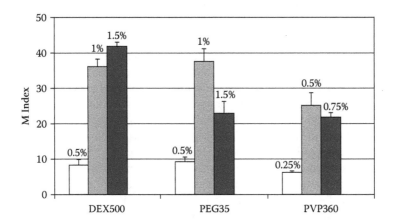

FIGURE 4.33 M index measured with a Myrenne aggregometer for rat RBC suspended in various polymer solutions. The three polymer concentrations employed are shown above each bar. DEX500: dextran (500 kDa); PEG35: poly(ethylene glycol) (35 kDa); PVP360: polyvinyl-pyrrolidone (350 kDa). Reproduced with permission from Baskurt, O. K., M. Bor-Kucukatay, O. Yalcin, H. J. Meiselman, and J. K. Armstrong. 2000. "Standard Aggregating Media to Test the 'Aggregability' of Rat Red Blood Cells." *Clinical Hemorheology and Microcirculation* 22:161–166.

$$PF = \frac{Ag_{Pl} - Ag_{Dex}}{Ag_{Pl}} \qquad (4.8)$$

where Ag_{Pl} is the difference between the aggregation of pathological and control blood each in their own plasma and Ag_{Dex} is the corresponding difference of aggregation parameters measured in 0.5% dextran 500. This approach allowed them to deduce the contribution of cellular factors by subtracting the value of PF from 1.

4.4   INTERPRETATION OF THE RESULTS

The methods discussed in this chapter provide quantitative measures of various aspects of RBC aggregation, with these data useful for monitoring changes in RBC aggregation due to plasmatic and cellular factors (Rampling et al. 2004). Several of the methods discussed were used to compare data from control and experimental groups and to assess disease-related alterations as discussed in detail in Chapters 2 and 8. However, several points should be kept in mind during interpretation of aggregation data due to the special nature of the aggregation parameters.

It is obvious from the description of the methods that some of the aggregation parameters are dimensionless indexes (e.g., AI, SA) while others reflect the kinetics of aggregation and are dependent on the geometry of the measurement chamber. Therefore, parameters obtained using different instruments are not directly comparable even if they are derived using the same mathematical approaches (Baskurt et al. 2009c). Furthermore, even data obtained in different laboratories using instruments by the same manufacturer have a wide distribution range (Vaya et al. 2003), thus limiting direct comparisons. It should be noted that due to the lack of aggregation standards, aggregation parameters are not similar in nature to blood chemistry parameters, which have normal ranges valid worldwide. Each laboratory should produce its own control data and normal range using a particular instrument. Alternatively, a control group can be used in each clinical and experimental study. Of course, comparisons should be done using data obtained under the same conditions (e.g., sample preparation, temperature, suspending medium).

The methods and parameters discussed in this chapter are characterized by different sensitivities to alterations of RBC aggregation as well as by different levels of precision. Therefore, detecting changes in RBC aggregation due to experimental or clinical influences depends on the sensitivity and precision of each parameter. The value of an aggregation parameter in detecting an alteration is given by its *power*. Power analysis for aggregation parameters should be done at the initial stage of a study (Cohen 1988). The power of a parameter can be estimated by calculating the *standardized difference*, given by the difference between the mean values of the experimental or patient group and the control group divided by their common standard deviation (Stuart et al. 1989). A higher standardized difference denotes more power for a parameter to detect an alteration in aggregation. Instrument precision and biological variability for parameters are among the main determinants of this power. Instrument precision can be estimated in terms of the coefficient of variation of repeated measurements (e.g., $n = 10$) on the same sample. This should be smaller than 5% for an acceptable aggregation parameter. This approach can also be used to estimate the required sample size (i.e., number of subjects in each group compared) to achieve a certain power (Stuart et al. 1989). Taking into account these statistical concerns would certainly increase the validity of data obtained by aggregation measurements.

There are a number of other concerns that should be kept in mind during aggregation measurements that may influence the result and interpretation accordingly:

a. *RBC morphology:* Even slight deviations from the biconcave discoid shape of normal RBC may result in significant alterations in RBC aggregation. This is especially true for severe echinocyte formation in that such shape-transformed RBC exhibit very low or no aggregability. Aggregation results are thus not meaningful if there are extensive morphological alterations induced by sampling or preparation procedures.

b. *RBC deformability:* Decreased deformability results in decreased aggregation, even if other plasmatic and cellular factors favor enhanced aggregation. Information about RBC deformability may therefore be helpful in interpreting results.

c. *Suspending phase viscosity:* Increased suspending phase viscosity slows the process of aggregation. This should be kept in mind during measurements on samples containing high concentrations of macroglobulins or polymers in the suspending medium. A correction for the influence of plasma viscosity has been proposed by multiplying the measured aggregation value by the measured plasma viscosity (Neumann et al. 1987).

d. *Disaggregation:* Most of the modern methods for the measurement of RBC aggregation require complete disaggregation at the initial stage of the measurement. If the procedure fails to provide sufficient shear force for complete disaggregation, lower aggregation indexes may result when, in fact, samples are characterized with high aggregation intensities. This paradoxical alteration can be detected and corrected by applying higher levels of shear.

A number of hemorheological, morphological, and biochemical assessments are strongly recommended to be part of RBC aggregation studies. These include measurements of plasma and whole blood viscosity, RBC deformability, plasma fibrinogen concentration, RBC indices including mean cellular volume and mean cellular hemoglobin concentration, and assessment of RBC and aggregate morphology.

LITERATURE CITED

Aggelopoulos, E. G., E. Karabetsos, and D. Koutsouris. 1997. "*In Vitro* Estimation of Red Blood Cells Aggregation Using Ultrasound Doppler Techniques." *Clinical Hemorheology and Microcirculation* 17:107–115.

Agosti, R., A. Clivati, M. Dettorre, F. Ferrarini, R. Somazzi, and E. Longhini. 1988. "Hematocrit Dependence of Erythrocyte Aggregation." *Clinical Hemorheology* 8:913–924.

Alexy, T., E. Pais, and H. J. Meiselman. 2009. "A Rapid Method to Estimate Westergren Sedimentation Rates." *Review of Scientific Instruments* 80: 096102.

Alexy, T., R. B. Wenby, E. Pais, L. J. Goldstein, W. Hogenauer, and H. J. Meiselman. 2005. "An Automated Tube-Type Blood Viscometer: Validation Studies." *Biorheology* 42:237–247.

Almog, B., R. Gamzu, C. H. Almog, J. B. Lessing, I. Shapira, S. Berliner, D. Pauzner, S. Maslovitz, and I. Levin. 2005. "Enhanced Erythrocyte Aggregation in Clinically Diagnosed Pelvic Inflammatory Disease." *Sexually Transmitted Diseases* 32:484–486.

Antonova, N., and P. Riha. 2006. "Studies of Electrorheological Properties of Blood." *Clinical Hemorheology Microcirculation* 35:19–29.

Antonova, N., P. Riha, and I. Ivanov. 2008. "Time Dependent Variation of Human Blood Conductivity as a Method for an Estimation of RBC Aggregation." *Clinical Hemorheology and Microcirculation* 39:69–78.

Barnes, H. A. 1995. "A Review of the Slip (Wall Depletion) of Polymer Solutions, Emulsions, and Particle Suspensions in Viscometers: Its Cause, Character, and Cure." *Journal of Non-Newtonian Fluid Mechanics* 56:221–251.

Barshtein, G., D. Wajnblum, and S. Yedgar. 2000. "Kinetics of Linear Rouleaux Formation Studied by Visual Monitoring of Red Cell Dynamic Organization." *Biophysical Journal* 78:2470–2474.

Baskurt, O. K., M. Bor-Kucukatay, O. Yalcin, H. J. Meiselman, and J. K. Armstrong. 2000. "Standard Aggregating Media to Test the 'Aggregability' of Rat Red Blood Cells." *Clinical Hemorheology and Microcirculation* 22:161–166.

Baskurt, O. K., M. Boynard, G. R. Cokelet, P. Connes, B. M. Cooke, S. Forconi, F. Liao, et al. 2009a. "New Guidelines for Hemorheological Laboratory Techniques." *Clinical Hemorheology and Microcirculation* 42:75–97.

Baskurt, O. K., R. A. Farley, and H. J. Meiselman. 1997a. "Erythrocyte Aggregation Tendency and Cellular Properties in Horse, Human, and Rat: A Comparative Study." *American Journal of Physiology—Heart and Circulatory Physiology* 273:H2604–H2612.

Baskurt, O. K., and F. Mat. 2000. "Importance of Measurement Temperature in Detecting the Alterations of Red Blood Cell Aggregation and Deformability Studied by Ektacytometry: A Study on Experimental Sepsis in Rats." *Clinical Hemorheology and Microcirculation* 23:43–49.

Baskurt, O. K., and H. J. Meiselman. 1997. "Cellular Determinants of Low-Shear Blood Viscosity." *Biorheology* 34:235–247.

Baskurt, O. K., and H. J. Meiselman. 2009. "Red Blood Cell Aggregability." *Clinical Hemorheology Microcirculation* 43:353–354.

Baskurt, O. K., H. J. Meiselman, and E. Kayar. 1998. "Measurement of Red Blood Cell Aggregation in a 'Plate-Plate' Shearing System by Analysis of Light Transmission." *Clinical Hemorheology Microcirculation* 19:307–314.

Baskurt, O. K., A. Temiz, and H. J. Meiselman. 1997b. "Red Blood Cell Aggregation in Experimental Sepsis." *Journal of Laboratory and Clinical Medicine* 130:183–190.

Baskurt, O. K., M. Uyuklu, and H. J. Meiselman. 2010. "Time Course of Electrical Impedance during Red Blood Cell Aggregation in a Glass Tube: Comparison with Light Transmittance." *IEEE Transactions on Biomedical Engineering* 57:969–978.

Baskurt, O. K., M. Uyuklu, and H. J. Meiselman. 2009b. "Simultaneous Monitoring of Electrical Conductance and Light Transmittance during Red Blood Cell Aggregation." *Biorheology* 46:239–249.

Baskurt, O. K., M. Uyuklu, S. Ozdem, and H. J. Meiselman. 2011. "Measurement of Red Blood Cell Aggregation in Disposable Capillaries and Correlation with Erythrocyte Sedimentation Rate." *Clinical Hemorheology and Microcirculation* 47: 295–305.

Baskurt, O. K., M. Uyuklu, P. Ulker, M. Cengiz, N. Nemeth, T. Alexy, S. Shin, M. R. Hardeman, and H. J. Meiselman. 2009c. "Comparison of Three Instruments for Measuring Red Blood Cell Aggregation." *Clinical Hemorheology and Microcirculation* 43:283–298.

Baskurt, O. K., M. Uyuklu, M. R. Hardeman, and H. J. Meiselman. 2009d. "Photometric Measurements of Red Blood Cell Aggregation: Light Transmission versus Light Reflectance." *Journal of Biomedical Optics* 14, no. 5.

Baskurt, O. K., O. Yalcin, F. Gungor, and H. J. Meiselman. 2006. "Hemorheological Parameters as Determinants of Myocardial Tissue Hematocrit Values." *Clinical Hemorheology and Microcirculation* 35:45–50.

Baskurt, O. K., O. Yalcin, S. Ozdem, J. K. Armstrong, and H. J. Meiselman. 2004. "Modulation of Endothelial Nitric Oxide Synthase Expression by Red Blood Cell Aggregation." *American Journal of Physiology—Heart and Circulatory Physiology* 286:H222–H229.

Bauersachs, R. M., R. B. Wenby, and H. J. Meiselman. 1989. "Determination of Specific Red Blood Cell Aggregation Indices via an Automated System." *Clinical Hemorheology* 9:1–25.

Becton Dickinson Co. 1996. "Evacuated blood collection tube for erythrocyte sedimentation rate determination." In *Seditainer Product Sheet.* Franklin Lakes, NJ: Becton Dickinson Co.

Ben Ami, R., G. Barshtein, D. Zeltser, Y. Goldberg, I. Shapira, A. Roth, G. Keren, et al. 2001. "Parameters of Red Blood Cell Aggregation as Correlates of the Inflammatory State." *American Journal of Physiology—Heart and Circulatory Physiology* 280:H1982–H1988.

Bennish, M., M. O. Beem, and V. Ormiste. 1984. "C-Reactive Protein and Zeta-Sedimentation Ratio as Indicators of Bacteremia in Pediatric-Patients." *Journal of Pediatrics* 104:729–732.

Berliner, S., R. Ben-Ami, D. Samocha-Bonet, S. Abu-Abeid, V. Schechner, Y. Beigel, I. Shapira, S. Yedgar, and G. Barshtein. 2004. "The Degree of Red Blood Cell Aggregation on Peripheral Blood Glass Slides Corresponds to Inter-Erythrocyte Cohesive Forces in Laminar Flow." *Thrombosis Research* 114:37–44.

Berliner, S., O. Rogowski, S. Aharonov, T. Mardi, T. Tolshinsky, M. Rozenblat, D. Justo, et al. 2005. "Erythrocyte Adhesiveness/Aggregation: A Novel Biomarker for the Detection of Low-Grade Internal Inflammation in Individuals with Atherothrombotic Risk Factors and Proven Vascular Disease." *American Heart Journal* 149:260–267.

Beving, H., L. E. G. Eriksson, C. L. Davey, and D. B. Kell. 1994. "Dielectric Properties of Human Blood and Erythrocytes at Radio Frequencies (0.2–10 MHz): Dependence on Cell Volume Fraction and Medium Composition." *European Biophysics Journal* 23:207–215.

Biernacki, E. 1897. "Samoistna sedymentacja krwi jako naukowa, praktyczno-kliniczna metoda badania." *Gazeta Lekarska* 17:962–996.

Borawski, J., and M. Mysliwiec. 2001. "The hematocrit-Corrected Erythrocyte Sedimentation Rate Can Be Useful in Diagnosing Inflammation in Hemodialysis Patients." *Nephron* 89:381–383.

Boycott, A. E. 1920. "Sedimentation of Blood Corpuscles." *Nature* 104:532.

Boynard, M., and J. C. Lelievre. 1990. "Size Determination of Red Blood Cell Aggregates Induced by Dextran Using Ultrasound Backscattering Phenomenon." *Biorheology* 27:39–46.

Bridgen, M. L., and N. E. Page. 1993. "Three Closed-Tube Methods for Determining Erythrocyte Sedimentation Rate." *Laboratory Medicine* 24:97–102.

Brooks, D. E., J. W. Goodwin, and G. V. F. Seaman. 1974. "Rheology of Erythrocyte Suspensions: Electrostatic Factors in the Dextran-Mediated Aggregation of Erythrocytes." *Biorheology* 11:69–76.

Bull, B. S., and J. D. Brailsford. 1972. "The Zeta Sedimentation Ratio." *Blood* 40:550–559.

Bull, B. S., and G. Brecher. 1974. "Evaluation of Relative Merits of Wintrobe and Westergren Sedimentation Methods, Including Hematocrit Correction." *American Journal of Clinical Pathology* 62:502–510.

Bull, B. S., S. Chien, J. Dormandy, H. Kiesewetter, S. M. Lewis, G. D. O. Lowe, H. J. Meiselman, et al. 1989. "Guidelines on Selection of Laboratory Tests for Monitoring the Acute Phase Response." *Journal of Clinical Pathology* 41:1203–1212.

Bull, B. S., S. Chien, J. Dormandy, H. Kiesewetter, S. M. Lewis, G. D. O. Lowe, H. J. Meiselman, et al. 1986. "Guidelines for Measurement of Blood Viscosity and Erythrocyte Deformability." *Clinical Hemorheology* 6:439–453.

Cha, C. H., C. J. Park, Y. J. Cha, H. K. Kim, D. H. Kim, Honghoon, J. H. Bae, et al. 2009. "Erythrocyte Sedimentation Rate Measurements by TEST-1 Better Reflect Inflammation Than Do Those by the Westergren Method in Patients with Malignancy, Autoimmune Disease, or Infection." *American Journal of Clinical Pathology* 131:189–194.

Chen, S., G. Barshtein, B. Gavish, Y. Mahler, and S. Yedgar. 1994. "Monitoring of Red Blood Cell Aggregability in a Flow-Chamber by Computerized Image Analysis." *Clinical Hemorheology* 14:497–508.

Chen, S., A. Eldor, G. Barshtein, S. Zhang, A. Goldfarb, E. Rachmilewitz, and S. Yedgar. 1996. "Enhanced Aggregability of Red Blood Cells of Beta-Thalassemia Major Patients." *American Journal of Physiology—Heart and Circulatory Physiology* 270:H1951–H1956.

Chen, S., B. Gavish, S. Zhang, Y. Mahler, and S. Yedgar. 1995. "Monitoring of Erythrocyte Aggregate Morphology under Flow by Computerized Image Analysis." *Biorheology* 32:487–496.

Chien, S., and K. M. Jan. 1973. "Ultrastructural Basis of the Mechanism of Rouleaux Formation." *Microvascular Research* 5:155–166.

Chien, S., and L. A. Sung. 1987. "Physicochemical Basis and Clinical Implications of Red Cell Aggregation." *Clinical Hemorheology* 7:71–91.

Cloutier, G. 1999. "Characterization of Erythrocyte Aggregation with Ultrasound." *Biorheology* 36:443–446.

Cloutier, G., and Z. Qin. 1997. "Ultrasound Backscattering from Non-Aggregating and Aggregating Erythrocytes—A Review." *Biorheology* 34:443–470.

Cloutier, G., Z. Qin, L. G. Durand, and B. G. Teh. 1996. "Power Doppler Ultrasound Evaluation of the Shear Rate and Shear Stress Dependences of Red Blood Cell Aggregation." *IEEE Transactions on Biomedical Engineering* 43:441–450.

Cloutier, G., and K. K. Shung. 1993. "Study of Red Cell Aggregation in Pulsatile Flow from Ultrasonic Doppler Power Measurements." *Biorheology* 30:443–461.

Cloutier, G., X. D. Weng, G. O. Roederer, L. Allard, F. Tardif, and R. Beaulieu. 1997. "Differences in the Erythrocyte Aggregation Level between Veins and Arteries of Normolipidemic and Hyperlipidemic Individuals." *Ultrasound in Medicine and Biology* 23:1383–1393.

Cohen, J. 1988. *Statistical Power Analysis for the Behavioral Sciences*. Hillsdale, NJ: Lawrence Earlbaum Associates.

Connes, P., N. Nemeth, H. J. Meiselman, and O. K. Baskurt. 2009. "Effect of Tourniquet Application during Blood Sampling on Red Blood Cell Deformability and Aggregation: Is It Better to Keep It On?" *Clinical Hemorheology and Microcirculation* 42:297–302.

Dadgostar, H., G. N. Holland, X. Huang, A. Tufail, A. Kim, T. C. Fisher, W. G. Cumberland, et al. 2006. "Hemorheologic Abnormalities Associated with HIV Infection: *In Vivo* Assessment of Retinal Microvascular Blood Flow." *Investigative Ophthalmology & Visual Science* 47:3933–3938.

Deng, L. H., J. C. Barbenel, and G. D. O. Lowe. 1994. "Influence of Hematocrit on Erythrocyte Aggregation Kinetics for Suspensions of Red-Blood-Cells in Autologous Plasma." *Biorheology* 3:193–205.

Desjardins, C., and B. R. Duling. 1987. "Microvessel Hematocrit: Measurement and Implications for Capillary Oxygen Transport." *American Journal of Physiology—Heart and Circulatory Physiology* 252:H494–H503.

Dintenfass, L., and C. D. Forbes. 1973. "About Increase of Aggregation of RBC with an Increase in Temperature in Normal and Abnormal Blood (in Cancer)." *Biorheology* 10:383–391.

Dobashi, T., A. Idonuma, Y. Toyama, and A. Sakanishi. 1994. "Effect of Concentration on Enhanced Sedimentation Rate of Erythrocytes in an Inclined Vessel." *Biorheology* 31:383–393.

Dobbe, J. G. G., G. J. Streekstra, J. Strackee, M. C. M. Rutten, J. M. A. Stijnen, and C. A. Grimbergen. 2003. "Syllectometry: The Effect of Aggregometer Geometry in the Assessment of Red Blood Cell Shape Recovery and Aggregation." *IEEE Transactions on Biomedical Engineering* 50:97–106.

Donner, M., M. Siadat, and J. F. Stoltz. 1988. "Erythrocyte Aggregation: An Approach by Light Scattering Determination." *Biorheology* 25:367–375.

Engeset, J., A. L. Stalker, and N. A. Matheson. 1966. "Effects of Dextran 40 on Erythrocyte Aggregation." *Lancet* 1:1124–1127.

Fåhraeus, R. 1921. "The Suspension Stability of the Blood." *Acta Medica Scandinavica* 55:1–228.

Fåhraeus, R. 1929. "The Suspension Stability of the blood." *Physiological Reviews* 9:241–274.

Falco, C., A. Vaya, M. Simo, T. Contreras, M. Santaolaria, and J. Aznar. 2005. "Influence of Fibrinogen Levels on Erythrocyte Aggregation Determined with the Myrenne Aggregometer and the Sefam Erythro-Aggregometer." *Clinical Hemorheology and Microcirculation* 33:145–151.

Foresto, P., M. D'Arrigo, L. Carreras, R. E. Cuezzo, J. Valverde, and R. Rasia. 2000. "Evaluation of Red Blood Cell Aggregation in Diabetes by Computerized Image Analysis." *Medicina (Buenos Aires)* 60:570–572.

Freyburger, G., G. Janvier, S. Dief, and M. R. Boisseau. 1993. "Fibrinolytic and Hemorheologic Alterations during and after Elective Aortic Graft Surgery: Implications for Postoperative Management." *Anesthesia and Analgesia* 76:504–512.

Fricke, H. 1924. "A Mathematical Treatment of the Electrical Conductivity and Capacity of Disperse Systems." *Physical Review* 4:575–587.

Fusman, R., R. Rotstein, K. Elishkewich, D. Zeltser, S. Cohen, M. Kofler, D. Avitzour, et al. 2001. "Image Analysis for the Detection of Increased Erythrocyte, Leukocyte and Platelet Adhesiveness/Aggregation in the Peripheral Blood of Patients with Diabetes Mellitus." *Acta Diabetologica* 38:129–134.

Gaspar-Rosas, A., and G. B. Thurston. 1988. "Erythrocyte Aggregate Rheology by Transmitted and Reflected Light." *Biorheology* 25:471–487.

Gaw, R. L., B. H. Cornish, and B. J. Thomas. 2008. "The Electrical Impedance of Pulsatile Blood Flowing through Rigid Tubes: A Theoretical Investigation." *IEEE Transactions on Biomedical Engineering* 55:721–727.

Giavarina, D., S. Capuzzo, U. Pizzolato, and G. Soffiati. 2006. "Length of Erythrocyte Sedimentation Rate (ESR) Adjusted for the Hematocrit: Reference Values for the TEST 1 (TM) Method." *Clinical Laboratory* 52:241–245.

Groner, W., N. Mohandas, and M. Bessis. 1980. "New Optical Technique for Measuring Erythrocyte Deformability with the Ektacytometer." *Clinical Chemistry* 26:1435–1442.

Haider, L., P. Snabre, and M. Boynard. 2004. "Ultrasound Scattering from Concentrated Suspensions of Aggregated Red Cells in Shear Flow." *Clinical Hemorheology and Microcirculation* 30:345–352.

Hardeman, M. R., J. G. G. Dobbe, and C. Ince. 2001. "The Laser-Assisted Optical Rotational Cell Analyzer (LORCA) as Red Blood Cell Aggregometer." *Clinical Hemorheology and Microcirculation* 25:1–11.

Hardeman, M. R., P. T. Goedhart, J. G. G. Dobbe, and K. P. Lettinga. 1994. "Laser-Assisted Optical Rotational Cell Analyzer (lorca). 1. A New Instrument for Measurement of Various Structural Hemorheological Parameters." *Clinical Hemorheology* 14:605–618.

Hardeman, M. R., M. Levitus, A. Pelliccia, and A. A. Bouman. 2010a. "Test 1 Analyser for Determination of ESR. 1. Practical Evaluation and Comparison with the Westergren Technique." *Scandinavian Journal of Clinical and Laboratory Investigation* 70:21–25.

Hardeman, M. R., M. Levitus, A. Pelliccia, and A. A. Bouman. 2010b. "Test 1 Analyser for Determination of ESR. 2. Experimental Evaluation and Comparison with RBC Aggregometry." *Scandinavian Journal of Clinical and Laboratory Investigation* 70:26–32.

Hoetink, A. E., Th. J. C. Faes, K. R. Visser, and R. M. Heethaar. 2004. "On the Flow Dependency of the Electrical Conductivity of Blood." *IEEE Transactions on Biomedical Engineering* 51:1251–1261.

Holder, D. 2005. "Brief Introduction to Bioimpedance." In *Electrical Impedance Tomography*, 411–422. Boca Raton, FL: CRC Press.

Hynes, M., and L. E. H. Whitby. 1938. "Correction of Sedimentation Rate for Anemia." *Lancet* 2:249–255.

Iliev, B., A. Herbers, J. P. Donnely, N. Blijlevens, and G. Pop. 2008. "Measuring *in Vivo* Hemorheological Parameters On-Line with a Central Venous Catheter in the Right Atrium." *Biorheology* 45:80.

International Committee for Standardization in Haematology. 1977. "Recommendations for Measurement of Erythrocyte Sedimentation Rate of Human Blood." *American Journal of Clinical Pathology* 68:505–507.

International Committee for Standardization in Haematology (Expert Panel on Blood Rheology). 1993. "ICSH Recommendations for Measurement of Erythrocyte Sedimentation Rate." *Journal of Clinical Pathology* 46:198–203.

Ismailov, R. M., N. A. Shechuk, and H. Khusanov. 2005. "Mathematical Model Describing Erythrocyte Sedimentation Rate. Implications for Blood Viscosity Changes in Traumatic Shock and Crush Syndrome." *Biomedical Engineering Online* 4:24.

Jan, K. M. 1986. "Roles of Surface Electrochemistry and Macromolecular Adsorption in Heparin-Induced Red Blood Cell Aggregation." *Biorheology* 23:91–98.

Jayavanth, S., and M. Singh. 2004. "Computerized Analysis of Erythrocyte Aggregation from Sequential Video-Microscopic Images under Gravitational Sedimentation." *ITBM-RBM* 25:67–74.

Jovtchev, S., S. Stoeff, K. Arnold, and O. Zschornig. 2008. "Studies on the Aggregation Behaviour of Pegylated Human Red Blood Cells with the Zeta Sedimentation Technique." *Clinical Hemorheology and Microcirculation* 39:229–233.

Jung, F., A. Seegert, H. G. Roggenkamp, C. Mrowietz, H. P. Nuttgens, H. Kiesewetter, H. Zeller, and G. Muller. 1987. "Simultaneous Measurement of Hematocrit, Erythrocyte Aggregation and Erythrocyte Disaggregation—Method, Quality-Control and Reference Range." *Biomedizinische Technik* 32:117–125.

Kaliviotis, E., I. Ivanov, N. Antonova, and M. Yianneskis. 2010. "Erythrocyte Aggregation At Non-Steady Flow Conditions: A Comparison of Characteristics Measured with Electrorheology and Image Analysis." *Clinical Hemorheology and Microcirculation* 44:43–54.

Kaliviotis, E., and M. Yianneskis. 2008. "Fast Response Characteristic of Red Blood Cell Aggregation." *Biorheology* 45:639–649.

Kameneva, M. V., J. F. Antaki, H. S. Watach, H. S. Borovetz, and R. L. Kormos. 1994. "Heparin Effect on Red Cell Aggregation." *Biorheology* 51:297–304.

Karabetsos, E., C. Papaodysseus, and D. Koutsouris. 1999. "A New Method for Measuring Red Blood Cell Aggregation Using Pattern Recognition Techniques on Backscattered Ultrasound Doppler Signals." *Clinical Hemorheology and Microcirculation* 20:63–75.

Kavitha, A., and S. Ramakrishnan. 2007. "Assessment of Human Red Blood Cell Aggregation Using Image Processing and Wavelets." *Measurement Science Review* 7:43–51.

Kawakami, S., Y. Isogai, J. Yamamoto, T. Maeda, S. Ikemoto, and M. Kaibara. 1994. "Rheological Study on Erythrocyte Aggregation with Special Reference to ESR—Application to Quick Estimation of ESR Value." *Clinical Hemorheology* 14:509–518.

Keidan, A. J., S. S. Marwah, and J. Stuart. 1987. "Evaluation of Phosphate and HEPES Buffers for Study of Erythrocyte Rheology." *Clinical Hemorheology* 7:627–635.

Kesler, A., Y. Yatziv, I. Shapira, S. Berliner, and E. Ben Assayag. 2006. "Increased Red Blood Cell Aggregation in Patients with Idiopathic Intracranial Hypertension—A Hitherto Unexplored Pathophysiological Pathway." *Thrombosis and Haemostasis* 96:483–487.

Kiesewetter, H., H. Radtke, R. Schneider, K. Mussler, A. Scheffler, and H. Schmid-Schönbein. 1982. "The Mini Erythrocyte Aggregometer: A New Apparatus for the Rapid Quantification of the Extent of Erythrocyte Aggregation." *Biomedical Technology (Berlin)* 28:209–213.

Kim, S., A. S. Popel, M. Intaglietta, and P. C. Johnson. 2005. "Aggregate Formation of Erythrocytes in Postcapillary Venules." *American Journal of Physiology—Heart and Circulatory Physiology* 288:584–590.

Kitamura, H., and S. Kawasaki. 1997. "Detection and Clinical Significance of Red Cell Aggregation in the Human Subcutaneous Vein Using a High-Frequency Transducer (10 MHz): A Preliminary Report." *Ultrasound in Medicine & Biology* 23:933–938.

Kitamura, H., J. Sigel, J. Machi, and E. J. Feleppa. 1995. "Roles of Hematocrit and Fibrinogen in Red Cell Aggregation Determined by Ultrasonic Scattering Properties." *Ultrasound in Medicine & Biology* 21:827–832.

Lacombe, C., and J. C. Lelievre. 1987. "Interpretation of Rheograms for Assessing RBC Aggregation and Deformability." *Clinical Hemorheology* 7:47–61.

Lademann, J., H. G. Weigmann, W. Sterry, A. Roggan, A. V. Muller, A. V. Priezzhev, and N. N. Firsov. 1999. "Investigation of the Aggregation and Disaggregation Properties of Erythrocytes by Light Scattering Measurements." *Laser Physics* 9:357–362.

Lamb, H. 1994. *Hydrodynamics*. Cambridge: Cambridge University Press.

Lee, B. K., T. Alexy, R. B. Wenby, and H. J. Meiselman. 2007. "Red Blood Cell Aggregation Quantitated via Myrenne Aggregometer and Yield Shear Stress." *Biorheology* 44:29–35.

Lim, H-J., J-H. Nam, Y-J. Lee, and S. Shin. 2009. "Measurement of The Temperature-Dependent Threshold Shear-Stress of Red Blood Cell Aggregation." *Review of Scientific Instruments* 80:096101.

Maeda, N., M. Seike, and T. Shiga. 1987. "Effect of Temperature on the Velocity of Erythrocyte Aggregation." *Biochimica Biophysica Acta* 13:319–329.

Maharshak, N., Y. Arbel, I. Shapira, S. Berliner, R. Ben-Ami, S. Yedgar, G. Barshtein, and I. Dotan. 2009. "Increased Strength of Erythrocyte Aggregates in Blood of Patients with Inflammatory Bowel Disease." *Inflammatory Bowel Diseases* 15:707–713.

Maharshak, N., I. Shapira, R. Rotstein, J. Serov, S. Aharonov, T. Mardi, A. Twig, et al. 2002. "The Erythrocyte Adhesiveness/Aggregation Test for the Detection of an Acute Phase Response and for the Assessment of Its Intensity." *Clinical and Laboratory Haematology* 24:205–210.

Manley, R. W. 1957. "The Effect of Room Temperature on Erythroycte Sedimentation Rate and Its Correction." *Journal of Clinical Pathology* 10:354–357.

Martinez M., A. Vaya, C. Lopez-Camacho, E. Coll-Sangrona, Y. Mira, and J. Aznar. 2000. "High and Low Molecular Weight Heparins Do Not Modify Red Blood Cell Aggregability *in Vitro*." *Clinical Hemorheology and Microcirculation* 23:67–70.

Mayer, J., Z. Popisil, and J. Litzman. 1992. "The Mechanism of Erythrocyte Sedimentation in Westergren's Examination." *Biorheology* 29:261–271.

Mchedlishvili, G., N. Beritashvili, D. Lominadze, and B. Tsinamdzvrishvili. 1993. "Technique for Direct and Quantitative-Evaluation of Erythrocyte Aggregability in Blood-Samples." *Biorheology* 30:153–161.

Miller, A., M. Green, and D. Robinson. 1983. "Simple Rule for Calculating Normal Erythrocyte Sedimentation Rate." *British Medical Journal* 286:266.

Morris, M. W., R. S. Pinals, and D. A. Nelson. 1977. "Zeta-Sedimentation-Ratio (ZSR) and Activity of Disease in Rheumatoid-Arthritis." *American Journal of Clinical Pathology* 68:760–762.

Morris, M. W., Z. Skrodzki, and D. A. Nelson. 1975. "Zeta Sedimentation Ratio (ZSR), a Replacement for the Erythrocyte Sedimentation Rate (ESR)." *American Journal of Clinical Pathology* 64:254–256.

Nam, J-H., Y. Yang, S. Chung, and S. Shin. 2010. "Comparison of Light Transmission and Backscattering Methods in the Measurement of RBC Aggregation." *Journal of Biomedical Optics* 15: 027003.

Nemeth, N., O. K. Baskurt, H. J. Meiselman, and I. Miko. 2009. "Species-Specific Effects of Anticoagulants on Red Blood Cell Deformability." *Clinical Hemorheology and Microcirculation* 43:257–259.

Neu, B., and H. J. Meiselman. 2007. "Red Blood Cell Aggregation." In *Handbook of Hemorheology and Hemodynamics*, ed. O. K. Baskurt, M. R. Hardeman, M. W. Rampling, and H. J. Meiselman, 114–136. Amsterdam, Berlin, Oxford, Tokyo, Washington, DC: IOS Press.

Neumann, F-J., H. Schmid-Schönbein, and H. Ohlenbusch. 1987. "Temperature-Dependence of Red Cell Aggregation." *Pflügers Archives* 408:524–530.

Oancea, S. 2007. "A Quantitative Analysis of Red Blood Cell Aggregation from Bovine Blood." *Romanian Journal of Biophysics* 17:205–209.

Oka, S. 1985. "A Physical Theory of Erythrocyte Sedimentation." *Biorheology* 22:315–321.

Ozanne, P., R. B. Francis, and H. J. Meiselman. 1983. "Red Blood Cell Aggregation in Nephrotic Syndrome." *Kidney International* 23:519–525.

Ozdem, S., H. S. Akbas, L. Donmez, and M. Gultekin. 2006. "Comparison of TEST 1 with SRS 100 and ICSH Reference Method for the Measurement of the Length of Sedimentation Reaction in Blood." *Clinical Chemistry and Laboratory Medicine* 44:407–412.

Patton, W. N., P. J. Meyer, and J. Stuart. 1989. "Evaluation of Sealed Vacuum Extraction Method (Seditainer) for Measurement of Erythrocyte Sedimentation Rate." *Journal of Clinical Pathology* 42:313–317.

Pawlotsky, Y., J. Goasguen, P. Guggenbuhl, E. Veillard, C. Jard, M. Pouchard, A. Perdriger, J. Meadeb, and G. Chales. 2004. "Sigma ESR—An Erythrocyte Sedimentation Rate Adjusted for the Hematocrit and Hemoglobin Concentration." *American Journal of Clinical Pathology* 122:802–810.

Pearson, M. J., and H. H. Lipowsky. 2004. "Effect of Fibrinogen on Leukocyte Margination and Adhesion in Postcapillary Venules." *Microcirculation* 11:295–306.

Phear, D. 1957. "The Influence of Erythrocyte Factors on Their Sedimentation Rate." *Journal of Clinical Pathology* 10:357–359.

Pieragalli, D., S. Forconi, M. Guerrini, A. Acciavatti, C. Galigani, C. Del Bigo, and T. Diperri. 1985. "A Syllectrometric Method to Study Erythrocyte Aggregation." *La Ricerca in Clinica e in Laboratorio* 15:79–86.

Piva, E., M. C. Sanzari, G. Servidio, and M. Plebani. 2001. "Length of Sedimentation Reaction in Undiluted Blood (Erythrocyte Sedimentation Rate): Variations with Sex and Age and Reference Limits." *Clinical Chemistry and Laboratory Medicine* 39:451–454.

Plebani, M. And E. Piva. 2002. "Erythorocyte Sedimentation Rate. Use of Fresh Blood for Quality Control" *American Journal of Clinical Pathology* 117: 621-626.

Plebani, M., S. De Toni, M. C. Sanzari, D. Bernardi, and E. Stockreiter. 1998. "The TEST 1 Automated System—A New Method for Measuring the Erythrocyte Sedimentation Rate." *American Journal of Clinical Pathology* 110:334–340.

Poole, J. C. F., and G. A. C. Summers. 1952. "Correction of ESR in Anaemia. Experimental Study Based on Interchange of Cells and Plasma between Normal Anaemic Subjects." *British Medical Journal* 1: 353–356.

Pop, G. A. M., Z. Y. Chang, C. J. Slager, B. J. Kooij, E. D. van Deel, L. Moraru, J. Quak, G. C. M. Meijer, and D. J. Duncker. 2004a. "Catheter-Based Impedance Measurements in the Right Atrium for Continuously Monitoring Hematocrit and Estimating Blood Viscosity Changes: An *in Vivo* Feasibility Study in Swine." *Biosensors & Bioelectronics* 15:1685–1693.

Pop, G. A. M., T. L. M. de Backer, M. de Jong, P. C. Struijk, L. Moraru, Z. Chang, H. G. Goovaerts, C. J. Slager, and A. J. J. C. Bogers. 2004b. "On-Line Electrical Impedance Measurement for Monitoring Blood Viscosity during On-Pump Heart Surgery." *European Journal Surgical Research* 36:259–265.

Pop, G. A. M., W. J. Hop, M. van der Jagt, J. Quak, D. Dekkers, Z. Chang, F. J. Gijsen, D. J. Dunncker, and C. J. Slager. 2003. "Blood Electrical Impedance Closely Matches Whole Blood Viscosity as Parameter of Hemorheology and Inflammation." *Applied Rheology* 13:305–312.

Potron, G., D. Jolly, P. Nguyen, J. L. Mailliot, and B. Pignon. 1994. "Approach to Erythrocyte Aggregation through Erythrocyte Sedimentation-Rate—Application of a Statistical-Model in Pathology." *Nouvelle Revue Francaise D Hematologie* 36:241–247.

Pribush, A., L. Hatzkelson, D. Meyerstein, and N. Meyerstein. 2007. "A Novel Technique for Quantification of Erythrocyte Aggregation Abnormalities in Pathophysiological Situations." *Clinical Hemorheology and Microcirculation* 36:121–132.

Pribush, A., H. J. Meiselman, D. Meyerstein, and N. Meyerstein. 1999. "Dielectric Approach to the Investigation of Erythrocyte Aggregation: I. Experimental Basis of the Method." *Biorheology* 36:411–423.

Pribush, A., H. J. Meiselman, D. Meyerstein, and N. Meyerstein. 2000. "Dielectric Approach to Investigation of Erythrocyte Aggregation. II. Kinetics of Erythrocyte Aggregation-Disaggregation in Quiescent and Flowing Blood." *Biorheology* 37:429–441.

Pribush, A., D. Meyerstein, H. J. Meiselman, and N. Meyerstein. 2004a. "Conductometric Study of Shear-Dependent Processes in Red Cell Suspensions. II. Transient Cross-Stream Hematocrit Distribution." *Biorheology* 41:29–43.

Pribush, A., D. Meyerstein, and N. Meyerstein. 2004b. "Conductometric Study of Shear-Dependent Processes in Red Cell Suspensions. I. Effect of red Blood Cell Aggregate Morphology on Blood Conductance." *Biorheology* 41:13–28.

Priezzhev, A. V., O. M. Ryaboshapka, N. N. Firsov, and I. V. Sirko. 1999. "Aggregation and Disaggregation of Erythrocytes in Whole Blood: Study by Backscattering Technique." *Journal of Biomedical Optics* 4:76–84.

Raich, P. C., and N. Temperly. 1976. "Comparison of the Wintrobe Erythrocyte Sedimentation Rate with the Zeta Sedimentation Ratio." *American Journal of Clinical Pathology* 65:690–703.

Rampling, M. W. 1988. "Red Cell Aggregation and Yield Stress." In *Clinical Blood Rheology*, ed. G. D. O. Lowe, 45–64. Boca Raton, FL: CRC Press.

Rampling, M. W., and G. Martin. 1989. "A Comparison of the Myrenne Erythrocyte Aggregometer with Older Techniques for Estimating Erythrocyte Aggregation." *Biorheology* 9:41–46.

Rampling, M. W., H. J. Meiselman, B. Neu, and O. K. Baskurt. 2004. "Influence of Cell-Specific Factors on Red Blood Cell Aggregation." *Biorheology* 41:91–112.

Rampling, M. W., P. Whittingstall, and O. Linderkamp. 1984. "The Effects of Fibrinogen and Its Plasmin Degradation Products on the Rheology of Erythrocyte Suspensions." *Clinical Hemorheology* 4:533–543.

Raphael, S. S. 1983. Practice of Hematology *Lynch's Medical Laboratory Technology*, 672–713. Philadelphia: W. B. Saunders Co.

Razavian, M., M. T. Guillemin, R. Guillet, Y. Beuzard, and M. Boynard. 1991. "Assessment of Red Blood Cell Aggregation with Dextran by Ultrasonic Interferometry." *Biorheology* 28:89–97.

Roggan, A., M. Friebel, K. Dirschel, A. Hahn, and G. Muller. 1999. "Optical Properties of Circulating Human Blood in the Wavelength Range 400–2500 nm." *Journal of Biomedical Optics* 4:36–46.

Romero, A., M. Munoz, and G. Ramirez. 2003. "Length of Sedimentation Reaction in Blood: A Comparison of the Test 1 ESR System with the ICSH Reference Method and the Sedisystem 15." *Clinical Chemistry and Laboratory Medicine* 41:232–237.

Rotstein, R., R. Fusman, S. Berliner, D. Levartovsky, O. Rogowsky, S. Cohen, E. Shabtai, et al. 2001. "The Feasibility of Estimating the Erythrocyte Sedimentation Rate within a Few Minutes by Using a Simple Slide Test." *Clinical and Laboratory Haematology* 23:21–25.

Rouffiac, V., P. Peronneau, J. P. Guglielmi, M. Del-Pino, N. Lassau, and J. Levenson. 2003. "Comparison of New Ultrasound Index with Laser Reference and Viscosity Indexes for Erythrocyte Aggregation Quantification." *Ultrasound in Medicine & Biology* 29:789–799.

Rourke, M. D., and A. C. Ernstene. 1930. "A Method for Correcting the Erythrocyte Sedimentation Rate for Variations in the Cell Volume Percentage of Blood." *Journal of Clinical Investigation* 8:545–558.

Saleem, A., E. Jafari, and M. K. Yapit. 1977. "Comparison of Zeta Sedimentation Ratio with Westergren Sedimentation Rate." *Annals of Clinical and Laboratory Science* 7:357–360.

Schmid-Schönbein, H. 1979. Apparatus for measuring the aggregation rate of particles. US patent 4135819, and issued 1979.

Schmid-Schöenbein, H., J. von Gosen, L. Heinich, H. J. Klose, and E. Volger. 1973. "A Counter-Rotating 'Rheoscope Chamber' for the Study of the Microrheology of Blood Cell Aggregation by Microscopic Observation and Microphotometry." *Microvascular Research* 6:366–376.

Shehada, R. E. N., R. S. C. Cobbold, and L. Y. L. Mo. 1994. "Aggregation Effects in Whole Blood: Influence of Time and Shear Rate Measured Using Ultrasound." *Biorheology* 31:115–135.

Shiga, T., K. Imaizumi, N. Harada, and M. Sekiya. 1983. "Kinetics of Rouleaux Formation Using TV Image Analyzer. I. Human Erythrocytes." *American Journal of Physiology— Heart and Circulatory Physiology* 245:H252–H258.

Shin, S., M. S. Park, Y. H. Ku, and J. S. Suh. 2006. "Shear-Dependent Aggregation Characteristics of Red Blood Cells in a Pressure-Driven Microfluidic Channel." *Clinical Hemorheology and Microcirculation* 34:353–361.

Shin, S., Y. Yang, and J. S. Suh. 2007. "Microchip-Based Cell Aggregometer Using Stirring-Disaggregation Mechanism." *Korea-Australia Rheology Journal* 19:109–115.

Shin, S., Y. Yang, and J. S. Suh. 2009a. "Measurement of Erythrocyte Aggregation in a Microchip-Based Stirring System by Light Transmission." *Clinical Hemorheology and Microcirculation* 41:197–207.

Shin, S., J. H. Nam, J. X. Hou, and J. S. Suh. 2009b. "A Transient, Microfluidic Approach to the Investigation of Erythrocyte Aggregation: The Threshold Shear-Stress for Erythrocyte Disaggregation." *Clinical Hemorheology and Microcirculation* 42:117–125.

Shung, K. K., and J. M. Reid. 1979. "Ultrasonic Instrumentation for Hematology." *Ultrasonic Imaging* 1:280–294.

Shung, K. K., R. A. Sigelman, and J. M. Reid. 1976. "Scattering of Ultrasound by Blood." *IEEE Transactions on Biomedical Engineering* 23:460–467.

Shung, K. K., and G. A. Thieme. 1993. *Ultrasonic Scattering in Biological Tissues*. Boca Raton, FL: CRC Press.

Shvartsman, L. D., and I. Fine. 2003. "Optical Transmission of Blood: Effect of Erythrocyte Aggregation." *IEEE Transactions on Biomedical Engineering* 50:1026–1033.

Singh, M., and J-F. Stoltz. 2002. "Influence of Temperature Variation from 5°C to 37°C on Aggregation and Deformability of Erythrocytes." *Clinical Hemorheology and Microcirculation* 26:1–7.

Skalak, R., and C. Zhu. 1990. "Rheological Aspects of Red Blood Cell Aggregation." *Biorheology* 27:309–325.

Soutani, M., Y. Suzuki, N. Tateishi, and N. Maeda. 1995. "Quantificative Evaluation of Flow Dynamics of Erythrocytes in Microvessels: Influence of Erythrocyte Aggregation." *American Journal of Physiology—Heart and Circulatory Physiology* 268:H1959–H1965.

Stoeff, S., S. Jovtchev, I. Dikov, D. Kolarov, T. Galabova, N. Trifonova, S. Hadjieva, et al. 2008. "Quantitative Expediency Assessment of the Zeta Sedimentation Ratio and the Plasma Viscosity in Arterial Hypertension Research." *Clinical Hemorheology and Microcirculation* 39:381–384.

Stoltz, J-F., F. Paulus, and M. Donner. 1987. "Experimental Approaches to Erythrocyte Aggregation." *Clinical Hemorheology* 7:109–118.

Stuart, J. 1991. "Rheological Methods for Monitoring the Acute-Phase Response." *Revista Portuguesa Hemorreologica* 5:57–62.

Stuart, J., P. C. W. Stone, G. Freyburger, M. R. Boisseau, and D. G. Altman. 1989. "Instrument Precision and Biological Variability Determine the Number of Patients Required for Rheological Studies." *Clinical Hemorheology* 9:181–197.

Tomita, M. 1989. Apparatus for measuring aggregation rate of whole blood red blood cells. US patent 4822568, and issued 1989.

Tomita, M., F. Gotoh, N. Tanahashi, and P. Turcani. 1986. "Whole-Blood Red Blood Cell Aggregometer for Human and Feline Blood." *American Journal of Physiology—Heart and Circulatory Physiology* 251:H1205–H1210.

Trudnowski, R. J., and R. C. Rico. 1974. "Specific Gravity of Blood and Plasma at 4 and 37°C." *Clinical Chemistry* 20:615–616.

Urbach, J., O. Rogowski, I. Shapira, D. Avitzour, D. Branski, S. Schwartz, S. Berliner, and T. Mardi. 2005. "Automatic 3-Dimensional Visualization of Peripheral Blood Slides—A New Approach for the Detection of Infection/Inflammation at the Point of Care." *Archives of Pathology & Laboratory Medicine* 129:645–650.

Uyuklu, M., M. Cengiz, P. Ulker, T. Hever, J. Tripette, P. Connes, N. Nemeth, H. J. Meiselman, and O. K. Baskurt. 2009a. "Effects of Storage Duration and Temperature of Human Blood on Red Cell Deformability and Aggregation." *Clinical Hemorheology and Microcirculation* 41:269–278.

Uyuklu, M., H. J. Meiselman, and O. K. Baskurt. 2009b. "Effect of Hemoglobin Oxygenation Level on Red Blood Cell Deformability and Aggregation Parameters." *Clinical Hemorheology and Microcirculation* 41:179–188.

Vaya, A., C. Falco, P. Fernandez, T. Contreras, M. Valls, and J. Aznar. 2003. "Erythrocyte Aggregation Determined with the Myrenne Aggregometer at Two Modes (M-0, M-1) and at Two Times (5 and 10 sec)." *Clinical Hemorheology and Microcirculation* 29:119–127.

Westergren, A. 1921. "Studies of the Suspension Stability of the Blood in Pulmonary Tuberculosis." *Acta Medica Scandinavica* 54:247–282.

Wintrobe, M. M., and J. Landsberg. 1935. "A Standardized Technique for the Blood Sedimentation Test." *American Journal of Medical Sciences* 189:102–117.

Yalcin, O., M. Uyuklu, J. K. Armstrong, H. J. Meiselman, and O. K. Baskurt. 2004. "Graded Alterations of RBC Aggregation Influence *in Vivo* Blood Flow Resistance." *American Journal of Physiology—Heart and Circulatory Physiology* 287:H2644–H2650.

Yu, F. T. H., and G. Cloutier. 2007. "Experimental Ultrasound Characterization of Red Blood Cell Aggregation Using the Structure Factor Size Estimator." *Journal of Acoustic Society of America* 122:645–656.

Yu, F. T. H., E. Franceschini, B. Chayer, J. K. Armstrong, H. J. Meiselman, and G. Cloutier. 2009. "Ultrasonic Parametric Imaging of Erythrocyte Aggregation Using the Structure Factor Size Estimator." *Biorheology* 46:343–363.

Zhao, H., X. Wang, and J. F. Stoltz. 1999. "Comparison of Three Optical Methods to Study Erythrocyte Aggregation." *Clinical Hemorheology and Microcirculation* 21:297–302.

Zhao, T. X., and B. Jacobson. 1997. "Quantitative Correlations among Fibrinogen Concentration, Sedimentation Rate and Electrical Impedance of Blood." *Medical & Biological Engineering & Computing* 35:181–185.

Zijlstra, W. G. 1958. "Syllectometry, a New Method for Studying Rouleaux Formation of Red Blood Cells." *Acta Physiol Pharmacol Neerl* 7:153–154.

Zilberman, L., O. Rogowski, M. Rozenblat, I. Shapira, J. Serov, P. Halpern, I. Dotan, N. Arber, and S. Berliner. 2005. "Inflammation-Related Erythrocyte Aggregation in Patients with Inflammatory Bowel Disease." *Digestive Diseases and Sciences* 50:677–683.

5 Effect of Red Blood Cell Aggregation on *in Vitro* Blood Rheology

5.1 INITIAL CONSIDERATIONS

Mammalian blood is a rheologically complex fluid composed of cells (i.e., red blood cells [RBC], white blood cells, platelets) with differing mechanical properties suspended in a Newtonian aqueous solution of micro ions (e.g., calcium, potassium, sodium), and protein macromolecules such as albumin and fibrinogen (Rampling 2007). It is a crowded suspension, usually with a volume fraction of RBC, termed hematocrit, of about 0.40–0.45 l/l (i.e., 40–45%), and hence cell–cell interactions are the rule rather than the exception (Goldsmith 1968; Goldsmith et al. 1983). These cell–cell interactions are due to both shear forces and cell collisions as well as to attractive forces between cells (i.e., RBC aggregation). As a result, *in vitro* measurement of blood rheology can be difficult and requires attention to several potential artifacts. The following material describes systems for viscosity measurement, areas where problems may occur, and the rheology of blood with special emphasis on the effects of RBC aggregation.

5.2 VISCOMETRIC SYSTEMS

The instrumental requirements for measuring the rheological properties of blood and blood plasma depend strongly on the ultimate application for such measurements. In general, obtaining data for the design of mass transfer devices requires less sophisticated instruments than for information relevant to clinical diagnosis; obtaining basic information on the physical properties of blood is most demanding. This chapter is focused on the physical properties of blood, yet data obtained for this purpose are generally useful for other applications. Two basic classes of instruments are used in blood rheology studies—those in which blood is sheared by a moving surface (e.g., rotating cylinder, cone and plate) and those in which blood is flowed past stationary walls (e.g., tube viscometers). There are several literature sources describing instruments, rheological techniques, and data analysis methods for a wide variety of simple and complex fluids (e.g., Barnes 2000; Barnes et al. 1989; Cokelet and Meiselman 2007a; Coleman 1996; Larson 1998; Macosko 1994; Malkin and Isayev 2005; Metzger 2006; Wilkinson 1960). In addition, there are published guidelines for measurement of blood viscosity (ECSH Expert Panel on Blood Rheology

1986; Baskurt et al. 2009) and monographs and handbooks that are specific to blood rheology (Baskurt et al. 2007; Chien et al. 1987; Dintenfass 1971; Ehrly 1991; Stoltz et al. 1999); these sources should be consulted for additional details.

5.2.1 ROTATING CYLINDER VISCOMETERS

One viscometric system in which the sample is subjected to laminar and steady shear is the rotating cylinder viscometer, usually termed a *Couette viscometer* (Figure 5.1). The inner or outer cylinder is rotated at a given rate and the shear stress is transmitted through the fluid in the gap, with the resulting torque usually measured at the stationary cylinder. A useful feature of the Couette viscometer is that the total shear torque is constant across the gap so the shear stress is inversely related to the square of the radius to the point of measurement. The shear rate is not constant across the gap but for a simple Newtonian fluid (e.g., saline, oil) it also varies inversely with the square of the radius, with the variation proportional to the ratio of $(R_C^2 R_B^2) / (R_C^2 - R_B^2)$. The usual design goal is to make the radii large and the gap small, but for large gaps it is necessary to use the shear rate at the surface at which the shear stress is determined.

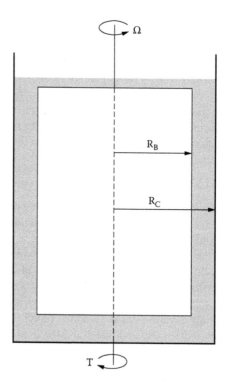

FIGURE 5.1 Cross-sectional schematic view of a concentric cylinder Couette viscometer. R_B is the radius of the inner cylinder (the bob), and R_C is the radius of the outer cylinder (the cup); the sample fills the space between the bob and cup. To reduce variation of shear rate within the gap between the bob and cup, both radii are made as large as practical and the gap between them is minimized.

Possible contributions of end and edge effects to the total torque measured in a Couette viscometer must be considered, with the usual experimental approach involving cylinders of constant diameter and identical shape but with different lengths. Torque measurements are made at several different lengths and a linear relation between torque and length indicates the absence of end effects. An alternative approach is the use of the Mooney configuration (Figure 5.2) in which the bottom of the inner cylinder has a conical end. The conical end is placed adjacent to the flat bottom of the cup with the separation adjusted so that, for simple shear, the shear rate in the cone-plate region is the same as that in the gap. Flow stability must also be considered, especially when the inner cylinder rotates, since Taylor vortices can develop above a critical Reynolds number; rotation of the outer cylinder stabilizes flow (i.e., no Taylor vortices) and allows much higher shear rates prior to the onset of turbulence.

Cone and plate viscometers are often used in hemorheology studies since their mechanical design and smaller sample size are somewhat less demanding than for coaxial cylinders. This type of instrument consists of a flat plate and a cone with a very obtuse angle (Figure 5.3). The apex of the cone just fails to touch the plate and the sample fills the gap formed by the cone and plate; an alternative design is to remove a very small portion of the tip of the cone, and then adjust the cone–plate

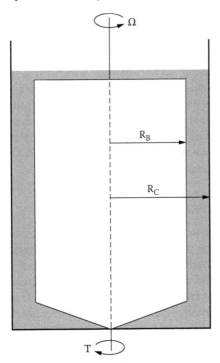

FIGURE 5.2 Cross-sectional schematic view of a concentric cylinder Couette viscometer with a cone-plate geometry at the bottom of the bob. This arrangement is referred to as a Mooney configuration and provides the same shear rate in the cone-plate region and in the gap between cylinders.

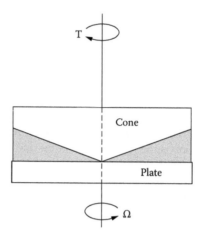

FIGURE 5.3 Cross-sectional schematic view of a cone-plate viscometer. The sample is placed between the cone and plate and is retained by surface tension forces. Usually the angle between the cone and plate is much smaller than shown (e.g., 1–2 degrees). This type of viscometer is particularly sensitive to sedimentation due to RBC aggregation since a cell-poor layer can form at the surface of the cone.

separation so that the apex of an "imaginary" cone just touches the plate. The cone or the plate can be rotated at desired speeds and the torque due to fluid drag measured on either element. Figure 5.3 shows the torque on the stationary cone being measured with the plate rotating; some cone-plate systems (e.g., Brookfield Engineering Laboratories, USA) rotate the cone and also measure torque on the cone. Surface tension of the sample is used to keep the sample in the gap, although at very high rotational speeds the fluid may be ejected. Note that since the gap at the apex is usually very small, gap width increases linearly with radial position and so does the linear velocity of the rotating element. Thus, if the flow is tangential, the shear rate in a Newtonian fluid is constant at all points in the gap.

5.2.2 Tube Viscometers

Tube viscometers, often termed *capillary viscometers*, are probably the most common method for studying the flow behavior of a fluid or suspension. At first glance, this measurement approach seems quite simple: measure flow at one or more pressure gradients or determine pressures necessary for required flow, then use these pressure–flow data to obtain a term reflecting viscosity. However, tube viscometry is often difficult to interpret, especially for non-Newtonian fluids. There are at least four potential problems in interpreting tube viscometer results: (1) the shear rate varies from zero at the center of the tube to a maximum at the tube wall; (2) entrance conditions can affect flow downstream from the inlet and, for suspensions, may affect particle concentration and distribution; (3) meniscus and surface tension artifacts can be important (see Section 5.3); (4) at very high flow rates, a portion of the driving force may remain in the fluid as kinetic energy rather than being dissipated by viscous friction.

Data analysis methods for tube viscometers are of two general classes: (1) use of pressure–flow results to calculate an apparent viscosity (i.e., flow resistance offered by a Newtonian fluid under the same flow conditions), and (2) detailed mathematical approach. The latter method calculates a wall shear stress based upon the pressure gradient, tube length, and tube diameter. It also calculates a wall shear rate using the gradient of normalized flow velocity (i.e., average velocity divided by tube diameter) versus wall shear stress. This approach, probably first developed by K. Weisenberg (Markovitz 1968), is limited to fluids for which the local shear stress is only a single-valued function of the local rate of deformation for all points in the tube. Note that although this approach for calculation of wall shear stress is generally applicable to all tube diameters, the Weisenberg approach for wall shear rate may not be valid for blood when using tube diameters of less than about 0.5 mm (Baskurt et al. 2009; Cokelet 1987; Cokelet and Meiselman 2007a; Cokelet et al. 1980).

5.3 POTENTIAL ARTIFACTS AFFECTING BLOOD VISCOMETRY

5.3.1 AIR–FLUID INTERFACE EFFECTS

Plasma proteins are surfactants and form a protein layer or film at fluid–air interfaces. This layer has mechanical strength, and since it is a semirigid film, it can transmit significant torque to or cause drag on the torque-measuring element. In general, this effect is a significant problem only for plasma, serum, and RBC–plasma or RBC–serum suspensions at very low hematocrits. At normal or high hematocrits, the film is of minor or no importance since the additional torque due to the film is only a small percentage of the total, especially if the radius of the shaft at the top of the inner cylinder is small (i.e., small torque arm). However, when measuring plasma, serum, or other protein solutions this film can markedly alter the results.

Figure 5.4 presents experimental data for normal human plasma measured in a cone-plate viscometer having a 0.8-degree cone with a radius of 2.4 cm. The viscometer (micro cone-plate, Brookfield Engineering Laboratories, USA) uses a stationary plate and a cone that rotates at various speeds and measures the torque acting on the cone. As is obvious, these results indicate that plasma is non-Newtonian with viscosity decreasing as shear rate increases. However, the non-Newtonian flow behavior in this figure is an artifact due to the resistance or drag of the film at the plasma–air interface; as the shear rate increases and the total torque increases, this drag becomes a smaller proportion of the total drag, and Newtonian flow is observed. In fact, plasma and serum are Newtonian fluids (Merrill 1965, p. 34): "it turns out after the artifacts had been carefully eliminated, that plasma is Newtonian. Serum is Newtonian, and every conceivable combination of the plasma proteins with and without lipid is Newtonian. But they must be measured in devices that eliminate the effects of surface layers ..."

Procedures to minimize or eliminate the influence of the protein film depend on the type of viscometer used for the measurements. Couette concentric-cylinder viscometers require the use of a cylindrical guard ring, which penetrates the film and prevents drag on the torque sensor by this film (Figure 5.5). It is important to note that the stationary guard ring can be applied only in systems in which the surface that

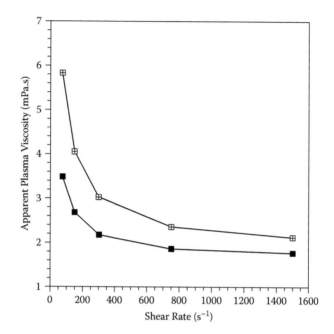

FIGURE 5.4 Plasma viscosity versus shear rate obtained using a viscometer without a guard ring to prevent surface film effects; the two plasma samples were obtained from different healthy adult donors. The decrease of viscosity with increasing shear rate is an artifact due to the film since plasma is a Newtonian fluid whose viscosity is shear rate independent. Adapted with permission from Baskurt, O. K., M. R. Hardeman, M. W. Rampling, and H. J. Meiselman, eds. 2007. *Handbook of Hemorheology and Hemodynamics*. Amsterdam, Berlin, Oxford, Tokyo, Washington, DC: IOS Press.

measures the torque (i.e., inner or outer cylinder) is held almost stationary (Cokelet and Meiselman 2007b; Merrill et al. 1963; Merrill 1965). Thus, in Figure 5.5, both the outer cylinder and the guard ring are stationary; the inner cylinder rotates but torque due to the film is not transmitted to the outer torque-measuring cylinder. An alternative approach is the use of two inner cylinders that are of different lengths and then employing a subtraction technique; this method assumes that the film at the interface is the same for both cylinders. A guard ring approach is also possible for cone-plate viscometers, although again the torque-measuring surface and the guard ring must be stationary or rotate at the same speed; a guard ring is not possible for cone-plate instruments (Figure 5.3) if the cone rotates and also senses torque.

Surface films are probably the most serious sources of error in tube viscometry of blood and plasma (Meiselman and Cokelet 1974) since there usually are two air–liquid interfaces to consider (i.e., advancing and receding menisci). In some situations, it is possible to resolve this source of error by using identical tubing diameters at both ends of the viscometer tube. However, experimental evidence indicates that this approach is not applicable for plasma and blood or even for clean distilled water (Meiselman and Cokelet 1974). In these situations, the generally accepted method for eliminating surface film artifacts is the use of a system in which the driving pressure

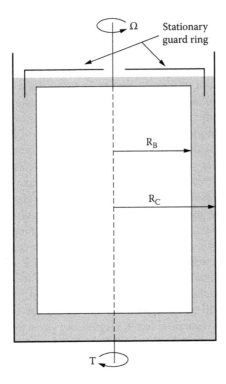

FIGURE 5.5 Cross-sectional schematic view of a concentric cylinder Couette viscometer equipped with a stationary guard ring. The guard ring penetrates the surface film thereby preventing torque due to the film from being sensed by the torque-measuring outside cylinder. When using a guard ring, plasma behaves as a Newtonian fluid.

gradient across the viscometer tube is measured through a fluid coupling, thereby eliminating any air–fluid interfaces. This approach has been successfully employed by Merrill and coworkers who observed Newtonian flow behavior for plasma when using a tube viscometer constructed so as to avoid air–plasma interfaces (Merrill et al. 1965).

5.3.2 Time-Dependent Effects

Essentially all rheological studies of blood are aimed at obtaining shear stress–shear rate data at steady state for a given shear stress or shear rate; blood is considered to be a time-independent fluid unless tested using dynamic experimental protocols (Cokelet 1987). However, RBC aggregation effects at low shear rates complicate obtaining correct values of shear stress.

RBC aggregation causes an effect termed red cell syneresis. This phenomenon is the inward movement of RBC as they aggregate; the inward movement leaves layers of cell-poor plasma at the walls of viscometers of all types (Cokelet et al. 2005) and is most evident in Couette systems. Although syneresis is usually considered to be a slow process (e.g., separation of fluid from a gel), the effects of syneresis are seen very soon after starting the viscometer. This has led to conflicting data for blood's rheological properties for shear rates below about 2–5 s^{-1}. For example, consider the events that occur when using a concentric cylinder viscometer, filled with well-mixed blood, upon starting the viscometer at a steady shear rate of 2–5 s^{-1} or less: (1) the torque-time record of the viscometer first rises from zero to a peak, (2) the torque decays in an exponential manner with time at a rate dependent on the rate of shear, and (3) the torque reaches a steady value after approximately 30 minutes. This time-dependent torque response is due to the startup of the fluid motion at very short times, syneresis of the erythrocytes away from the viscometer surfaces at all times, and finally, at very long times, sedimentation effects. Note that essentially all investigators agree that using the final steady-state value of torque is incorrect since this value is the result of fully completed syneresis plus some variable degree of sedimentation: its use results in shear stress and viscosity values that are too low.

Two different methods have been proposed for assigning the correct torque at shear rates below about 2 s–1: Chien's approach uses the peak torque value to calculate the apparent viscosity of blood (Chien 1970), while Merrill and coworkers extrapolate the exponentially decaying torque data back to zero time and use that torque value to calculate apparent viscosity (Cokelet et al. 1963). The rationale for the procedure by Merrill and coworkers is that during the time of the exponential torque decay, only the syneresis effect is a significant cause for the decreasing torque behavior. Thus extrapolation back to zero time assumes that the syneresis effect is the same in nature even in the earlier, prepeak time period. Comparison of the results of the two methods for a normal blood at a 0.4 l/l hematocrit indicate that they are in close agreement for shear rates greater than 0.5 s^{-1} but that they diverge at lower shear; at low shear the peak torque method of Chien markedly underestimates the true torque. It should be noted that the extent and "visibility" of torque-time effects depend critically on the dynamic behavior of a given viscometer: if the torque-sensing system is quick to respond (i.e., "stiff") then the effects may be seen, whereas as if the sensing system is slow (i.e., "soft") the peak and decay may be damped and blunted or even absent.

5.3.3 SEDIMENTATION EFFECTS

Since the density difference between the cell (1.096 gm/ml) and plasma (1.025 gm/ml) is small, RBC sedimentation is slow for individual cells. Consequently, the sedimentation rate of an individual erythrocyte in plasma is only about 0.13 µm/minute or 0.08 mm/h. If the suspending medium is serum and thus does not promote RBC aggregation, and if the RBC volume fraction (i.e., hematocrit) is 0.15 l/l or greater, the presence of neighboring red cells causes the settling to be hindered and hence slower. For hindered settling free of wall effects, the sedimentation rate varies with the fourth power of hematocrit (Cokelet and Meiselman 2007b).

RBC aggregation, as seen in plasma at stasis or slow flow, greatly increases the sedimentation rate since the settling rate of particles is dependent on the square of the particle size. A typical, steady, hindered settling rate for 0.4 l/l hematocrit normal human blood is about 200 µm/min or about 10 mm/h. Usually, sedimentation in viscometers is not a significant problem if the wall shear rate is above about several s^{-1}. In addition, Couette viscometers usually have a quantity of blood above the gap and space below the inner cylinder (Figure 5.1) and thus settling is problematic only if these two conditions are not present. However, at lower shear rates, the effect of erythrocyte sedimentation must be considered, especially for horizontal tube and cone-and-plate viscometers. While settling will have a minor effect in large tubes, even a 10 μm settling (i.e., about one RBC diameter) in small tubes elevates the hematocrit in the RBC phase (see Chapter 6, Section 6.5.5). Settling can also lead to phase separation in tubing used for pump and pressure sensor connections to the viscometer tube and give rise to two-phase flow in this tubing, and possibly to altered blood composition at the entrance to the tube. In cone-plate viscometers, RBC aggregation promotes settling away from the cone, thereby resulting in a two-phase fluid in the gap and a cell-poor layer adjacent to the cone. Obviously, this phase separation alters the torque on the measuring element, thereby vitiating the data; sedimentation plus surface film effects usually preclude very low shear measurements in cone-plate systems.

5.4   RHEOLOGICAL BEHAVIOR OF BLOOD

As mentioned previously and in Chapter 1, there are numerous published studies and monographs on the rheological behavior of blood, with the earliest dating from the 1846 paper by the French physician Jean-Leonard Poiseuille (Pfitzner 1976; Poiseuille 1846). Interestingly, it appears that Poiseuille began his investigations using animal blood but was unsuccessful due to blockages caused by blood clotting; he subsequently turned to simpler fluids such as water and oil. The majority of blood rheology studies deal with macroscopic, bulk rheology in which the geometry of the measuring system is much larger than the size of red or white blood cells. This macroscopic approach fails when tube diameters or viscometer gaps are less than about 500 μm (Cokelet 1980; Cokelet 1987; Cokelet and Meiselman 2007b). The following material deals only with the macroscopic flow behavior of blood and RBC suspensions, yet is focused on the rheological effects of cell–cell interactions (i.e., RBC aggregation); Chapter 6 presents information on blood flow in small tubes.

5.4.1   GENERAL FLOW BEHAVIOR

The most direct way of demonstrating the effects of RBC aggregation is to compare the flow behavior of RBC in plasma versus the same RBC washed and resuspended in a protein-free isotonic buffer. Figure 5.6 presents typical viscosity shear rate results for these two red cell suspensions at equal 0.4 l/l hematocrits. It is obvious that there is a marked effect of shear rate on viscosity for RBC in plasma and thus marked non-Newtonian flow behavior: over the shear rates shown, there is an 11-fold decrease in blood viscosity. Flow behavior becomes essentially Newtonian at high shear rates, although there is a continued small decrease until maximal RBC deformation is achieved; the transition to Newtonian-like flow depends on hematocrit and the strength of RBC aggregation (Merrill and Pelletier 1967). In contrast to the plasma suspension, the RBC-buffer suspension is much less shear dependent with less than a 2-fold decrease over the same shear rate range.

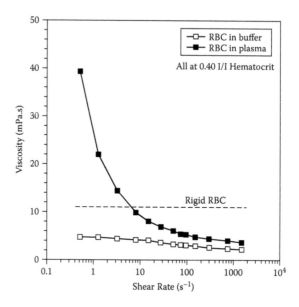

FIGURE 5.6 Representative viscosity versus shear rate data for three types of RBC suspensions at 0.4 l/l hematocrit: RBC in plasma (i.e., blood), RBC in protein-free isotonic buffer, and RBC made rigid by aldehyde fixation also suspended in isotonic buffer. RBC aggregation increases viscosity at low shear for RBC in plasma with cell deformation occurring at higher shear. RBC or rigid RBC in buffer do not aggregate and rigid cells do not deform at high shear. This figure was adapted from the original, Chien, S. 1970. "Shear Dependence of Effective Cell Volume as a Determinant of Blood Viscosity." *Science* 168:977–978, and is from Baskurt, O. K., M. R. Hardeman, M. W. Rampling, and H. J. Meiselman, eds. 2007. *Handbook of Hemorheology and Hemodynamics.* Amsterdam, Berlin, Oxford, Tokyo, Washington, DC: IOS Press, with permission.

The reason for the plasma-buffer differences in Figure 5.6 is, in fact, the subject of this monograph: normal RBC in plasma undergo reversible aggregation while those in buffer do not (see Chapter 2), and thus at constant hematocrit, RBC aggregation is the primary determinate of low-shear blood viscosity. With increasing shear forces, the RBC aggregates in the plasma suspension become progressively dispersed and the two suspensions approach each other. Since plasma has a higher viscosity than buffer, the two curves do not coincide at higher shear but would do so if suspending viscosities were equalized (Chien 1975). Note that RBC suspensions do not exhibit normal stresses, and for the usual time scales encountered in macrorheology studies, the rheological properties are history and time independent (i.e., no evidence of thixotropy or rheopexy).

Also shown in Figure 5.6 are data for RBC made rigid by chemical treatment with an aldehyde, thereby yielding cells that are unable to deform in response to shear forces and unable to form RBC aggregates (Chien 1975; Hardeman et al. 1994). Rigid RBC in either plasma or buffer exhibit Newtonian flow behavior and, at comparable hematocrits, are more viscous at any shear rate.

Based upon the data presented in Figure 5.6, it can be concluded that normal human blood behaves as a shear-thinning non-Newtonian fluid with flow behavior affected by two mechanisms: (1) RBC aggregates that are formed at stasis or very low shear are dispersed by increasing shear forces with complete dispersion observed in the range of 100–120 s^{-1} and (2) red cells deform and align with flow at higher shear rates. Thus, in general, increases of low-shear viscosity usually indicate enhanced aggregation while increases at high shear reflect RBC deformability.

5.4.2 Possible Artifact at Low Shear Rates

The data presented in Figure 5.6 confirm that increased low-shear viscosity indicates enhanced RBC aggregation and that increased viscosity at high shear indicates reduced cell deformability. In most cases this is correct. However, for RBC with only moderately reduced deformability, increases of low-shear viscosity *and* enhanced non-Newtonian flow behavior can occur in the *absence* of RBC aggregation! Such a situation is illustrated in Figure 5.7 for RBC suspended in buffer containing several concentrations of dinitrophenol (DNP). This chemical alters RBC shape (i.e., biconcave disc to crenated sphere) and reduces cell deformability in a dose-dependent manner so that the cells are poorly deformable at low shears but will deform normally at sufficiently high shear (Meiselman 1978; Meiselman 1981).

The results shown in Figure 5.7 imply enhanced RBC aggregation with increasing concentrations of DNP. However, microscopic examination of these cells in buffer indicates a total absence of aggregate formation, with all cells existing as individuals. Similar observations have been reported for RBC treated with hydrogen peroxide or very low levels of glutaraldehyde (see Figure 4.7 in Chapter 4) or heated to 48°C: either unaltered or slight decreases of aggregation in plasma yet markedly increased low-shear viscosity (Baskurt and Meiselman 1997). These results thus suggest the importance of microscopic observations as well as rheological measurements when considering blood rheology data.

5.4.3 Hematocrit Effects

Since blood is a suspension of particles in a liquid, basic principles suggest that its rheological behavior is markedly affected by the volume fraction or percentage of these particles; since RBC are by far the most numerous, it is their concentration (i.e., hematocrit) that is the important determinant (Baskurt and Meiselman 2003). Figures 5.8 and 5.9 present viscosity shear rate and viscosity-hematocrit data for normal human blood at four levels of hematocrit. Inspection of these results indicates that hematocrit affects both viscosity and the degree of non-Newtonian behavior: (1) at the lowest shear rate of 0.1 s^{-1}, a hematocrit change from 0.2 to 0.5 l/l causes a 9-fold increase of viscosity while at the highest shear shown, the same hematocrit change causes only a 3-fold increase; (2) deviation from Newtonian behavior is also hematocrit dependent such that at 0.2 l/l hematocrit there is a 6-fold decrease in viscosity as the shear rate is increased from 0.1 to 20 s^{-1} while over a 20-fold decrease is observed at 0.5 l/l hematocrit. Not surprisingly, interpretation of *in vitro* blood rheology data in terms of RBC aggregation requires careful attention to hematocrit levels.

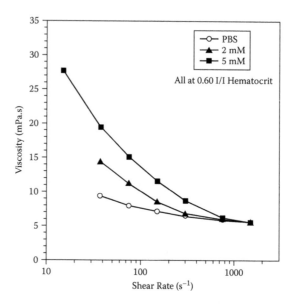

FIGURE 5.7 Viscosity versus shear rate data for RBC suspended in isotonic phosphate-buffered saline (PBS) as a function of dinitrophenol (DNP) concentration. DNP causes a slight dose-dependent increase of RBC rigidity yet cells are able to deform normally at sufficiently high shear forces. The DNP data are suggestive of enhanced aggregation, yet like control cells in PBS, DNP-treated RBC do *not* aggregate in buffer. Adapted from Baskurt, O. K., M. R. Hardeman, M. W. Rampling, and H. J. Meiselman, eds. 2007. *Handbook of Hemorheology and Hemodynamics.* Amsterdam, Berlin, Oxford, Tokyo, Washington, DC: IOS Press, with permission.

5.4.4 YIELD SHEAR STRESS

While there had been previous reports indicating the non-Newtonian flow behavior of blood (see reviews by Dintenfass 1971; Dintenfass 1985), the 1961 study by Wells and Merrill (1961) seems to be the first that explicitly examined the effects of RBC aggregation on low-shear viscosity. In an effort to quantitate low shear rate–shear stress data and to determine the strength of blood as an elastic solid, Cokelet, Merrill, and coworkers (Cokelet et al. 1963; Merrill et al. 1963) employed a rheological model initially proposed by Casson (1959) for printing ink. In this model, particles are suspended in a Newtonian fluid and form rodlike aggregates at low shear rates, with the length of these aggregates an inverse function of shear rate. In practice, the square root of shear stress is plotted versus the square root of shear rate and the low-shear straight line portion of the data is extrapolated to zero shear rate. The resulting intercept on the stress axis provides a value for the yield shear stress, which is the stress below which deformation or flow do not occur. Figure 5.10 shows a Casson plot for the data presented in Figure 5.8, where it can be seen that, over the range of hematocrit and shear rate employed, there are straight lines that extrapolate to nonzero intercepts (i.e., to the square root of the yield shear stress). Yield stress versus hematocrit data from Figure 5.10 can be fitted via power law regression

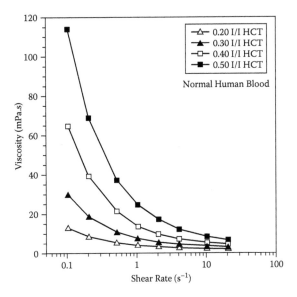

FIGURE 5.8 Viscosity versus shear rate for normal human blood at various hematocrits. Both viscosity and the degree of non-Newtonian behavior are markedly affected by hematocrit. Adapted from Baskurt, O. K., M. R. Hardeman, M. W. Rampling, and H. J. Meiselman, eds. 2007. *Handbook of Hemorheology and Hemodynamics.* Amsterdam, Berlin, Oxford, Tokyo, Washington, DC: IOS Press, with permission.

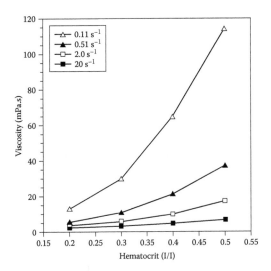

FIGURE 5.9 Viscosity versus hematocrit for normal human blood at various shear rates. Note that the effects of hematocrit changes are much greater at lower shear rates. Adapted from Baskurt, O. K., M. R. Hardeman, M. W. Rampling, and H. J. Meiselman, eds. 2007. *Handbook of Hemorheology and Hemodynamics.* Amsterdam, Berlin, Oxford, Tokyo, Washington, DC: IOS Press, with permission.

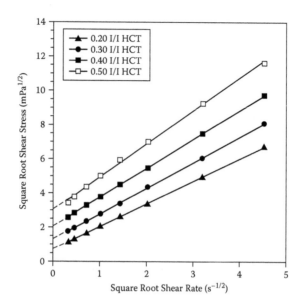

FIGURE 5.10 Square root of shear stress versus square root of shear rate, often termed a Casson plot, for blood at various hematocrits. Over the range of shear rates shown (i.e., 20 s^{-1} and lower) the data are well fit by straight lines obtained using linear regression; the intercepts on the ordinate are the square roots of yield shear stress. Data are from Figure 5.8. Adapted from Baskurt, O. K., M. R. Hardeman, M. W. Rampling, and H. J. Meiselman, eds. 2007. *Handbook of Hemorheology and Hemodynamics.* Amsterdam, Berlin, Oxford, Tokyo, Washington, DC: IOS Press, with permission.

(Figure 5.11), resulting in a value of 3.1 for the hematocrit exponent (r = 0.99, p < 0.01); this cubic relation is consistent with the findings of Merrill et al. (1963).

In addition to the Casson extrapolation method, the yield shear stress for blood can be demonstrated directly by measuring the residual pressure gradient across a tube. In this method, the pressure gradient is measured using liquid-coupled connections to a differential pressure transducer and blood flow through the tube is initiated manually. Subsequently, the tube-transducer system is hydrodynamically isolated and the flow is allowed to decay to zero. For nonaggregating suspensions, such as RBC in buffer, the pressure gradient also decays to zero, while at zero flow there is a finite residual pressure for RBC in plasma. Yield stress values obtained via this method are in agreement with those obtained via extrapolation of viscometer data (Merrill et al. 1965).

At constant hematocrit and temperature, the yield shear stress reflects the strength of RBC aggregation in the suspension under study (Lee et al. 2007; Rampling 1988). This association between RBC aggregation and yield stress has been shown by varying levels of the strongly pro-aggregating plasma protein fibrinogen by removal from plasma (Merrill et al. 1963), by addition to isotonic buffer (Merrill et al. 1966), and by analysis of native blood samples from over 100 healthy subjects (Merrill et al. 1968). Figure 5.12 presents results for both RBC in native plasma and for RBC suspended in isotonic buffer containing added fibrinogen. Interestingly, there is a

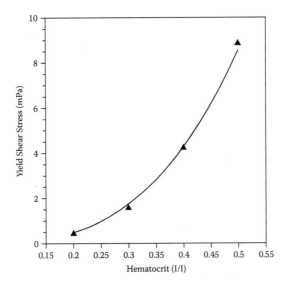

FIGURE 5.11 Yield shear stress versus hematocrit data obtained from the results presented in Figure 5.10. The curved line was obtained via power law regression: yield shear stress = 74 x (hematocrit)$^{3.1}$, $r = 0.99$, $p < 0.01$. Adapted from Baskurt, O. K., M. R. Hardeman, M. W. Rampling, and H. J. Meiselman, eds. 2007. *Handbook of Hemorheology and Hemodynamics.* Amsterdam, Berlin, Oxford, Tokyo, Washington, DC: IOS Press, with permission..

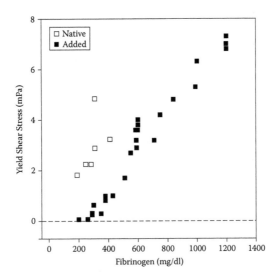

FIGURE 5.12 Yield shear stress versus fibrinogen concentration for blood from six normal donors (open squares) and for RBC in buffer with various levels of added fibrinogen (closed squares). Note the concentration threshold for added fibrinogen and the differences between native and added at the same concentration. Data from Merrill, E. W., E. R. Gilliland, T. S. Lee, and E. W. Salzman. 1966. "Blood Rheology: Effect of Fibrinogen Deduced by Addition." *Circulation Research* 18:437–446.

threshold effect for added fibrinogen and there are differences of yield stress levels at identical concentrations, presumably due to the nature of the freeze-dried protein added to the buffer; a similar threshold effect for added fibrinogen has been reported by Rampling and Whittingstall (1986). Associations between polymer-induced and fibrinogen-induced RBC aggregation and yield shear stress have also been reported (Meiselman et al. 1967; Morris et al. 1989).

Although the concept of a yield shear stress for suspensions exhibiting RBC aggregation seems straightforward, its measurement can be problematic: (1) the square root–square root extrapolation method is only valid for shear rates near zero and not greater than about 10 s^{-1}; (2) application of the Casson approach to data obtained at higher shear rates is possible, but extrapolation to zero shear will result in incorrect values of yield stress since there is a transition toward Newtonian flow at high shear (Merrill and Pelletier 1967); (3) both the Casson extrapolation method and the determination of residual pressure in a tube viscometer are made going from low shear to zero shear or from low flow to zero flow, with measurements made starting at zero easily affected by artifacts including RBC settling or syneresis in tubes; (4) determination of a yield stress for blood involves considering the time of observation since if sufficient time is allowed, plasma flow through aggregated cells may allow the stress to decay to zero.

5.4.5 TEMPERATURE EFFECTS

Not surprisingly, the rheological properties of blood depend strongly on temperature. Figure 5.13 presents typical viscosity-temperature results for human blood at a hematocrit of 0.4 l/l, and Figure 5.14 shows the same data plotted on a log-log axis system. Blood viscosity at all shear rates decreases with increasing temperature, yet both Figures 5.13 and 5.14 indicate that the effect of temperature is shear rate dependent: (1) over the range of 10 to 37°C, viscosity decreases by 80% at 0.15 s^{-1} whereas there is only a 20% decrease at 95 s^{-1} (Figure 5.13); (2) at 95 s^{-1} the dependence on temperature is essentially identical to that for water, whereas blood viscosity has a greater sensitivity at lower rates of shear (Figure 5.14). Square root (i.e., Casson) extrapolation of the data in Figure 5.13 over the range of 0.51 to 1.3 s^{-1} can be utilized to obtain yield shear stress as a function of temperature. Figure 5.15 presents yield stress data and also a curve for the viscosity of water multiplied by 10 in order to use the same vertical scale. The difference in sensitivity is obvious: over the range of 37 to 10°C there is a 3-fold increase of yield shear stress versus a 2-fold increase of the viscosity of water. A similar reciprocal relation between temperature and the shear stress necessary to disperse RBC aggregates has been reported by Lim and coworkers (Lim et al. 2010).

Figures 5.14 and 5.15 thus indicate that the strength of RBC aggregation, as indexed by low-shear viscosity and yield stress, increases with decreasing temperature. The mechanisms involved in this inverse temperature dependence are not, as yet, clear. Based upon the depletion model for RBC aggregation (see Chapter 3), the depletion interaction energy, and hence the difference between the chemical potential of the solvent in polymer solution and in polymer-free solution should increase with temperature. An increase of depletion interaction energy due to increased temperature

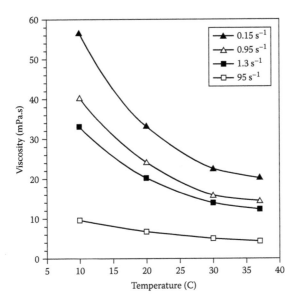

FIGURE 5.13   Viscosity versus temperature at various shear rates for blood at 0.4 l/l hematocrit. The effects of temperature increase with decreasing shear rate.

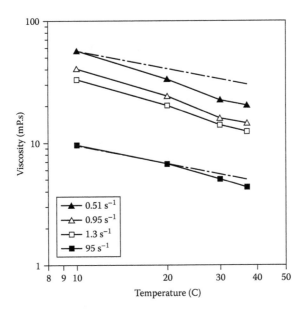

FIGURE 5.14   Viscosity versus temperature at various shear rates for blood at 0.4 l/l hematocrit shown on a log-log coordinate system. The dashed lines indicate the behavior of water and are scaled to be coincident with 10°C blood viscosity. Note that at low shear rates, blood viscosity has a greater sensitivity to temperature than water.

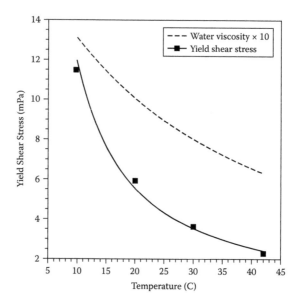

FIGURE 5.15 Yield shear stress versus temperature for blood at 0.4 l/l hematocrit. Also shown, as a dashed line, is the viscosity of water multiplied by a factor of 10. Note that RBC aggregation, as indexed by the yield stress, increases with decreasing temperature and has a greater sensitivity than water. (unpublished data).

would then be expected to enhance aggregation, yet yield stress and the stability of RBC aggregates in flow do not support this temperature-based expectation.

5.4.6 POLYMER-INDUCED RBC AGGREGATION

In addition to altering plasma levels of the protein fibrinogen or other pro-aggregant macromolecules (see Section 5.4.4 above), it is possible to alter RBC aggregation and the rheological behavior of blood by the presence of water-soluble polymers of appropriate size. The literature in this area is large, with much of the early work related to various molecular mass fractions of the polyglucose dextran (e.g., Chen et al. 1989; Chien et al. 1977; Chien and Sung 1987; Derrick and Guest 1971; Eiseman and Bosomworth 1963). For water-soluble nonionic polymers, there now seems to be general agreement as to the importance of the polymer's hydrated size rather than its specific chemistry (Armstrong et al. 2004). Below a critical size, the polymer may reduce aggregation of RBC in plasma or have no effect if it is dissolved in buffer, whereas above this size there is an increase of aggregation (Chien 1975; Chien and Sung 1987; Neu et al. 2008; Neu et al. 2001; Neu and Meiselman 2002) until the molecular size becomes too large (Neu et al. 2008) (See Chapter 2, Section 2.2).

Representative viscosity shear rate and viscosity-concentration data are shown in Figures 5.16 and 5.17 for a 500 kDa fraction of dextran added to whole blood. This molecular mass of dextran is a strong pro-aggregant (Armstrong et al. 2004; Baskurt et al. 2000; Chien 1975; Neu et al. 2008). As anticipated, increased RBC aggregation

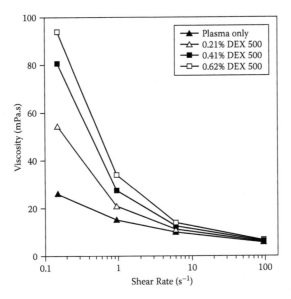

FIGURE 5.16 Viscosity versus shear rate for 0.4 l/l hematocrit blood containing various concentrations of 500 kDa dextran. This dextran fraction is a strong pro-aggregant and over the range shown increases low-shear viscosity and the degree of non-Newtonian flow behavior in a dose-dependent manner. At the high-shear rate, RBC aggregates are dispersed and viscosities are similar.

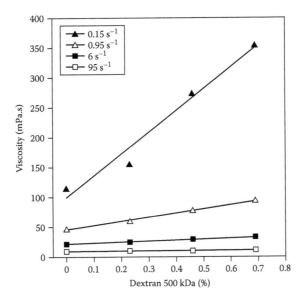

FIGURE 5.17 Viscosity versus 500 kDa dextran concentration at four shear rates; data are from Figure 5.16. Note the very strong concentration dependence at low shear (0.15 s^{-1}) and a greatly reduced sensitivity at the highest shear rate.

induced by increasing concentrations of this polymer enhances both low-shear viscosity and the degree of non-Newtonian flow behavior. Interestingly, the change of viscosity with dextran concentration is shear-rate dependent: increasing dextran concentration to 0.7 g/dL results in a 3-fold increase of viscosity at 0.15 s^{-1} yet only a 20% increase at 95 s^{-1}. RBC aggregation, induced by either specific plasma proteins or polymers of appropriate molecular size, thus has its greatest effect at low levels of shear rate, whereas aggregates are dispersed at higher shear and do not affect blood or RBC suspension rheology.

LITERATURE CITED

Armstrong, J. K., R. B. Wenby, H. J. Meiselman, and T. C. Fisher. 2004. "The hydrodynamic radii of macromolecules and their effect on red blood cell aggregation." *Biophysical Journal* 87:4259–4270.

Barnes, H. A. 2000. *A Handbook of Elementary Rheology*. Aberystwyth: University of Wales.

Barnes, H. A., J. F. Hutton, and K. Walters. 1989. *An Introduction to Rheology*. Amsterdam: Elsevier Science.

Baskurt, O. K., M. Bor-Kucukatay, O. Yalcin, H. J. Meiselman, and J. K. Armstrong. 2000. "Standard Aggregating Media to Test the 'Aggregability' of Rat Red Blood Cells." *Clinical Hemorheology and Microcirculation* 22:161–166.

Baskurt, O. K., M. Boynard, G. R. Cokelet, P. Connes, B. M. Cooke, S. Forconi, F. Liao, et al. 2009. "New Guidelines for Hemorheological Laboratory Techniques." *Clinical Hemorheology and Microcirculation* 42: 75–97.

Baskurt, O. K., M. R. Hardeman, M. W. Rampling, and H. J. Meiselman, eds. 2007. *Handbook of Hemorheology and Hemodynamics*. Amsterdam, Berlin, Oxford, Tokyo, Washington, DC: IOS Press.

Baskurt, O. K., and H. J. Meiselman. 1997. "Cellular Determinants of Low Shear Blood Viscosity." *Biorheology* 34:235–247.

Baskurt, O. K., and H. J. Meiselman. 2003. "Blood Rheology and Hemodynamics." *Seminars in Thrombosis and Hemostasis* 29:435–450.

Casson, N. 1959. "A Flow Equation for Pigment-Oil Suspension of the Printing Ink Type." In *Rheology of Disperse Systems*, ed. C. C. Mills, 84–104. New York: Pergamon Press.

Chen, R. Y., R. D. Carlin, S. Simchon, K. M. Jan, and S. Chien. 1989. "Effects of Dextran-induced Hyperviscosity on Regional Blood Flow and Hemodynamics in Dogs." *American Journal of Physiology—Heart and Circulatory Physiology* 256:H898–H905.

Chien, S. 1970. "Shear Dependence of Effective Cell Volume as a Determinant of Blood Viscosity." *Science* 168:977–978.

Chien, S. 1975. "Biophysical Behaviour of Red Cells in Suspensions." In *The Red Blood Cell*, ed. D. M. Surgenor, 1031–1133. New York: Academic Press.

Chien, S., J. Dormandy, E. Ernst, and A. Matrai. 1987. *Clinical Hemorheology*. Dordrecht/Boston/Lancaster: Martinus Nijhoff Publ.

Chien, S., S. Simchon, R. E. Abbott, and K. M. Jan. 1977. "Surface Adsorption of Dextrans on Human Red Blood Cell Membrane." *Journal of Colloid and Interface Science* 62:461–470.

Chien, S., and L. A. Sung. 1987. "Physicochemical Basis and Clinical Implications of Red Cell Aggregation." *Clinical Hemorheology* 7:71–91.

Cokelet, G. R. 1980. "Rheology and Hemodynamics." *Annual Review of Physiology* 42:311–324.

Cokelet, G. R. 1987. "Rheology of Tube Flow of Blood." In *Handbook of Engineering*, ed. R. Skalak and S. Chien, 14.1–14.17. New York: McGraw-Hill Book Co.

Cokelet, G. R., J. R. Brown, S. O. Codd, and J. D. Seymour. 2005. "Magnetic Resonance Microscopy–Determined Velocity and Hematocrit Distributions in a Couette Viscometer." *Biorheology* 42:385–399.

Cokelet, G. R., and H. J. Meiselman. 2007a. "Basic Aspects of Hemorheology." In *Handbook of Hemorheology and Hemodynamics*, ed. O. K. Baskurt, M. R. Hardeman, M. W. Rampling and H. J. Meiselman, 3–33. Amsterdam, Berlin, Oxford, Tokyo, Washington, DC: IOS Press.

Cokelet, G. R., and H. J. Meiselman. 2007b. "Macro- and Micro-Rheological Properties of Blood." In *Handbook of Hemorheology and Hemodynamics*, ed. O. K. Baskurt, M. R. Hardeman, M. W. Rampling, and H. J. Meiselman, 45–71. Amsterdam, Berlin, Oxford, Tokyo, Washington, DC: IOS Press.

Cokelet, G. R., H. J. Meiselman, and D. E. Brooks. 1980. *Erythrocyte Mechanics and Blood Flow*. New York: A.R. Liss.

Cokelet, G. R., E. W. Merrill, E. R. Gilliland, H. Shin, A. Britten, and R. E. Wells. 1963. "The Rheology of Human Blood-Measurement near and at Zero Shear Rate." *Transactions of Society of Rheology* 7:303–317.

Coleman, B. D. 1996. *Viscometric Flows of Non-Newtonian Fluids*. New York: Springer-Verlag.

Derrick, J. R., and M. M. Guest. 1971. *Dextrans: Current Concepts of Basic Actions and Clinical Applications*, 1–222. Springfield, IL: Charles C. Thomas.

Dintenfass, L. 1971. *Blood Microrheology*. New York: Appleton-Century-Crofts.

Dintenfass, L. 1985. *Hyperviscosity and Hyperviscosemia*. Lancaster, UK: MTP Press.

ECSH Expert Panel on Blood Rheology. 1986. "Guidelines for Measurement of Blood-Viscosity and Erythrocyte Deformability." *Clinical Hemorheology* 6:439–453.

Ehrly, A. M. 1991. *Therapeutic Hemorheology*. Berlin: Springer-Verlag.

Eiseman, B., and P. Bosomworth. 1963. *Evaluation of Low Molecular Weight Dextran in Shock*, 1–135. Washington, DC: National Academy of Sciences.

Goldsmith, H. L. 1968. "The Microrheology of Red Blood Cell Suspensions." *Journal of General Physiology* 52:5s–28s.

Goldsmith, H. L., K. Takamura, and D. Bell. 1983. "Shear-Induced Collisions between Human Blood Cells." *Annals of NY Academy of Sciences* 416:299–318.

Hardeman, M. R., P. T. Goedhart, and N. H. Schut. 1994. "Laser-Assisted Optical Rotational Cell Analyzer (LORCA); II Red Blood Cell Deformability: Elongation Index Versus Cell Transit Time." *Clinical Hemorheology* 14:619–630.

Larson, R. G. 1998. *The Structure and Rheology of Complex Fluids*. Oxford: Oxford University Press.

Lee, B. K., T. Alexy, R. B. Wenby, and H. J. Meiselman. 2007. "Red Blood Cell Aggregation Quantitated via Myrenne Aggregometer and Yield Shear Stress." *Biorheology* 44:29–35.

Lim, H. J., Y. J. Lee, J. H. Nam, S. Chung, and S. Shin. 2010. "Temperature-Dependent Threshold Shear Stress of Red Blood Cell Aggregation." *Journal of Biomechanics* 43:546–550.

Macosko, C. W. 1994. *Rheology: Principals, Measurements and Application*. New York: Wiley-VCH.

Malkin, A. Y., and A. I. Isayev. 2005. *Rheology: Concepts, Methods and Applications*. Norwich, NY: ChemTec Publishing.

Markovitz, H. 1968. "The Emergence of Rheology." *Physics Today* 21:23–30.

Meiselman, H. J. 1978. "Rheology of Shape-Transformed Human Red Cells." *Biorheology* 15:225–237.

Meiselman, H. J. 1981. "Morphological Determinants of Red Cell Deformability." *Scandinavian Journal of Clinical and Laboratory Investigation* 41:27–34.

Meiselman, H. J., and G. R. Cokelet. 1974. "Blood Rheology: Instrumentation and Techniques." In *Flow: Its Measurement and Control in Science and Industry*, ed. R. B. Dowdell, 1337–1346. Pittsburgh, PA: Instrument Society of America.

Meiselman, H. J., E. W. Merrill, E. Salzman, E. R. Gilliland, and G. A. Pelletier. 1967. "The Effect of Dextran on the Rheology of Human Blood: Low Shear Viscosity." *Journal of Applied Physiology* 22:480–486.

Merrill, E. W. 1965. "Rheology of Human Blood and Some Speculations on Its Role in Vascular Hemostasis." In *Biophysical Mechanisms in Vascular Hemostasis and Intravascular Thrombosis*, ed. P. H. Sawyer, 31–42. New York: Appleton-Century-Crofts.

Merrill, E. W., A. M. Benis, E. R. Gilliland, T. K. Sherwood, and E. W. Salzman. 1965. "Pressure-Flow Relations of Human Blood in Hollow Fibers at Low Flow Rates." *Journal of Applied Physiology* 20:954–967.

Merrill, E. W., C. S. Cheng, and G. A. Pelletier. 1968. "Yield Stress of Normal Human Blood as a Function of Endogenous Fibrinogen." *Journal of Applied Physiology* 26:1–3.

Merrill, E. W., G. R. Cokelet, A. B. Britten, and R. E. Wells. 1963. "Non-Newtonian Rheology of Human Blood-Effect of Fibrinogen Deduced by Subtraction." *Circulation Research* 13:48–55.

Merrill, E. W., E. R. Gilliland, G. R. Cokelet, H. Shin, A. Britten, and R. E. Wells. 1963. "Rheology of Human Blood near and at Zero Flow." *Biophysical Journal* 3:199–213.

Merrill, E. W., E. R. Gilliland, T. S. Lee, and E. W. Salzman. 1966. "Blood Rheology: Effect of Fibrinogen Deduced by Addition." *Circulation Research* 18:437–446.

Merrill, E. W., and G. A. Pelletier. 1967. "Viscosity of Human Blood: Transition from Newtonian to Non-Newtonian." *Journal of Applied Physiology* 23:178–182.

Metzger, T. G. 2006. *The Rheology Handbook*. New York: Elsevier Science.

Morris, C. L., D. L. Rucknagel, R. Shuka, R. A. Gruppo, C. M. Smith, and P. L. Blackshear. 1989. "Evaluation of the Yield Stress of Normal Blood as a Function of Fibrinogen Concentration and Hematocrit." *Microvascular Research* 37:323–338.

Neu, B., J. K. Armstrong, T. C. Fisher, and H. J. Meiselman. 2001. "Aggregation of Human RBC in Binary Dextran-PEG Polymer Mixtures." *Biorheology* 38:53–68.

Neu, B., and H. J. Meiselman. 2002. "Depletion-Mediated Red Blood Cell Aggregation in Polymer Solutions." *Biophysical Journal* 83 (5):2482–2490.

Neu, B., R. Wenby, and H. J. Meiselman. 2008. "Effects of Dextran Molecular Weight on Red Blood Cell Aggregation." *Biophysical Journal* 95:3059–3065.

Pfitzner, J. 1976. "Poiseuille and His Law." *Anesthesia* 31:273–275.

Poiseuille, J. L. M. 1846. "Recherches experimentales sur la mouvement des liquides dan les tubes trespetits diametres." *Academia Royal de Sciences Paris* 9: 433–453.

Rampling, M. W. 1988. "Red Cell Aggregation and Yield Stress." In *Clinical Blood Rheology*, ed. G. D. O. Lowe, 45–64. Boca Raton, FL: CRC Press.

Rampling, M. W. 2007. "Compositional Properties of Blood." In *Handbook of Hemorheology and Hemodynamics*, ed. O. K. Baskurt, M. R. Hardeman, M. W. Rampling, and H. J. Meiselman, 34–44. Amsterdam, Berlin, Oxford, Tokyo, Washington, DC: IOS Press.

Rampling, M. W., and P. Whittingstall. 1986. "A Comparison of Five Methods for Estimating Red Cell Aggregation." *Klinische Wochenschrift* 64:1084–1088.

Stoltz, J. F., M. Singh, and P. Riha. 1999. *Hemorheology in Practice*. Amsterdam, Berlin, Oxford, Tokyo, Washington, DC: IOS Press.

Wells, R. E., and E. W. Merrill. 1961. "Shear Rate Dependence of the Viscosity of Whole Blood and Plasma." *Science* 133:763–764.

Wilkinson, W. L. 1960. *Non-Newtonian Fluids*. London: Pergamon Press.

6 Effect of Red Blood Cell Aggregation on Tube Flow

Understanding blood flow in the circulatory system has attracted scientific interest for several centuries, with perhaps the most notable early work being the 1628 publication of *An Anatomical Study of the Motion of the Heart and of the Blood in Animals* by William Harvey. However, due to technical difficulties in conducting appropriate experiments in the vasculature of living organisms, blood flow studies until early in the twentieth century were mostly based on observations in cylindrical tubes. Therefore, observations on blood flow in narrow tubes have provided the majority of basic information on blood flow dynamics, including the influence of red blood cell (RBC) properties. These early studies on blood flow in cylindrical tubes are briefly summarized in this chapter, with more detailed information available elsewhere (Goldsmith et al. 1989).

6.1 HISTORICAL PERSPECTIVES

Jean-Leonard Poiseuille is considered to be the pioneer of hydrodynamics and is famous for his studies of the flow of various fluids in narrow tubes. His work led to the Poiseuille equation describing the relationship between volumetric flow rate, pressure gradient, the geometry of the flow system, and the fluid viscosity. His original work did not explain the behavior of blood in narrow tubes, but did stimulate extensive research done in the second half of the nineteenth century and the first several decades of the twentieth century (Goldsmith et al. 1989). Hess showed that the flow behavior of blood in glass tubes could be predicted by the Poiseuille equation only if the flow rate was sufficiently high. The apparent viscosity of blood was observed to increase at lower flow rates, with this dependence on flow rate mainly attributed to the elastic deformation of RBC (Hess 1915). Interestingly, Walter Rudolph Hess was a Swiss physiologist who originally was interested in hemodynamics, but is widely known as a neurophysiologist, most likely due to the 1949 Nobel Prize awarded to him "for his discovery of the functional organization of the interbrain as a coordinator of the activities of the internal organs" (Nobelprize.org 2010).

Early reports indicated that blood flow in the peripheral vasculature (i.e., locations other than cardiac valves and aorta) is laminar (Hahn et al. 1945; Hess 1917; Ralston and Taylor 1945); these findings significantly influenced the understanding of blood flow dynamics in tube flow. It was also suggested by Thoma in the early 1900s that the formation of a marginal cell-poor layer during blood flow in narrow tubes may contribute to the flow dependence of the apparent viscosity of blood (Goldsmith et al. 1989).

Although RBC aggregation during flow was observed as early as the eighteenth century, Robin Fåhraeus was the first scientist who indicated the importance of rouleaux formation in tube flow (Fåhraeus 1921). He described two phenomena, the Fåhraeus effect and the Fåhraeus-Lindqvist effect, both of which influence the resistance to blood flow in narrow tubes. These effects depend strongly on the distribution of blood cells across the tube diameter during flow, and are discussed in more detail below in Sections 6.3 and 6.4 of this chapter. Importantly, he recognized that RBC aggregation can promote the axial accumulation of RBC and amplify these effects and demonstrated that the thickness of the peripheral cell-poor fluid layer increases with RBC aggregation. Figure 6.1 shows his observations using human and bovine blood flowing in a 95 μm diameter glass tube. The wider peripheral cell-poor layer for human blood compared to bovine blood is obvious and is due to the more intense RBC aggregation in human blood (Fåhraeus 1958).

6.2   TUBE FLOW

6.2.1   VELOCITY PROFILE

The radial distribution of flow velocity in a cylindrical tube with laminar flow is well known for a Newtonian fluid. The velocity profile can be predicted based on the Navier-Stokes equation, with the velocity at a given radial position (u_r) given by:

$$u_r = u_{max} \left[1 - \left(\frac{r}{R} \right)^2 \right]$$

(6.1)

where u_{max} is the maximum (centerline) velocity, R is the tube radius, and r is the radial distance from the center of the tube (Chatzimavroudis 2002). As is obvious

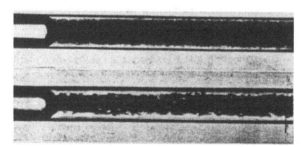

FIGURE 6.1   Bovine and human blood flowing in glass tubes of 95 μm diameter at a flow rate of 3.3 mm/s. The upper panel shows the flow of bovine blood, while the lower panel shows the flow of human blood, which exhibits more intense red blood cell aggregation. The figure is reproduced from the original publication of Fåhraeus, R. 1958. "the Influence of the Rouleauz Formation of the Erythrocytes on the Rheology of the Blood." *Acta Medica Scandinavica* 161:151-165.

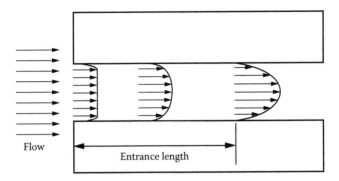

FIGURE 6.2   Development of parabolic velocity profile in a cylindrical tube within a certain distance from the entrance which is termed the entrance length.

from Equation (6.1), the velocity at the center of the tube where $r = 0$ is equal to u_{max} and approaches zero at the tube wall where $r = R$; the velocity distribution thus has a parabolic profile (Figure 6.2) (Fung 1997). This velocity profile will fully develop within a certain distance from the entrance of the tube (i.e., entrance length), with the distance determined by tube diameter and the Reynolds number (Chatzimavroudis 2002). The velocity profile can also be influenced by the presence of nonsteady pulsatile flow, curvature, and branching in the tube system.

The properties of the flowing fluid may also influence the shape of the velocity profile, especially for blood flow in cylindrical tubes. It has been shown that velocity profiles are blunted in tubes perfused with blood (Figure 6.3) and that the degree of blunting becomes more pronounced in tubes with diameters less than ~130 μm and for samples having higher hematocrit (Gaehtgens et al. 1970). Velocity profiles are also influenced by RBC mechanical properties, including deformability (Gaehtgens et al. 1970) and aggregation (Bishop et al. 2002; Bishop et al. 2001). Section 6.5.1 provides additional velocity distribution details for blood flowing in tubes > 150 μm.

There are a number of obvious consequences of a velocity profile in tube flow: (1) Due to the curved velocity profile, the difference in the velocities of adjacent flow streamlines close to the tube axis is smaller than the difference near the tube wall. The difference in the velocity of adjacent streamlines defines the local shear rate, and thus shear rate varies from a minimum (e.g., zero for a parabolic profile) at the axis to a maximum at the wall, (2) The viscosity of non-Newtonian fluids is not a constant within a tube but rather changes over the cross section of the tube as a result of the change in shear rate. However, based on the Poiseuille equation, an effective viscosity value can be calculated using pressure-flow data and tube geometry (Goldsmith 1989). This value may not always correspond to a viscosity measured at a given shear rate using a viscometer (see Section 6.5.1). (3) The suspended particles (e.g., blood cells) moving at different axial positions have different velocities and hence different transit times through a tube. If the distribution of such particles is not homogenous over the cross section of the tube, the suspending phase and particles may have different average velocities. This difference may also exist among different types of particles if their distribution across the tube is not uniform.

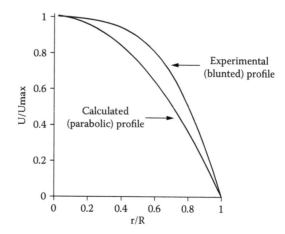

FIGURE 6.3 Velocity profiles are calculated using Equation (6.1) and experimentally measured in an 80 μm diameter cylindrical tube for RBC in plasma at a nominal shear rate of 15.4 s⁻¹. Note the blunting of the profile for the RBC suspension. R = tube radius, r = radial position from axis, U = velocity at position r, and Umax = the centerline (maximum) velocity. The experimental data are redrawn from Gaehtgens, P., H. J. Meiselman, and H. Wayland. 1970. "Velocity Profiles of Human Blood at Normal and Reduced Hematocrit in Glass Tubes up to 130 μm Diameter." *Microvascular Research* 2:13–23.

6.2.2 AXIAL MIGRATION

Due to simple exclusion considerations, usually referred to as the Vand effect, RBC concentration very near the tube wall (i.e., ~1–3 μm from the wall) is slightly less than in the rest of the suspension (Maude 1967). This slight reduction of local hematocrit occurs because the cells pack less efficiently at a solid surface; an imperfect analogy is the packing of spheres near a solid surface and the spaces between adjacent spheres and the wall. However, an "active" process occurs during tube flow: suspended particles (e.g., RBC) tend to have a net lateral movement toward the center of the tube (Goldsmith 1989). This phenomenon has been investigated widely, not only for blood cells but also for different types of particles in suspensions or emulsions (Brenner 1966; Leal 1980). An earlier hypothesis regarding the mechanism of this lateral movement was related to a radial pressure gradient across the diameter of the tube as estimated based on the Bernouilli Theorem and the density or size of the particles (Goldsmith et al. 1989). However, later and more detailed analysis of the problem revealed that Bernouilli's Theorem does not predict a radial pressure gradient, but rather that the tube wall, together with the inertia of the fluid, results in the axial migration of particles; Goldsmith et al. (1989) discuss this phenomenon in detail. The newer theory, with slightly different mechanisms, is applicable to the migration of both deformable and rigid particles. The migration velocity varies with the third power of the ratio of particle to tube radius and depends on radial position: migration velocity is maximum near the tube wall and decreases sharply as the particle moves toward the center. Given the sensitivity to the particle size to tube diameter ratio, axial migration of suspended particles is favored in smaller diameter tubes; the consequences of this migration in smaller tubes is discussed in Sections 6.3 and 6.4 of this chapter.

The obvious consequences of wall exclusion and axial migration of RBC are as follows: (1) the average velocity of RBC during flow in a cylindrical tube is higher than the average velocity of the suspending medium (e.g., plasma); (2) a particle-poor, or even a particle-free, fluid layer is formed near the tube wall, tending to decrease the frictional resistance between the flowing fluid and the wall and hence to decrease hydrodynamic resistance. Axial migration mechanisms are expected to influence all particles, although there may be different effects due to variations of their physical properties as well as the behavior of other particles. Nobis et al. (1985) demonstrated that white blood cells (WBC) were accumulated in the central flow zone of small tubes at high shear rates. However, at flow rates allowing RBC aggregation, margination of WBCs was enhanced in aggregating RBC suspensions, while WBCs accumulated in the central flow region in nonaggregating RBC suspensions (Nobis et al. 1985).

6.3   FÅHRAEUS EFFECT

At sufficiently high flow rates and at steady state, the concentration of suspended particles (i.e., hematocrit for RBC in suspension) in a small-sized tube (tube hematocrit, H_T) is lower than the concentration in the larger diameter feed tube (feed hematocrit, H_F). This reduction of concentration is a direct consequence of the two important phenomena characterizing tube flow discussed in Section 6.2: (1) migration of particles (e.g., RBC) toward the axial flow zone, and (2) higher velocity of this axial fluid zone relative to the marginal fluid zone close to the tube wall. This decrease of particle concentration within the tube as a function of tube diameter is known as the *Fåhraeus effect* (Fåhraeus 1929).

The Fåhraeus effect occurs in cylindrical tubes with diameters smaller than 300 μm. The ratio of H_T to H_F decreases with tube diameter, approaching a minimum at tube diameters in the size range of the suspended particles (Figure 6.4). It should be noted that a similar difference between H_T and H_F may exist in small tubes if particle entry into the tube is hindered; this limitation to particle entry is referred to as *screening*. Obviously, if screening occurs, the particle concentration in the fluid discharged from the small tube (i.e., H_D, discharge hematocrit in case of RBC suspensions) would be lower than H_F; H_F is equal to H_D in the absence of screening (Gaehtgens et al. 1978).

If several types of particles (e.g., RBC, WBC, platelets) are suspended in the fluid flowing in a tube with a diameter <300 μm, it might be expected that these particles would have different radial distribution patterns and thus different average velocities. Accumulation of RBC in the axial flow zone displaces WBCs and platelets to the marginal flow zone, resulting in a lower average velocity for these cells with respect to RBC and plasma. The concentration of a given type of particle in flow is inversely proportional to its velocity: particles with higher velocity (i.e., particles moving in the axial flow zone) would have a lower concentration in the tube (H_T, hematocrit in case of RBC) compared to that in the feed tube. Conversely, particles with lower velocity (i.e., particles moving in the marginal flow zone) would have a higher concentration than in the feed tube. This increase of particle concentration within the tube occurs for WBCs and platelets and is known as the *inverse Fåhraeus effect* (Goldsmith et al. 1999).

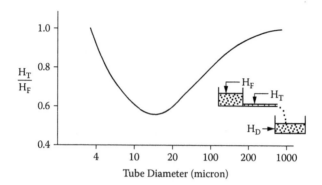

FIGURE 6.4 Ratio of tube hematocrit (H_T) to feed hematocrit (H_F) as a function of tube diameter. H_F should be equal to discharge hematocrit (H_D) in the absence of entrance (i.e., screening) effects.

6.4   FÅHRAEUS–LINDQVIST EFFECT

It has been recognized for more than a century that the effective viscosity of suspensions (e.g., paints, clay) during flow in a cylindrical tube becomes smaller as the tube radius decreases and flow rate increases (Bingham and Green 1919). This phenomenon was investigated for blood flow in small tubes by Fåhraeus and Lindqvist (1931). Figure 6.5 presents the relative viscosity of RBC suspensions as the tube radius decreases from ~1000 µm; relative viscosity is obtained by dividing the effective viscosity calculated using the Poiseuille equation by the suspending phase viscosity. The data shown in Figure 6.5 are for human RBC suspensions with hematocrits in the range of 0.4–0.45 l/l and were obtained at flow rates sufficiently high to prevent RBC aggregation (Goldsmith et al. 1989). In agreement with the original observations (Fåhraeus and Lindqvist 1931), the Fåhraeus–Lindqvist effect is operative for blood flow in cylindrical tubes with diameters smaller than 300 µm.

The decrease in relative viscosity of flowing RBC suspensions with tube diameter can usually be explained by the reduction of hematocrit with tube diameter (Fåhraeus effect, Figure 6.4). Axial migration of RBC during tube flow, as discussed in Section 6.2.2, generates a plasma-rich flow zone near the vessel wall with a thickness on the order of the particle (i.e., RBC) size; viscosity in this marginal zone may approach that of plasma. Formation of this fluid layer significantly decreases the energy dissipation during flow and contributes to the reduction in viscosity (Goldsmith et al. 1989). Using tubes with diameters of 28 to 101 µm perfused at various flow rates with human blood, Alonso et al. (1993) showed an inverse relation between the width of this marginal zone relative to tube diameter versus relative viscosity (Figure 6.6).

While the previously mentioned formation of a cell-poor zone near the tube wall is useful in explaining tube diameter effects, Barbee and Cokelet tested the assumption regarding the decrease of viscosity with tube diameter: the Fåhraeus effect explains the Fåhraeus–Lindqvist effect. They evaluated this explanation by: (1) determining pressure-flow and tube hematocrit (H_T) data in tubes as small as 29 µm, and (2)

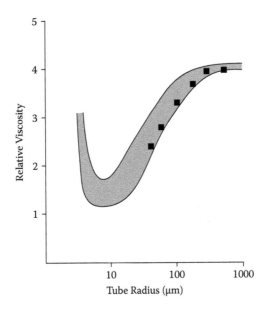

FIGURE 6.5   Relative viscosity of human RBC suspensions with hematocrits in the range of 0.40–0.45 1/1 during flow in tubes with diameters of fewer than 1,000 microns. The flow rate was sufficiently high to prevent RBC aggregation. Relative viscosity was calculated by dividing effective viscosity by the suspending phase viscosity. The shaded area represents published data reported by various investigators, and the black squares are data points published by Fåhraeus, R., and T. Lindqvist. 1931. "The Viscosity of the Blood in Narrow Capillary Tubes." *American Journal of Physiology* 96:562–568. The figure is redrawn from Goldsmith, H. L., G. R. Cokelet, and P. Gaehtgens. 1989. "Robin Fåhraeus: Evolution of His Concepts in Cardiovascular Physiology." *American Journal of Physiology—Heart and Circulatory Physiology* 257:H1005–H1015. With permission.

measuring shear stress–shear rate relations in a Couette viscometer (see Chapter 5) for the same suspensions at hematocrits spanning the range of H_T found in their tube studies (Barbee and Cokelet 1971a; Barbee and Cokelet 1971b). The viscometric data, at appropriate levels of hematocrit, were then used to predict RBC suspension pressure–flow relations for various tube diameters: there was excellent agreement between actual pressure-flow results and those predicted based on viscometric data. In brief, knowing small tube hematocrit H_T and shear stress–shear rate relations at the same hematocrit allows prediction of small tube flow behavior.

6.5   EFFECT OF RED BLOOD CELL AGGREGATION ON TUBE FLOW

6.5.1   Flow Behavior in Larger Tubes

Tubes with diameters sufficiently large to minimize or eliminate the Fåhraeus and Fåhraeus–Lindqvist effects (i.e., >200–300 μm) and employed in systems designed to avoid sedimentation (see Section 6.5.2) have frequently been used to determine pressure–flow relations for various RBC suspensions. There are several approaches to

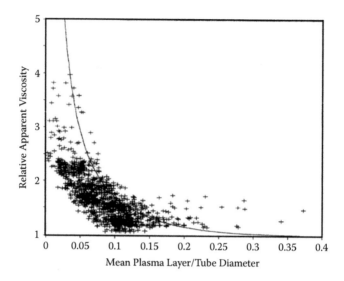

FIGURE 6.6 Relative apparent viscosity plotted against the ratio of mean plasma-layer thickness to tube diameter. Data points are values measured in tubes with diameters between 28 and 101 μm for the flow of human blood at various flow rates and at various time points following the start of flow. The solid line is the prediction of the same relationship from a model study. (Reproduced from Alonso, C., A. R. Pries, and P. Gaehtgens. 1993. "Time-Dependent Rheological Behavior of Blood at Low Shear in Narrow Vertical Tubes." *American Journal of Physiology—Heart and Circulatory Physiology* 265:H553–H561. With permission.)

the use of such tubes as viscometers, including applying steady flow and measuring the resulting pressure drop, applying a steady pressure and determining the resulting flow, and applying a constantly declining pressure gradient and determining changes of flow with pressure (Merrill et al. 1965; Alexy et al. 2005). Given the non-Newtonian flow behavior of human blood, analysis of the resulting data in terms of apparent viscosity as a function of shear rate can be complex and often requires use of the Mooney-Rabinowitz-Weissenberg equation (Cokelet and Meiselman 2007). However, early work by Merrill et al. (1965) greatly aided the analysis process. Their approach involved a pseudo shear rate Ũ, calculated as average flow velocity in the tube divided by tube diameter; Ũ is one-eighth the wall shear rate for Newtonian fluid and an index to this shear rate for more complex fluids (e.g., human blood). They also employed an empirical equation developed by Casson (1959) for the functional relation between shear stress and Ũ. In this equation, shear stress and Ũ (i.e., shear rate) have a linear relation on a square root coordinate system and there can be a finite intercept on the stress axis (i.e., the fluid exhibits a yield shear stress, see Figures 5.10 and 5.11 in Chapter 5). At constant hematocrit, the yield shear stress increases with enhanced RBC aggregation and is the stress level below which the fluid does not deform but rather acts as a solid and above which the fluid has viscous behavior.

Knowing the wall shear stress and the yield shear stress, together with the Casson equation, allows prediction of velocity profiles during tube flow for various ratios of wall to yield stress. As shown in Figure 6.7, the profile is strongly dependent on this

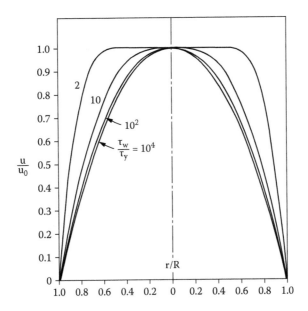

FIGURE 6.7 Predicted velocity profiles for blood flow in a tube based on the Casson shear stress–shear rate relation. The vertical axis is velocity relative to centerline velocity and the horizontal axis is the radial position where r is the distance from center and R is the tube radius. The variable is the ratio of wall shear stress to yield shear stress; 50% of unsheared blood at a ratio of 2 and a parabolic profile when the ratio is 10^4. Reproduced from Merrill, E. W., A. M. Benis, E. R. Gilliland, T. K. Sherwood, and E. W. Salzman. 1965. "Pressure-Flow Relations of Human Blood in Hollow Fibers at Low Flow Rates." *Journal of Applied Physiology* 20:945–957 with permission.

ratio: when the ratio is 2, there is a flat profile (i.e., plug flow) of unsheared blood over 50% of the diameter, whereas the profile is essentially parabolic when the ratio is 10^4. Thus, as expected, increased shear forces reduce RBC aggregation and hence the influence of the yield stress on the flow pattern.

Flow data for tubes with diameters between 288 and 850 µm are presented in Figure 6.8 as a plot of log wall shear stress versus log $\bar{U}$; the data were obtained for 0.39 1/1 hematocrit normal human blood at 20°C. Inspection of these results indicates: (1) in general, the data for all of the tubes collapse onto a single line; (2) the line is not straight but tends to curve upward at lower values of $\bar{U}$; (3) at higher $\bar{U}$ values, the line is parallel to the dashed line with a slope of unity but diverges from this slope at lower $\bar{U}$ values. All three of these observations are consistent with a fluid where shear stress is only a function of shear rate, the flow is steady and laminar, and there is no slip at the wall (Merrill et al. 1965). The curvature of the line and its divergence from a slope of unity at lower values of $\bar{U}$ indicate that, as expected, blood is behaving as a non-Newtonian, shear thinning fluid (see Chapter 5): Newtonian fluids have a slope of unity regardless of shear stress or shear rate. Note that if the yield stress is known and $\bar{U}$ is used as the shear rate term, the curved line shown in Figure 6.8 can be accurately predicted using the Casson equation (Merrill et al. 1965).

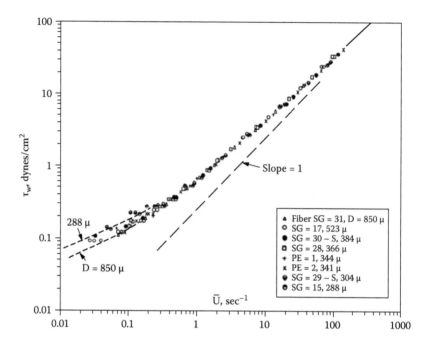

FIGURE 6.8   Wall shear stress versus Ũ (i.e., average flow velocity divided by tube diameter) for 0.39 l/l hematocrit normal human blood at 20°C. Tube diameters range from 288 to 850 μm and are fabricated from either soft glass (SG) or polyethylene (PE); all tubes were horizontal. Dashed line with slope of unity indicates Newtonian flow behavior. The divergence of the data at low values of Ũ most likely reflects sedimentation effects in the smaller tubes (see Section 6.5.5). Reproduced from Merrill, E. W., A. M. Benis, E. R. Gilliland, T. K. Sherwood, and E. W. Salzman. 1965. "Pressure-Flow Relations of Human Blood in Hollow Fibers at Low Flow Rates." *Journal of Applied Physiology* 20:945–957 with permission.

6.5.2   FLOW BEHAVIOR IN SMALLER TUBES

Flow in smaller tubes can be markedly affected by a nonhomogenous distribution of RBC across the diameter. RBC aggregation promotes the accumulation and compaction of RBC in the central flow zone and improves the effectiveness of axial migration and related mechanisms (Alonso et al. 1993; Cokelet and Goldsmith 1991). It has been experimentally observed that blood with higher RBC aggregation exhibits higher degrees of axial migration and related phase separation (Figure 6.1). Based on such observations, aggregation is expected to reduce the effective viscosity during tube flow by promoting the Fåhraeus effect and by increasing the relative thickness of the lower-viscosity marginal fluid layer (see Figure 6.9). However, aggregation is also well known to enhance RBC sedimentation (Fabry 1987; Raphael 1983). The influence of sedimentation on axial accumulation and phase separation thus needs to be considered, especially in relation to the orientation of the tube with respect to gravity. This aspect of the role of RBC aggregation in tube flow is discussed in Section 6.5.5.

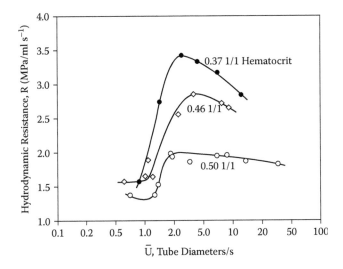

FIGURE 6.9 Hydrodynamic resistance as a function of average flow velocity, expressed as tube diameters per second (Ū), measured in a 170 μm diameter by 12.2 cm long tube. Human RBC were suspended in an isotonic medium that promotes aggregation (1.5% solution of 110 kDa dextran in phosphate-buffered saline [PBS]). Data are for three RBC suspensions with hematocrits of 0.37 l/l, 0.46 l/l, and 0.50 l/l. Reproduced from Cokelet, G. R., and H. L. Goldsmith. 1991. "Decreased Hydrodynamic Resistance in the Two-Phase Flow of Blood through Small Vertical Tubes at Low Flow Rates." *Circulation Research* 68:1–17 with permission.

6.5.3 IMPORTANCE OF FLOW RATE

The effect of aggregation on the radial distribution of RBC is shear rate and shear stress dependent, and follows from the well-known influence of shear forces on RBC aggregation: RBC aggregation is minimized or abolished at sufficiently high shear forces while aggregation is favored at low shear (i.e., low flow rate) or at stasis (see Chapter 5). Figure 6.9 shows the influence of tube flow velocity Ū, expressed as tube diameters per second, on hydrodynamic resistance in a 170 μm diameter tube. In this study, human RBC were suspended in an isotonic medium that promotes aggregation (i.e., 1.5% dextran 110 kDa in PBS) (Cokelet and Goldsmith 1991). The three curves in Figure 6.9 were obtained at different hematocrits and exhibit a similar pattern of dependence on flow rate, although the value of flow resistance at a given flow rate was hematocrit dependent. Interestingly, resistance values approach each other in the low flow rate region, which is characterized by a significant degree of RBC aggregation. The flow behavior shown in Figure 6.9 has been observed for various aggregating RBC suspensions, whereas nonaggregating RBC suspensions in similar flow systems have a higher hydrodynamic resistance at low flow rates (Cokelet and Goldsmith 1991; Reinke et al. 1987).

As shown in Figure 6.10, the relative viscosity of RBC suspensions in nonaggregating media (e.g., saline, serum) is greater than for aggregating systems (e.g., plasma, high-molecular-mass dextran) *if* the flow rate is low enough to allow red cell

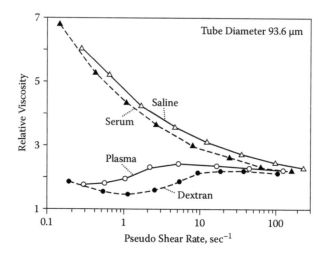

FIGURE 6.10 Relative viscosity obtained using pressure-flow data from a 93.6 μm tube during flow of aggregating (plasma or 4.5% dextran 250 kDa in buffer) and nonaggregating (saline or serum) RBC suspensions. Reproduced from Reinke, W., P. Gaehtgens, and P. C. Johnson. 1987. "Blood Viscosity in Small Tubes: Effect of Shear Rate, Aggregation, and Sedimentation." *American Journal of Physiology—Heart and Circulatory Physiology* 253:H540–H547 with permission.

aggregation and radial migration. The large difference in the viscosities of aggregating and nonaggregating RBC suspensions at low shear rates, together with the convergence at high shear rates, clearly illustrates that RBC aggregation is an important determinant of flow dynamics in small tubes.

6.5.4 Time Dependence of Aggregation Effect on Tube Flow

Several studies have shown that the aggregation-related decrease in viscosity during tube flow (Figures 6.9 and 6.10) develops with a time course determined by the tube geometry and orientation, flow rate, and suspension properties (e.g., intensity of RBC aggregation) (Alonso et al. 1995; Alonso 1993; Alonso et al. 1989). Figure 6.11 presents the time course of relative apparent viscosity for normally aggregating human RBC suspensions in vertical cylindrical tubes with diameters of 28 to 101 μm (Alonso et al. 1993). Relative apparent viscosity decreased with time after the start of flow when the wall shear stress was low (i.e., 20 mPa) and approached a steady-state minimum in ~60 s for all tubes (Figure 6.11a); the greatest change with time occurred with the largest diameter tube (101 μm). An important observation in these experiments was that the relative apparent viscosity did not change with time if the flow rate was high enough to prevent RBC aggregation (Figure 6.11b). Note that the steady-state viscosities in each tube were less at low shear stress (Figure 6.11a) compared to those measured for the same tube at higher stress (i.e., 100 mPa, Figure 6.11b), thus indicating again that RBC aggregation can reduce hydrodynamic resistance. Interestingly, the characteristic time course for reducing hydrodynamic resistance due to radial migration is similar to that required for completion of RBC

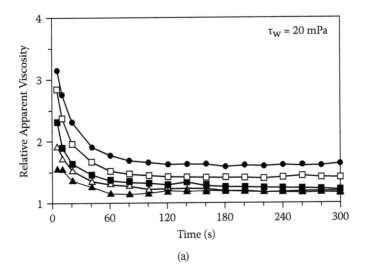

(a)

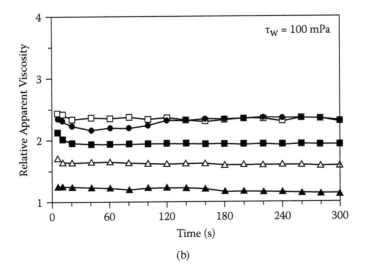

(b)

FIGURE 6.11   Time course of relative apparent viscosity in vertical cylindrical tubes with various diameters for the flow of normal human blood at 0.44 l/l hematocrit: (a) measured at a wall shear stress of 20 mPa, which allowed RBC aggregation; (b) measured at a wall shear stress of 100 mPa, which abolishes RBC aggregation. Tube diameters: ▲ 28 μm; △ 41 μm; ■ 59 μm; □ 82 μm; ● 101 μm. Reproduced from Alonso, C., A. R. Pries, and P. Gaehtgens. 1993. "Time-Dependent Rheological Behavior of Blood at Low Shear in Narrow Vertical Tubes." *American Journal of Physiology—Heart and Circulatory Physiology* 265:H553–H561 with permission.

aggregation (see Chapter 4, Section 4.1.6.2.2). Alonso et al. (1993) also showed that the parameters associated with reduced hydrodynamic resistance (e.g., ratio of core radius to tube radius) also follow a similar time course.

6.5.5   VERTICAL VERSUS HORIZONTAL TUBES

Aggregation promotes RBC sedimentation since the rate of sedimentation is proportional to the square of the size (e.g., radius) of the settling particle (see Chapter 4, Section 4.1.1.2). The steady-state settling velocity is determined by the balance between the frictional drag on the cell or aggregate and a force due to the difference between the weight of the cell or aggregate and its buoyancy. RBC aggregation, and hence sedimentation, are prevented by high shear forces (i.e., high flow rates), resulting in only a very thin (i.e., ~2–4 μm) marginal cell-poor layer, which is due to wall exclusion (Alonso et al. 1995). The resulting radial distribution is symmetric, regardless of the orientation of the flow system with respect to gravity (Figure 6.12). At lower flow rates that allow RBC aggregates to form, cells can begin to sediment resulting in an asymmetric distribution. Figure 6.12 shows the development of asymmetric phase separation in a horizontal tube in which RBC have settled on the dependent, lower side (Alonso et al. 1995). This marked asymmetry does not develop in a vertically oriented tube, although irregularities in the marginal and axial flow zones are observed.

The consequences of sedimentation on the dependent side of the tube are not merely limited to the loss of beneficial hydrodynamic effects due to RBC axial migration, but can actually increase flow resistance and hence apparent viscosity in horizontal tubes (Alonso et al. 1995; Alonso et al. 1989; Cokelet and Goldsmith 1991; Murata 1987). This effect of RBC aggregation–mediated sedimentation in horizontal tubes is in contrast with that observed in vertical tubes (Alonso et al. 1995; Alonso et al. 1993; Cokelet and Goldsmith 1991; Gaehtgens 1987; Murata and Secomb 1989). Figure 6.13 demonstrates this contrasting behavior for normal human blood in a 100 μm tube: RBC aggregation at lower flow rates (i.e., lower shear forces) leads to decreased relative effective viscosity in a vertical tube, while a very prominent increase occurs if the tube is oriented horizontally (Alonso et al. 1995). Note that the divergence of pressure-flow data shown in Figure 6.8 at low Ũ values is most likely due to sedimentation in horizontal tubes: the shear stress results for the 288 μm tube are higher than for the 850 μm tube, a finding that is to be expected since for a given settling distance, the effect increases with decreasing tube diameter.

Cokelet and Goldsmith (1991) compared the hydrodynamic resistance in a 170-μm diameter vertical tube for RBC suspensions flowing upward and downward (Figure 6.14). The dependence of flow resistance on the flow rate exhibited similar patterns for both orientations with a marked reduction at lower flow rates. However, resistance values measured during upward flow were higher than those obtained during downward flow; the difference became larger as flow rate decreased, indicating that the difference reflects the influence of RBC aggregation. The basis for the higher flow resistance during upward flow is related to the opposing effect of RBC sedimentation (i.e., downward movement of RBC under the influence of gravity), whereas sedimentation has a synergistic effect with downward flow.

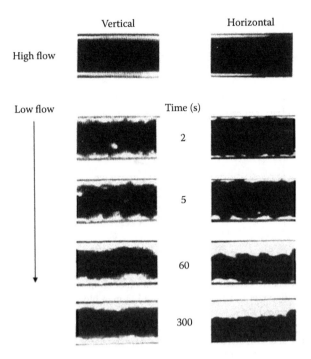

FIGURE 6.12 Microscopic photographs of vertically and horizontally oriented 60 μm tubes during flow of 0.44 l/l hematocrit human blood with normal aggregation. The *high flow* images at a wall stress of 30 P_a show RBC to be uniformly distributed with only a very narrow ~2 to 4 μm marginal cell-poor layer. After flow was abruptly reduced to a wall shear stress of 10 mPa, a cell-free, nearly symmetric plasma layer develops in the vertical tube within 60 s, whereas in the horizontal tube radial asymmetry develops with the formation of a wide plasma layer. Reproduced from Alonso, C., A. R. Pries, K. D. Lerche, and P. Gaehtgens. 1995. "Transient Rheological Behavior of Blood in Low-Shear Tube Flow: Velocity Profiles and Effective Viscosity." *American Journal of Physiology—Heart and Circulatory Physiology* 268:H25–H32 with permission.

6.6   CONCLUSION

Experimental data obtained using cylindrical tubes with diameters <300 μm indicates that RBC aggregation promotes blood flow via decreasing hydrodynamic resistance. However, these results may not be directly applicable to *in vivo* blood flow conditions due to several significant limitations, including: (1) Cylindrical tubes are oversimplified models of the vasculature. The geometry of the circulatory system is extremely complex and is characterized by a very wide range of tube dimensions from several μm to several cm, extensive branching, and active regulation of the vessel diameter especially in vessels with dimensions in the range of the tubes discussed in this chapter, (2) *In vitro* experiments usually utilized tubes whose length is many times greater than their diameter, and thus their length-to-diameter ratio is very far from being applicable to the vessel dimensions in the circulatory system, (3) The hydrodynamic mechanisms related to RBC aggregation may require several minutes to fully develop.

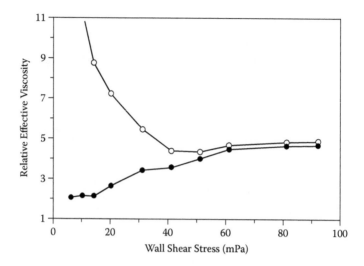

FIGURE 6.13 Relative effective viscosity as a function of wall shear stress in cylindrical tubes of 100 μm diameter during flow of normal human blood. ● vertically oriented tube; ○ horizontally oriented tube. Reproduced from Alonso, C., A. R. Pries, K. D. Lerche, and P. Gaehtgens. 1995. "Transient Rheological Behavior of Blood in Low-Shear Tube Flow: Velocity Profiles and Effective Viscosity." *American Journal of Physiology—Heart and Circulatory Physiology* 268:H25–H32 with permission.

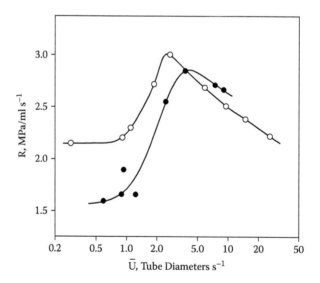

FIGURE 6.14 Hydrodynamic resistance as a function of tube flow rate (Ŭ) in a 170 μm diameter tube during upward (○) and downward (●) flow of RBC suspensions. Reproduced from Cokelet, G. R., and H. L. Goldsmith. 1991. "Decreased Hydrodynamic Resistance in the Two-Phase Flow of Blood through Small Vertical Tubes at Low Flow Rates." *Circulation Research* 68:1–17 with permission..

This time scale significantly exceeds the transit times of RBC through unbranched lengths of blood vessels with diameters <300 μm. The *in vivo* implications of the hydrodynamic mechanisms presented in this chapter are considered in Chapter 7.

LITERATURE CITED

Alexy, T., R. B. Wenby, E. Pais, and L. J. Goldstein, W. Hogenauer, and H. J. Meiselman. 2005. "An Automated Tube-Type Blood Viscometer: Validation Studies." *Biorheology* 42:511–520.

Alonso, C., A. R. Pries, and P. Gaehtgens. 1989. "Time-Dependent Rheological Behavior of Blood Flow at Low Shear in Narrow Horizontal Tubes." *Biorheology* 26:229–246.

Alonso, C., A. R. Pries, and P. Gaehtgens. 1993. "Time-Dependent Rheological Behavior of Blood at Low Shear in Narrow Vertical Tubes." *American Journal of Physiology—Heart and Circulatory Physiology* 265:H553–H561.

Alonso, C., A. R. Pries, K. D. Lerche, and P. Gaehtgens. 1995. "Transient Rheological Behavior of Blood in Low-Shear Tube Flow: Velocity Profiles and Effective Viscosity." *American Journal of Physiology—Heart and Circulatory Physiology* 268:H25–H32.

Barbee, J. H., and G. R. Cokelet. 1971a. "The Fåhraeus Effect." *Microvascular Research* 3: 6–16.

Barbee, J. H., and G. R. Cokelet. 1971b. "Prediction of Blood Flow in Tubes with Diameters as Small as 29 Microns." *Microvascular Research* 3: 17–21.

Bingham, E. C., and H. Green. 1919. "Paint, a Plastic Material and Not a Viscous Liquid:The Measurement of Its Mobility and Yield Value." *Proceedings of American Society for Testing Materials* 19:640–664.

Bishop, J. J., P. R. Nance, A. S. Popel, M. Intaglietta, and P. C. Johnson. 2001. "Effect of Erythrocyte Aggregation on Velocity Profiles in Venules." *American Journal of Physiology—Heart and Circulatory Physiology* 280:H222–H236.

Bishop, J. J., A. S. Popel, M. Intaglietta, and P. C. Johnson. 2002. "Effect of Aggregation and Shear Rate on the Dispersion of Red Blood Cells Flowing in Venules." *American Journal of Physiology—Heart and Circulatory Physiology* 283:H1985–H1996.

Brenner, H. 1966. "Hydrodynamic Resistance of Particles at Small Reynolds Numbers." In *Advances in Chemical Engineering*, ed. T. B. Drew, J. W. Hoopes, and T. Vermuelen, 287–438. New York: Academic Press.

Casson, N. 1959. "A Flow Equation for Pigment-Oil Suspension of the Printing Ink Type." In: *Rheology of Disperse Systems,* ed. C.C. Mills, 84–104. New York: Pergamon Press.

Chatzimavroudis, G. P. 2002. "Blood flow dynamics." In *Fluid Flow Handbook*, ed. J. Saleh, 1–26. New York: McGraw-Hill.

Cokelet, G. R., and H. L. Goldsmith. 1991. "Decreased Hydrodynamic Resistance in the Two-Phase Flow of Blood through Small Vertical Tubes at Low Flow Rates." *Circulation Research* 68:1–17.

Cokelet, G. R., and H. J. Meiselman. 2007. "Basic Aspects of Hemorheology." In *Handbook of Hemorheology and Hemodynamics*, ed. O. K. Baskurt, M. Hardeman, M. W. Rampling, and H. J. Meiselman, 21–33. Amsterdam, Berlin, Oxford, Tokyo, Washington, DC: IOS Press.

Fabry, T. L. 1987. "Mechanism of Erythrocyte Aggregation and Sedimentation." *Blood* 70:1572–1576.

Fåhraeus, R. 1921. "The Suspension Stability of the Blood." *Acta Medica Scandinavica* 55:1–228.

Fåhraeus, R. 1929. The Suspension Stability of the Blood. *Physiological Reviews.* 9:241-274.

Fåhraeus, R. 1958. "The Influence of the Rouleaux Formation of the Erythrocytes on the Rheology of the Blood." *Acta Medica Scandinavica* 161:151–165.

Fåhraeus, R., and T. Lindqvist. 1931. "The Viscosity of the Blood in Narrow Capillary Tubes." *American Journal of Physiology* 96:562–568.

Fung, Y. C. 1997. *Biomechanics of Circulation*, 114–118. New York: Springer.

Gaehtgens, P. 1987. "Tube Flow of Human Blood at near Zero Shear." *Biorheology* 24:267–276.

Gaehtgens, P., K. H. Albrecht, and F. Kreutz. 1978. "Fåhraeus Effect on Cell Screening during Tube Flow of Human Blood. I. Effect of Variation of Flow Rate." *Biorheology* 15:147–154.

Gaehtgens, P., H. J. Meiselman, and H. Wayland. 1970. "Velocity Profiles of Human Blood at Normal and Reduced Hematocrit in Glass Tubes up to 130 μm Diameter." *Microvascular Research* 2:13–23.

Goldsmith, H. L., D. N. Bell, S. Spain, and F. A. McIntosh. 1999. "Effect of Red Blood Cells and Their Aggregates on Platelets and White Cells in Flowing Blood." *Biorheology* 36:461–468.

Goldsmith, H. L., G. R. Cokelet, and P. Gaehtgens. 1989. "Robin Fåhraeus: Evolution of His Concepts in Cardiovascular Physiology." *American Journal of Physiology—Heart and Circulatory Physiology* 257:H1005–H1015.

Hahn, P. F., W. D. Donald, and R. C. Grier Jr. 1945. "The Physiological Bilaterality of the Portal Circulation: Streamline Flow of Blood into the Liver as Shown by Radioactive Phosphorus." *American Journal of Physiology* 143:105–107.

Hess, W. R. 1915. "Gehorcht das Blut dem allgemeinen Stromungsgesetz der Flussigkeiten?" *Pflügers Archieves* 164:603–644.

Hess, W. R. 1917. "Über die periphere Regulierung der Blutzirkulation." *Pflügers Archieves* 168:439–490.

Leal, L. G. 1980. "Particle Motions in a Viscous Fluid." *Annual Review of Fluid Mechanics* 12:435–476.

Maude, A. D. 1967. "Theory of the Wall Effect in the Viscosity of Suspensions." *British Journal of Applied Physiology* 18:1193–1197.

Merrill, E. W., A. M. Benis, E. R. Gilliland, T. K. Sherwood, and E. W. Salzman. 1965. "Pressure-Flow Relations of Human Blood in Hollow Fibers at Low Flow Rates." *Journal of Applied Physiology* 20:945–957.

Murata, T. 1987. "Effects of Sedimentation of Small Red Blood Cell Aggregates on Blood Flow in Narrow Horizontal Tubes." *Biorheology* 33:267–283.

Murata, T., and T. W. Secomb. 1989. "Effects of Aggregation on the Flow Properties of Red Blood Cell Suspensions in Narrow Vertical Tubes." *Biorheology* 26:247–259.

Nobelprise.org. 2010. The Nobel Prize in Physiology or Medicine 1949. http://nobelprize.org/nobel_prizes/medicine/laureates/1949/. Accessed on November 6, 2010.

Nobis, U., A. R. Pries, G. R. Cokelet, and P. Gaehtgens. 1985. "Radial Distribution of White Cells during Blood Flow in Small Tubes." *Microvascular Research* 29:295–304.

Ralston, H. J., and A. N. Taylor. 1945. "Streamline Flow in the Arteries of the Dog and Cat: Implications for the Work of the Heart and the Kinetic Energy of Blood Flow." *American Journal of Physiology* 144:706–710.

Raphael, S. S. 1983. *Lynch's Medical Laboratory Technology*. Philadelphia, PA: W. B. Saunders Co.

Reinke, W., P. Gaehtgens, and P. C. Johnson. 1987. "Blood Viscosity in Small Tubes: Effect of Shear Rate, Aggregation, and Sedimentation." *American Journal of Physiology—Heart and Circulatory Physiology* 253:H540–H547.

7 *In Vivo* Hemodynamics and Red Blood Cell Aggregation

7.1 BASIC APPROACH TO *IN VIVO* BLOOD FLOW

Maintenance of adequate blood flow to tissues is of crucial importance for proper physiological function. The properties of circulatory system and blood tissue have evolved in higher, multicellular organisms, in order to adapt to the requirements of tissue perfusion. The circulatory system consists of a pump (heart) and a very extensive network of blood vessels, which are tubes with diameters varying from several micrometers to several centimeters in vertebrates. A detailed discussion of the anatomy of the circulatory system is beyond the scope of this book; general morphological and functional features of the mammalian circulatory system are briefly described below.

The mammalian heart is a four-chambered organ that can be perceived as two separate pumps, namely the left and right hearts. Each pump is characterized by a smaller chamber accepting the blood returning to the heart (right and left atria) and a pumping chamber (right and left ventricles). The right heart accepts blood returning from the body (systemic circulation) and pumps into the pulmonary circulation perfusing the alveoli of the lungs for oxygenation. The left heart accepts the oxygenated blood returning from the lungs and pumps it to the systemic circulation serving all organs and tissues, except lung alveoli. The heart is a pulsatile pump; the contraction of the heart muscle (myocardium) increases the pressure in the ventricle, ejecting ~60% of the blood initially filling the ventricles to the pulmonary or systemic circulation. This ejection period is known as *systole*. Each systole is followed by the relaxation of myocardium (*diastole*), allowing venous blood to return to the ventricles to be ejected during the next systole. The left heart is a significantly more powerful pump generating a mean pressure of ~100 mmHg in most mammals under physiological conditions, whereas the right ventricle is characterized by a less-powerful muscle pump and the mean pressure at its outlet is ~17 mmHg.

Blood is carried away from the heart by the aorta, which branches into arteries of smaller diameter. Branching in the arterial tree is extensive, giving rise to smaller diameters but a larger number of daughter vessels at each level. This branching pattern results in a very significant increment in the total cross-sectional area as blood flows into smaller blood vessels. It should be noted that blood leaves the left heart via a single, large-diameter blood vessel (i.e., aorta) while there is an increment in the blood vessel number on the order of 10^7 at the microcirculatory level. This results in an increase of total cross-sectional area on the order of 800-fold and a reduction of blood velocity of the same order as blood approaches the microcirculation. The microcirculation is the site for exchange of metabolites between the blood and tissues. Arterioles are the endpoints of the arterial system and connect to capillaries with diameters around 3–5 microns in most mammals. Blood is collected from capillaries by venules and flows into the venous system, which returns the blood to right heart.

Blood flow from the heart to the microcirculation and back to the heart is basically determined by the general rules of hydrodynamics. That is, the magnitude of blood flow in a given blood vessel under a given pressure gradient is determined by the hydraulic resistance, which in turn depends on the geometry of the blood vessel and flow properties of blood. This relationship can be formulated by the Poiseuille equation:

$$Q = \frac{P \cdot \pi \cdot D^4}{128 \cdot \eta \cdot L} \qquad (7.1)$$

where Q is the volumetric flow, P the pressure gradient, D and L are the diameter and length of the blood vessel, respectively, and η is blood viscosity. It follows from the Equation (7.1) that flow resistance in a given blood vessel is determined by its geometry (diameter and length) and blood viscosity. It has been well known since the nineteenth century that the majority of flow resistance in the systemic circulation is in the microcirculation, mainly due to geometry (Pries et al. 1994). The microcirculation is thus the site where the largest pressure drop occurs, due to the high hydrodynamic resistance related to the geometry of the blood vessels; obviously, blood viscosity also makes significant contributions to flow resistance.

Blood viscosity is a shear-dependent parameter (see Chapter 5), ranging between ~5 mPa.s at shear rates above ~100 sec^{-1} and several hundred mPa.s at shear rates <0.1 sec^{-1}. This range corresponds to the *ex vivo* measurements done using rotational or tube viscometers. It follows from Chapters 2 and 5 that red blood cell (RBC) aggregation is strongly influenced by shear forces. Therefore, aggregation is an important determinant of blood viscosity if the flow conditions are characterized by low shear forces. This shear dependency of blood viscosity does not permit the direct application of simple hydrodynamic principles, including the Poiseuille equation, to *in vivo* blood flow since blood viscosity is a variable determined by the shearing conditions.

The shear rate in the vasculature is proportional to the flow rate in a given vessel, and therefore also varies within the circulatory system. Table 7.1 presents mean flow velocity and calculated shear rate in various segments of the human circulatory system. Note that shear rate varies from 40 sec^{-1} in venous vessels to several thousand sec^{-1} in large arteries (Table 7.1).

Shear stress in a vascular segment is proportional to the pressure drop across that segment, and therefore high shear stresses are expected in microcirculatory vessels where the pressure changes are greatest (i.e., the arterial side of the microcirculation).

TABLE 7.1
Shear Rate Values Estimated for Various Levels of the Human Circulatory System

Vessel Type	Diameter (cm)	Mean Flow Velocity (cm/sec)	Wall Shear Rate (sec⁻¹)
Aorta	1.6–3.2	6^0	150–300
Artery	0.2–0.6	15–50	200–2,000
Arteriole	0.004	0.5	1,000
Capillary	0.0005–0.001	0.05–0.1	400–1,600
Post-capillary venule	0.002	0.2	800
Venule	0.01	0.5	400
Vein	0.5–1	15–20	120–320
Vena cava	2	10.15	40–60

Sources: Lipowsky, H. H. 1995. Shear Stress in the Circulation. *In Flow Dependent Regulation of Vascular Function.* J. Bevan and G. Kaley, eds. New York: Oxford University Press. 28-45; McDonald, D.A. 1974. *Blood Flow in Arteries.* Baltimore: The Williams and Wilkins Co. 356-378.

Note: Shear rate is calculated as 8(Q/d), where Q is the mean flow velocity, and d is the diameter of the vessel and assumes a parabolic profile.

However, it has been argued that wall shear stress is sensed by blood vessels and that vessel diameters are adjusted to maintain a relatively constant stress level throughout the arterial system (Secomb and Pries 2007). The level of shear stress maintained depends on the size of the mammal, and is higher for small rodents (e.g., mouse, rat); in the human arterial system, wall shear stress is maintained below 1 Pa, whereas for smaller animals it approaches 10 Pa (Greve et al. 2006).

There are additional *in vivo* hemodynamic mechanisms that result in significant deviations of *in vivo* pressure–flow relationships from those dictated by basic hydrodynamic rules. These include axial migration of RBC, the Fåhraeus-Lindqvist and Fåhraeus effects (see Chapter 6) and their consequences such as plasma skimming, lower microcirculatory hematocrit, and other mechanisms discussed in Section 7.4 of this chapter. Many of these mechanisms are prominently influenced by RBC aggregation, and thus a full understanding of the influence of RBC aggregation on blood flow *in vivo* can only be achieved by studies in living organisms.

7.2   *IN VIVO* VERSUS *IN VITRO* BLOOD VISCOSITY

Blood viscosity can be ideally measured using a rotational viscometer to characterize its shear-dependent flow behavior by measurements under various shear rates (or stresses) within well-defined shearing conditions. The viscosity of a fluid can also be estimated based on the pressure–flow relationship obtained in cylindrical tubes by applying Poiseuille's Law. This approach works well for Newtonian fluids (e.g., water, plasma) if the other special conditions of the equation (e.g., steady, laminar flow in a cylindrical, rigid tube) are also fulfilled (Secomb and Pries 2007). The calculated viscosity would not be dependent on the flow rate and tube geometry as long as the fluid flow is laminar. However, for a non-Newtonian fluid (e.g., blood), the calculated values of viscosity using the previously mentioned approach depend on the tube geometry and flow rate. Therefore, blood viscosity calculated by this approach only has a meaning under the specific measurement conditions and is termed *apparent viscosity* (Berne and Levy 2001). The apparent (or effective, see Chapter 6) viscosity of blood can be estimated by evaluating pressure–flow relationships in living blood vessels, vascular networks, and even the entire circulatory system. The analysis may also include comparisons with data obtained using cell-free suspending media (e.g., plasma) in the same flow system, yielding *relative apparent viscosity*.

Studies based on the concept of measuring apparent viscosity of blood in the living vasculature of experimental animals have frequently been termed *in vivo viscometry*. One of the most frequently cited examples of such studies is the paper published in 1933 by Whittaker and Winton. They simultaneously studied the apparent viscosity of blood in the isolated-perfused dog hind limb and a glass tube (0.93 mm id, 303 mm length). They used a complicated system termed a *triple pump-lung perfusion apparatus* to perfuse the isolated hind limb and glass tube with blood at hematocrit values ranging between 0.025 and 0.79 l/l under well-controlled conditions (Whittaker and Winton 1933). Figure 7.1 indicates the main finding of their study for the apparent viscosity of blood measured in a living vascular network and the glass tube.

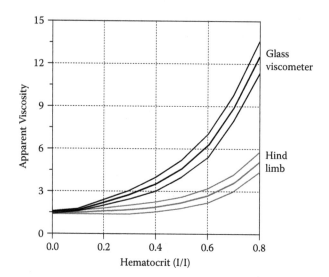

FIGURE 7.1 Apparent viscosity of blood with hematocrit values ranging between 0.025 and 0.79 l/l measured simultaneously in isolated dog hind limb and a glass tube (0.93 mm id and 303 mm length), using the triple pump-lung perfusion apparatus. The middle lines for each data set represent the mean of 11 experiments, whereas bordering lines mark the "probable error." Whittaker, S. R. F., and F. R. Winton. 1933. "The Apparent Viscosity of Blood Flowing in the Isolated Hindlimb of the Dog and Its Variation with Corpuscular Concentration." *Journal of Physiology London* 78:339–368.

Figure 7.1 shows that there is a very significant difference between the apparent viscosity of blood measured in the isolated hind limb preparation and the glass tube. This difference was especially prominent at higher hematocrits: ~100% higher value in the glass tube at ~0.4 l/l hematocrit and ~140% greater at the highest hematocrit level (~0.8 l/l). This interesting observation of the discrepancy between the *in vivo* and *in vitro* findings stimulated active research studies and discussions regarding the hemorheological and hemodynamic mechanisms leading to the enhanced fluidity of blood flowing in the living vascular systems (Benis et al. 1970; Djojosugito et al. 1970; Eliassen et al. 1973). Studies investigating the flow properties of blood included *in vivo* experiments with modified hematocrit (Baskurt et al. 1991; Benis et al. 1970; Fan et al. 1980), plasma viscosity (Chen et al. 1989), RBC deformability (Simchon et al. 1987) and RBC aggregation (Cabel et al. 1997; Charansonney et al. 1993; Durussel et al. 1998; Soutani et al. 1995; Yalcin et al. 2004).

The main conclusion from the previously mentioned *in vivo* studies is that the hydrodynamic behavior of blood in the living vasculature, and especially in the microcirculation, is very complex. This complexity relates to factors specific to the mechanical behavior of blood and blood elements (mainly RBC), heterogeneities in the distribution of blood elements and their rheological behavior, and special features of the vascular network (Pries et al. 1996). The following discussion presents factors relevant to the *in vivo* effects of RBC aggregation.

7.3 EXPERIMENTAL STUDIES INVESTIGATING *IN VIVO* EFFECTS OF RED BLOOD CELL AGGREGATION

The hemodynamic effects of altered RBC aggregation have been investigated experimentally by studying pressure–flow relationships in the vascular systems of various experimental models. These studies can be classified according to the methods employed to study the pressure–flow relationship and the approach to modify RBC aggregation. The results of studies employing different methodologies are frequently in conflict with each other; methodologies are briefly discussed in the following text to provide the basis for the detailed discussion of these conflicting results. The reader is referred to the cited publications for detailed information on each method.

7.3.1 Methods for Studying the Pressure–Flow Relationship

A number of studies have employed intravital microscopy for the quantification of blood flow in the microcirculation (e.g., arterioles, capillaries, and postcapillary venules). Pressure–flow relationships *in vivo* have also been investigated using various techniques including direct measurements of blood pressure and flow for whole organs (e.g., isolated heart preparation, isolated hind limbs) or at certain circulatory levels (e.g., venous blood vessels).

7.3.1.1 Intravital Microscopy

Blood flow in microcirculation can be quantified using microscopic image analysis. This approach has been used to study blood flow in nail fold capillaries of human subjects (Lawall and Angelkort 1999; Maricq 2006). Tissues of laboratory animals (e.g., mouse, rat, hamster, rabbit) that can be exteriorized and spread on specially designed microscopic stages are frequently used for such investigations. These tissues include mesentery, cremaster muscle, spinotrapezius muscle, and hamster cheek pouch (Hester 2006; Svensjo 2006). Chronically implanted dorsal skin chambers have also been used in microcirculatory studies (Figure 7.2) (Fukumura 2006).

A typical intravital microscopy system includes an inverted microscope equipped with suitable illuminating systems and video camera, recording and image analysis systems, micromanipulators, hydrostatic pressure monitors, and suitable perfusion

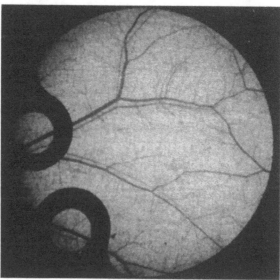

FIGURE 7.2 Rat dorsal skin chamber used for microcirculatory studies. Fukumura, D. 2006. *Window Models. in Microvascular Research*, D. Shepro, ed. Burlington, San Diego, London: Elsevier Academic Press. 151-163. With permission.

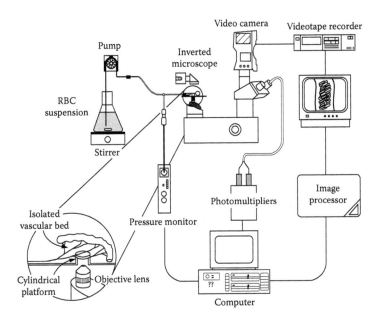

FIGURE 7.3 Typical intravital microscopy system used for *in vivo* hemorheology studies. Reproduced from Soutani, M., Y. Suzuki, N. Tateishi, and N. Maeda. 1995. "Quantificative Evaluation of Flow Dynamics of Erythrocytes in Microvessels: Influence of Erythrocyte Aggregation." *American Journal of Physiology—Heart and Circulatory Physiology* 268:H1959–H1965 with permission.

equipment (Figure 7.3). Microscope systems may include attachments for fluorescence measurement. Tissues under investigation are spread on a transparent platform to allow proper illumination, with this spreading of the tissue usually resulting in all microvessels oriented horizontally with respect to gravity. Maintenance of a constant, physiological temperature is essential; heated tissue chambers, as well as enclosures to contain the microscope, tissue under investigation, and perfusion systems have been used for this purpose.

Blood flow is usually quantified in terms of RBC velocity, with the calculations often performed offline using video recordings. Various methods for on-line calculation of RBC velocity in microvessels have been developed (Sapuppo et al. 2007; Wayland and Johnson 1967). Laser Doppler velocimetry has also been used for measuring RBC velocity in microvessels (Durussel et al. 1998; Seki 1990).

Intravital microscopy techniques can be used for the assessment of important hemodynamic phenomena such as phase separation and cell-poor layer formation in microvessels (Kim, Kong, et al. 2006; Soutani et al. 1995). Additionally, formation of RBC aggregates in postcapillary venules can be monitored and the time course of aggregation can be assessed using video images obtained by intravital microscopy (Kim et al. 2005; Kim, Zhen, et al. 2007). Such recordings have also been used to estimate aggregate size by applying image analysis techniques (Bishop et al. 2004) and functional capillary density under the influence of altered RBC aggregation (Kim, Popel, et al. 2006).

7.3.1.2 Organ Perfusion Studies

Isolated-perfused whole-organ preparations have been used to study the influence of RBC aggregation on blood flow resistance. These studies include isolated heart (Charansonney et al. 1993; Yalcin et al. 2005), liver (Rogausch 1987), skeletal muscle (Cabel et al. 1997), and complete hind limb preparations (Yalcin et al. 2004). This approach has been used for investigating flow dynamics at selected levels of the circulatory system such as the venous blood vessels (Cabel et al. 1997). Alternatively, pressure–flow relationships can be studied in the complete vascular network of the organ, including arterial, microcirculatory, and venous vessels (Yalcin et al. 2004). Blood flow was monitored by either electromagnetic or Doppler flow meters, and hydrostatic pressure monitors were placed in line with the perfusion circuits of these preparations. Perfusion systems can be driven by simple gravity-dependent pressurizing systems or by various types of pumps including pressure servo–controlled equipment. Figure 7.4 shows the perfusion system used by Cabel et al. (1997) to study venous hemodynamics and the influence of altered RBC aggregation.

Figure 7.5 shows a schematic of a servo-controlled pressure perfusion system used for investigating pressure–flow relationships in various isolated organs (e.g., isolated

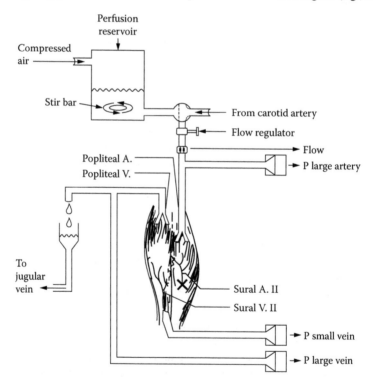

FIGURE 7.4 Perfusion system used by Cabel et al. for studying the effects of RBC aggregation on venous hemodynamics in a cat gastrocnemius muscle preparation. Reproduced from Cabel, M., H. J. Meiselman, A. S. Popel, and P. C. Johnson. 1997. "Contribution of Red Blood Cell Aggregation to Venous Vascular Resistance in Skeletal Muscle." *American Journal of Physiology—Heart and Circulatory Physiology* 272:H1020–H1032 with permission.

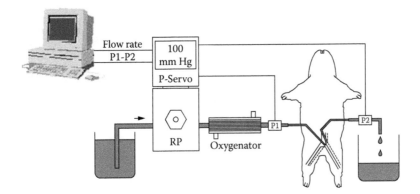

FIGURE 7.5 Pressure servo-control system used for perfusion of isolated organs and monitoring pressure–flow relationships. RP: roller pump; P-Servo: Pressure monitor and the servo control unit. Pressure gradient across the femoral artery and femoral vein (P1-P2) and the flow rate are monitored on an external computer.

guinea pig heart, rat hind limb) (Yalcin et al. 2004; Yalcin et al. 2005). Such a system consists of a servo control unit, which can be set to maintain a constant pressure or pressure gradient across the perfused circulatory network (e.g., pressure differences between arterial and venous catheters in femoral artery and femoral vein, Figure 7.5) by alternating the speed of a roller pump. The servo control signal can also be used to monitor the flow rate, thus providing an accurate method for obtaining blood flow data during perfusion of the organ at a constant pressure. Alternatively, the system can be set to maintain a constant flow rate and monitor the pressure gradient.

7.3.2 METHODS TO MODIFY RED BLOOD CELL AGGREGATION

There are two main approaches to modify RBC aggregation in experimental studies: (1) modifying suspending phase properties in terms of aggregating macromolecules, and (2) modifying RBC surface properties to alter aggregability. Most experimental studies employ the first method by introducing aggregating macromolecules (e.g., fibrinogen, high-molecular-mass dextrans) as a model of enhanced *in vivo* RBC aggregation. These macromolecules can be included in the RBC suspensions used in perfusion studies or solutions of these substances can be infused into the circulation of the organs/tissues under investigation. (See Chapter 2 for a detailed discussion on the influence of macromolecules on RBC aggregation.)

It should be noted that introducing high-molecular-mass polymers or proteins into the suspending phase of RBC suspensions (e.g., plasma) has dual effects: (1) enhanced RBC aggregation, and (2) increased suspending phase viscosity. High-molecular-mass dextrans, frequently used for modifying the aggregation behavior of RBC suspensions, have significant effects on suspending phase viscosity (Figure 7.6). Therefore, experiments using this approach reflect the influence of both enhanced aggregation and increased suspending phase viscosity.

Experimental methods based on modifications of cellular properties affecting aggregation in a defined suspending medium (i.e., RBC aggregability) allow

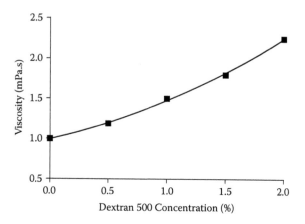

FIGURE 7.6 Viscosity of solutions containing 0–2% of 500 kDa dextran measured at 25°C indicating a 140% increase at the highest concentration.

investigation of altered RBC aggregation without the influence of suspending phase properties. A recently developed method of surface coating of RBC with specially designed copolymers (poloxamers) enables preparation of RBC suspensions with various degrees of aggregation in the same medium (e.g., autologous plasma) (Armstrong et al. 2005).

Poloxamer coating of RBC has been used to gradually modify aggregation (Figure 7.7). This was achieved by reacting the cells at poloxamer concentrations between 0.00125 and 0.5 mg/mL: RBC aggregation was increased by 60 to 200% in a dose-dependent manner (Yalcin et al. 2004). Aggregation of these poloxamer-coated RBC is characterized by strong adhesive interactions between adjacent cells and thus higher shear forces are required for disaggregation. Additionally, aggregate morphology can differ from that arising from physiological aggregation, especially at high poloxamer concentrations.

Given these differences, the method employed to modify *in vivo* RBC aggregation may be an important determinant of experimental results and may contribute to conflict among literature reports.

7.3.3 Comparison of the Results of Experimental Studies

Evidence for the effects of RBC aggregation on *in vivo* hemodynamics is based on observations of flow resistance as a function of flow rate (Cabel et al. 1997; House and Johnson 1986; Soutani et al. 1995; Thulesius and Johnson 1966). It can be assumed that RBC aggregation should become more intense at lower flow rates (i.e., lower shear forces), an assumption that might be expected to lead to increased flow resistance considering *in vitro* viscometric measurements (See Chapter 5). This assumption was strongly supported by the observations of Cabel et al. (1997) made in the venous vessels of cat gastrocnemius muscle (see Figure 7.4). Figure 7.8 presents venous vascular resistance as a function of blood flow, with normally aggregating cat blood showing a typical inverse relationship. The curves

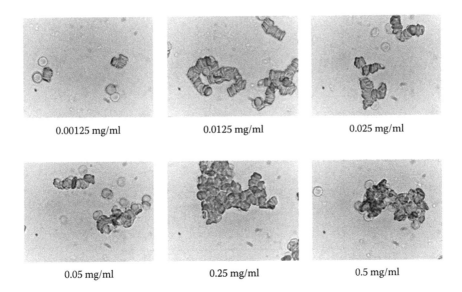

0.00125 mg/ml   0.0125 mg/ml   0.025 mg/ml

0.05 mg/ml   0.25 mg/ml   0.5 mg/ml

FIGURE 7.7 Aggregation of human red blood cells coated with an increasing amount of poloxamer (Pluronic F98) achieved by reacting the cells at concentrations between 0.00125 and 0.5 mg/mL. All coated cells are resuspended in the same autologous plasma. Modified from Yalcin, O., M. Uyuklu, J. K. Armstrong, H. J. Meiselman, and O. K. Baskurt. 2004. "Graded Alterations of RBC Aggregation Influence *in Vivo* Blood Flow Resistance." *American Journal of Physiology—Heart and Circulatory Physiology* 287:H2644–H2650 with permission.

for experiments with RBC suspended in dextran 40 kDa where RBC aggregation was absent exhibited almost no dependence of flow resistance on flow rate. An interesting observation of Cabel et al. shown in Figure 7.8 is the reduced dependence of flow resistance on flow rate with increased levels of RBC aggregation induced by the addition of dextran 250 kDa to the blood used for perfusion. This important experimental study underlines the importance of physiological levels of RBC aggregation for maintaining normal venous hemodynamics (Cabel et al. 1997), which in turn are important determinants of capillary flow dynamics, tissue perfusion, and function. It should be noted that the results by Cabel et al. are limited to flow dynamics in venous blood vessels and may not be applicable to parts of the circulatory network with higher shear forces (i.e., arterial or capillary blood vessels). The 1997 study by Cabel and coworkers prompted further investigations of venous flow dynamics under the influence of RBC aggregation and possible mechanisms of its influence.

Using intravital microscopy techniques, Bishop et al. (2001a, 2002) studied the effects of RBC aggregation enhanced by dextran 500 kDa infusions on velocity profiles in ~50 µm diameter venules of rat spinotrapezius muscle. They traced and measured the velocity of fluorescent-labeled RBC for various perfusion pressures. A blunted velocity profile during perfusion with strongly aggregating blood was observed, but only under "slow-flow" conditions following the reduction of perfusion pressure to 50 ± 14 mmHg from the control level of 123 ± 11 mmHg (Figure 7.9)

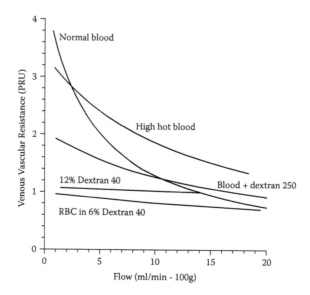

FIGURE 7.8 Venous vascular resistance as a function of blood flow measured in an isolated-perfused cat gastrocnemius muscle preparation (see Figure 7.4). Red blood cell aggregation was absent in dextran 40 kDa solutions while it was significantly enhanced in blood containing dextran 250 kDa. Reproduced from Cabel, M., H. J. Meiselman, A. S. Popel, and P. C. Johnson. 1997. "Contribution of Red Blood Cell Aggregation to Venous Vascular Resistance in Skeletal Muscle." *American Journal of Physiology—Heart and Circulatory Physiology* 272:H1020–H1032 with permission.

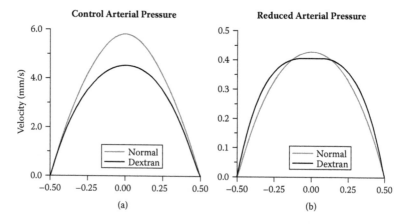

FIGURE 7.9 Red blood cell velocity profiles in ~50 μm diameter venules of rat spinotrapezius muscle. Arterial pressure for control conditions was 123 ± 11 mmHg (a) and was reduced to 50 ± 14 mmHg to achieve the "slow flow" conditions (b). Aggregation was enhanced by dextran 500 kDa infusions. Modified from Bishop, J. J., A. S. Popel, M. Intaglietta, and P. C. Johnson. 2002. "Effect of Aggregation and Shear Rate on the Dispersion of Red Blood Cells Flowing in Venules." *American Journal of Physiology—Heart and Circulatory Physiology* 283:H1985–H1996 with permission.

(Bishop et al. 2002). These results support the data of Cabel et al. and provide a hydrodynamic explanation for the observed dependence of venous flow resistance on RBC aggregation (Bishop et al. 2001a).

Formation of RBC aggregates in venules has been visually demonstrated by Bishop et al. (2004) using a high-speed camera and an image analysis system to determine aggregate size. They showed an inverse relationship between aggregate size and shear rate (Bishop et al. 2004); aggregate size was increased with dextran 500 kDa infusions in agreement with prior results of blunting of velocity profiles in the same size blood vessels (Bishop et al. 2002). They also observed that the aggregate size was greater in the central region of the vessel (Bishop et al. 2004). Kim et al. (2005), using a much faster camera (i.e., 4,450 frames/sec), demonstrated the time course of aggregate formation in postcapillary venules (Figure 7.10). About 50% of RBC were found to be associated with an aggregate within 300 msec after blood entered a postcapillary venule (Figure 7.11) and thus within about 40 μm distance from the end of the capillary (Kim et al. 2005).

The time course of *in vivo* aggregation presented in Figure 7.11 (Kim et al. 2005) is of interest in that it exhibits two phases with different time constants. This behavior is similar to the time course of RBC aggregation monitored *ex vivo*, which can be characterized by a double-exponential equation (see Chapter 4, Section 4.1.6.2.2).

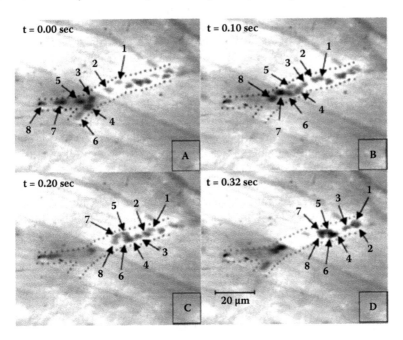

FIGURE 7.10 Formation of red blood cell aggregates in a postcapillary venule of rat spinotrapezius muscle. Panels A–D demonstrate the same venule at different time points (*t* = 0 to 0.32 sec). Red blood cells are labeled with numbers (1–8) to show their position in the resulting aggregate. Reproduced from Kim, S., A. S. Popel, M. Intaglietta, and P. C. Johnson. 2005. "Aggregate Formation of Erythrocytes in Postcapillary Venules." *American Journal of Physiology—Heart and Circulatory Physiology* 288:H584–H590 with permission.

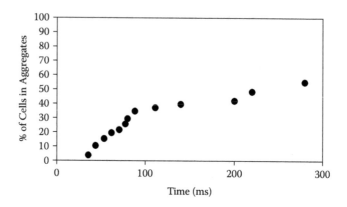

FIGURE 7.11   Time course of red blood cell aggregation in postcapillary venules of rat spinotrapezius muscle. Reproduced from Kim, S., A. S. Popel, M. Intaglietta, and P. C. Johnson. 2005. "Aggregate Formation of Erythrocytes in Postcapillary Venules." *American Journal of Physiology—Heart and Circulatory Physiology* 288:H584–H590 with permission.

However, the time constants for the aggregation in a postcapillary venule of ~15 μm was about 10 times smaller (i.e., faster time courses) than those observed in aggregometers with measurement chambers using ~300 μm or larger blood film thicknesses (see Chapter 4, Section 4.1.6). This very significant difference of aggregation time constants *in vivo* should be considered when attempting to extrapolate *ex vivo* measurements of RBC aggregation to *in vivo* flow conditions.

Kim, Zhen, et al. (2007) also demonstrated that the time course of aggregate formation is determined by the collision frequency at the very beginning of the venous system as the blood leaves the capillaries. However, they also observed that all collisions of RBC did not result in aggregate formation, and therefore collision efficiency is also an important determinant of this time course (Kim, Zhen, et al. 2007). RBC aggregability (see Chapter 2), which varies within an RBC population, is an important factor contributing to collision efficiency; RBC with higher aggregability may participate in an aggregate in the proximal part of the venule, leading to a decreased percentage of this type of cell and hence reduced collision efficiency in the more distal parts of the vessel.

Soutani et al. (1995) studied the influence of altered RBC aggregation induced by dextran 70 kDa on pressure–flow relationships in 25 to 35 μm-diameter microvessels of the rabbit mesentery when perfused with human RBC suspensions. The exact nature of the microvessels studied (i.e., arterial or venous side) was not mentioned but is probably not an important factor since flow dynamics were studied using an isolated perfusion approach (Soutani et al. 1995). Dextran 70 kDa had a dose-dependent effect on RBC aggregation: in agreement with the general pattern reported by others (Baskurt et al. 1997), the highest degree of aggregation was observed at 2.5 g/dl with further increments of concentration resulting in decreased RBC aggregation. The relationship between dextran 70 concentration and flow resistance was also not monotonic (Figure 7.12a): there was a sharp increase in flow resistance as dextran concentration increased up to ~1.5 g/dl, followed by a concentration range

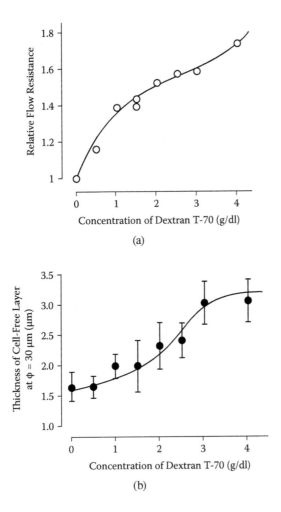

FIGURE 7.12 Relative flow resistance (a) and cell-free layer thickness (b) as a function of dextran 70 kDa concentration in microvessels of rabbit mesentery perfused by human RBC suspensions at 0.3 l/l hematocrit. Reproduced from Soutani, M., Y. Suzuki, N. Tateishi, and N. Maeda. 1995. "Quantificative Evaluation of Flow Dynamics of Erythrocytes in Microvessels: Influence of Erythrocyte Aggregation." *American Journal of Physiology— Heart and Circulatory Physiology* 268:H1959–H1965 with permission.

of 1.5–3 g/dl having a lower dependence on concentration. This is then followed by a second region of increased dependence on dextran concentration, which corresponds to decreased RBC aggregation compared to lower concentrations. They also studied the thickness of the cell-free layer near the vessel wall (Figure 7.12b), which demonstrated an opposite pattern of dependence on dextran concentration: (1) the lower range of dextran concentration corresponding to lower degrees of RBC aggregation showed a low degree of dependence of the cell-free layer thickness, (2) the dependence was more prominent over the concentration range in which RBC aggregation reaches its maximum, and (3) further increases of dextran concentration in the region where there is decreased aggregation resulted in a blunted dependence of thickness on dextran concentration. This inverse relationship between the dependencies of flow resistance and cell-free layer thickness on dextran 70 concentration, and hence on the extent of RBC aggregation, may provide an interesting ground for the discussion of specific hemodynamic mechanisms related to RBC aggregation. These mechanisms are further discussed in Section 7.4.1 of this chapter.

Soutani et al. also investigated the homogeneity of flow in 8 to 9 μm-diameter microvessels by recording light intensity and analyzing the power spectrum, following transformation of the data to the frequency domain, by a fast-Fourier algorithm (Soutani et al. 1995). They reported that the power spectrum shifted to lower frequencies with increased dextran 70 concentrations and interpreted this as indicating nonhomogeneous flow due to enhanced aggregation. This nonhomogeneous flow was attributed to irregular disaggregation of RBC aggregates at bifurcations and other geometric irregularities, leading to temporary capillary blockage in the downstream region. Kim, Popel, et al. (2006) have provided more direct information regarding the perfusion of capillaries in the presence of enhanced RBC aggregation. They determined the percentage of capillaries with moving RBC (i.e., functional capillary density) in a rat spinotrapezius muscle preparation. Dextran 500 kDa infusions decreased the functional capillary density significantly, with the effect

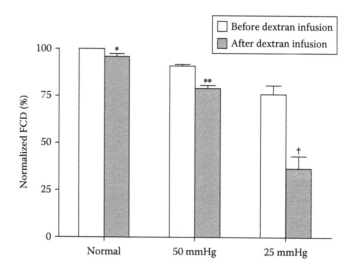

FIGURE 7.13 Functional capillary density (FCD) normalized to the value before dextran infusion at normal arterial pressure (126 ± 4 mmHg), before and after the infusion of dextran 500 kDa to induce enhanced red blood cell aggregation. Dextran infusion significantly affected functional capillary density at normal pressure (*, $p < 0.05$) and the effect was more prominent at reduced arterial pressures of 50 mmHg (**, $p < 0.005$) and 25 mmHg (†: $p < 0.0001$). Reproduced from Kim, S., A. S. Popel, M. Intaglietta, and P. C. Johnson. 2006. "Effect of Erythrocyte Aggregation at Normal Human Levels on Functional Capillary Density in Rat Spinotrapezius Muscle." *American Journal of Physiology—Heart and Circulatory Physiology* 290:H941–H947 with permission.

being much more prominent if arterial pressure was reduced (Figure 7.13). They also observed that the reduction in functional capillary density in preparations with enhanced aggregation and reduced arterial flow was compensated for by increased flow rate in those capillaries with flow. They suggested that the decreased functional capillary density should have resulted from enhanced plasma skimming but not from capillary plugging. Mchedlishvili et al. (1993) also reported decreased functional capillary density in rat mesentery following dextran 500 infusions; however, they observed capillaries filled with nonmoving RBC and their explanation for this effect was different than suggested by Kim, Popel, et al. (2006).

It should be noted that all experimental studies investigating the effect of enhanced RBC aggregation on *in vivo* blood flow dynamics discussed previously have employed the same approach: introducing high-molecular-mass dextrans with a molecular mass of 70 kDa or greater into the suspension used for perfusion. However, these high-molecular-mass dextrans cause significantly increased plasma or suspending medium viscosity as well as enhanced RBC aggregation (see Figure 7.6). This viscosity increase may have important hemodynamic-vascular effects (see Section 7.4.3), and most investigators are aware of this problem and usually discuss it as a factor potentially influencing their results.

Durussel et al. (1998) followed a different approach to study the *in vivo* effects of RBC aggregation. They studied the influence of infusing dextran 500 solutions into rat mesentery using intravital microscopy, and they compared blood flow indexes normalized to flow before dextran 500 infusions in two groups of animals, both of which received dextran infusions. In one group they inhibited dextran-mediated RBC aggregation by the use of a pharmacological agent, Troxerutine; this agent is hypothesized to attach to the RBC glycocalyx and interfere with aggregating molecules (Durussel et al. 1998). Therefore, both groups had elevated plasma viscosity, but the group that received Troxerutine injection had lower RBC aggregation. Their results indicate a significantly lower blood flow (i.e., higher flow resistance) in the group with both enhanced aggregation and increased plasma viscosity compared to the group with only increased plasma viscosity (Figure 7.14). Their data thus support the findings of most intravital microscopy studies by confirming that enhanced RBC aggregation is associated with increased microvascular flow resistance. However, the study may be criticized due to the use of dextran infusions; the resulting higher plasma viscosity may influence vascular mechanisms that may interfere with the hemodynamic effects observed (see Section 7.4.3).

It is obvious from the previous discussion that almost all studies employing intravital microscopy indicate that enhanced RBC aggregation is associated with increased blood flow resistance. The majority of these studies focused on flow patterns in venous blood vessels or capillaries. However, studies reporting impaired capillary perfusion associated with enhanced RBC aggregation also strongly suggest effects on the arterial side of the microcirculation (Kim, Popel, et al. 2006; Mchedlishvili et al. 1993; Soutani et al. 1995). This suggestion is supported by the observed decrease of functional capillary density (i.e., increased number of capillaries without moving RBC), since this alteration can be explained by hemodynamic alterations in the arterial vessels prior to capillaries (Kim, Popel, et al. 2006).

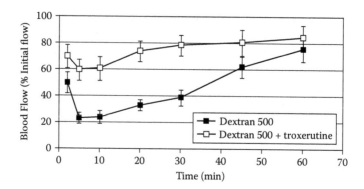

FIGURE 7.14 Blood flow normalized to the initial flow before dextran 500 infusions in rat mesentery; dextran was infused with or without troxerutine. Troxerutine reduces dextran-induced red blood cell aggregation but does not alter the elevated suspending phase viscosity due to dextran. Modified from Durussel, J. J., M. F. Berthault, G. Guiffant, and J. Dufaux. 1998. "Effects of Red Blood Cell Hyperaggregation on the Rat Microcirculation Blood Flow." *Acta Physiologica Scandinavica* 163, no. 1:25–32.

Several investigators have studied the influence of RBC aggregation on flow resistance in arterial blood vessels. Mchedlishvili et al. (1993) observed significant decrements in blood flow velocity in arterioles of rat mesentery following increases in RBC aggregation induced by dextran 500 kDa infusions. Dextran 500 infusions also resulted in significantly increased blood pressure in rats (Mchedlishvili et al. 1993). Vicaut et al. (1994) studied flow velocity in arterioles of rat cremaster muscle following the infusion of dextrans with molecular mass of 40 kDa, 70 kDa, and 480 kDa. They observed that flow velocity and functional capillary density decreased with the infusion of dextran solutions with 70 and 480 kDa, but not with 40 kDa (Vicaut et al. 1994). Dextran with molecular mass of 40 kDa does not induce RBC aggregation.

The influence of RBC aggregation on the flow resistance of the complete circulatory network, including the arterial blood vessels, can be studied via whole-organ perfusion. Unlike the studies using intravital microscopy, the results of these whole-organ perfusion studies are not always in accordance. Verkeste et al. (1992) observed no alteration in uteroplacental blood flow with enhanced RBC aggregation. Comparisons of the flow resistance in an isolated-perfused liver preparation during perfusion with aggregating and nonaggregating RBC suspensions indicated that the apparent viscosity of aggregating suspensions was higher compared to those without aggregation (Rogausch 1987). Both studies used dextrans to induce RBC aggregation. Gustafsson et al. (1981) reported no significant alteration of apparent viscosity measured *in vivo* during the perfusion of the dog hind limb by RBC suspensions in high molecular mass dextran solutions; in this study, blood vessels in the isolated-perfused preparation were maximally dilated by papaverin infusions (Gustafsson et al. 1981). In contrast with these two studies, Baskurt et al. (1999) demonstrated that blood flow resistance was lower in isolated-perfused guinea pig hind limbs perfused with aggregating suspensions of RBC in a 3% dextran 70 kDa solution compared to nonaggregating suspensions of cells in a viscosity-matched dextran 40 (Mm: 40

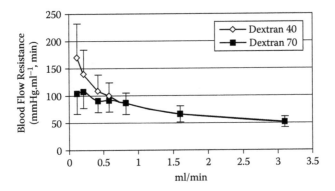

FIGURE 7.15 Blood flow resistance measured in isolated-perfused hind limbs of guinea pigs during perfusion with aggregating (in 3% dextran 70 solution) and nonaggregating (in viscosity-matched dextran 40 solution) RBC suspensions. Reproduced from Baskurt, O. K., M. Bor-Kucukatay, and O. Yalcin. 1999. "The Effect of Red Blood Cell Aggregation on Blood Flow Resistance." *Biorheology* 36:447–452 with permission.

kDa) solution (Figure 7.15) (Baskurt et al. 1999). In these studies the isolated vasculature of the hind limbs was chemically fixed by perfusion of 0.4% glutaraldehyde solution prior to perfusion with RBC suspensions, and therefore vascular compliance was significantly reduced and vasomotor control mechanisms were abolished. The difference in flow resistance during perfusion with aggregating and nonaggregating suspensions were observed at flow rates below 0.5 ml/min, corresponding to perfusion pressures below 50 mmHg, but flow resistance values measured above 0.5 ml/min were almost identical (Figure 7.15).

Charansonney et al. (1993) reported an interesting study performed on isolated-perfused rat hearts. They perfused isolated hearts via the Langendorf method using nonaggregating RBC suspensions prepared in Krebs-albumin solution or RBC suspensions prepared in dextran 70 kDa solutions (Charansonney et al. 1993). Their results are summarized in Figure 7.16.

Figure 7.16a presents coronary blood flow normalized to the value measured at 80 mmHg perfusion pressure during perfusion with nonaggregating RBC suspensions in Krebs-albumin solution. Note that the flow was higher if the isolated heart preparation was perfused with the suspension containing the proaggregant 1% dextran 70. Furthermore, when the normalized blood flow indexes were corrected for the higher viscosity of the dextran suspension, the difference between aggregating and nonaggregating RBC suspensions became even greater. This pattern is clearly in contrast with the whole-organ studies mentioned previously (Rogausch 1987; Verkeste et al. 1992), which reported increased flow resistance or no effect with aggregating RBC suspensions.

Interestingly, perfusion of isolated heart preparations with suspension containing 2% dextran 70, which exhibit higher degrees of aggregation, resulted in decreased blood flow index compared to nonaggregating RBC suspensions (Figure 7.16b). However, the blood flow indexes, after correction for the increased suspending medium viscosity, were not different from those measured during perfusion with

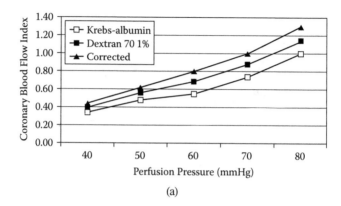

(a)

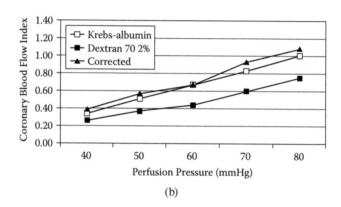

(b)

FIGURE 7.16 Coronary blood flow index measured in isolated-perfused rat heart preparations perfused by nonaggregating (Krebs-albumin) and aggregating (dextran 70 solutions) RBC suspensions: (a) dextran 70 concentration of 1%, (b) dextran 70 concentration of 2%. Values corrected for the increased viscosity of the dextran 70 solutions are also shown as "corrected." Redrawn from Charansonney, O., S. Mouren, J. Dufaux, M. Duvelleroy, and E. Vicaut. 1993. "Red Blood Cell Aggregation and Blood Viscosity in an Isolated Heart Preparation." *Biorheology* 30:75–84 with permission.

nonaggregating RBC suspensions. These findings indicate that the influence of RBC aggregation on *in vivo* blood flow resistance depends on the level of aggregation. It has thus been suggested that RBC aggregation may have opposite effects at various levels of the circulatory network depending on the degree of aggregation (Vicaut 1995).

Further evidence for the dependence of *in vivo* hemodynamic effects on the level of RBC aggregation was provided by Yalcin et al. (2004). Hind limbs of guinea pigs were perfused via the femoral artery and vein using the servo-controlled perfusion system shown in Figure 7.5 and described in Section 7.3.1.2. RBC aggregation was modified by surface coating RBC with the poloxamer F-98 at various concentrations, thereby resulting in a gradual enhancement of aggregation of 60–200% of control

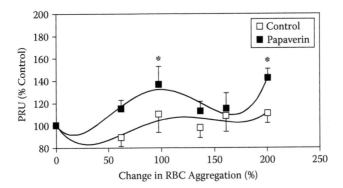

FIGURE 7.17 Blood flow resistance in isolated-perfused guinea pig hind limb preparations perfused with suspensions of normal and poloxamer F98-coated RBC. The poloxamer concentration during the coating procedure was varied between 0.00125 and 0.5 mg/mL resulting in a 60–200% enhancement of aggregation. Flow resistance (R) is expressed as a percentage of the values measured with normal, noncoated suspensions (Control). Experiments were conducted in two series of organ preparations with and without inhibiting vessel tone by papaverin. The asterisk (*) indicates a significant difference from perfusion experiment using normal RBC suspensions. Redrawn from Yalcin, O., M. Uyuklu, J. K. Armstrong, H. J. Meiselman, and O. K. Baskurt. 2004. "Graded Alterations of RBC Aggregation Influence *in Vivo* Blood Flow Resistance." *American Journal of Physiology—Heart and Circulatory Physiology* 287:H2644–H2650.

levels (See Section 7.3.2). Perfusion of the isolated hind limbs with RBC suspensions having a 100% increment in aggregation resulted in significantly increased flow resistance (Figure 7.17). Interestingly, flow resistance returned to control levels with further enhancement in RBC aggregation up to 150%. Finally, perfusion with RBC suspensions with 200% enhanced aggregation level resulted in significantly increased flow resistance (Yalcin et al. 2004). These results also confirm that the level of RBC aggregation determines the overall influence of aggregate formation on *in vivo* hemodynamic alterations. However, significant alterations of resistance during perfusions using suspensions with 100% and 200% enhanced aggregation could only be detected in the isolated hind-limb preparations after pretreatment with papaverin; this agent inhibits vascular smooth muscle tone and hence abolishes vasomotor control. The perfusion experiments with the same protocol in hind limb preparations with intact vasomotor tone (not treated with papaverin) did not result in any significant changes in blood flow resistance (Figure 7.17). This finding underlines the importance of vasomotor control mechanisms in maintaining adequate blood flow, despite the changes in hemorheological factors (including RBC aggregation). This is an important factor that should be kept in mind in evaluating the results of *in vivo* hemorheology experiments, further considered in Section 7.3.3.1.2.

The results summarized in Figure 7.17 differ from previously described studies in that a different method was used to modify the aggregation properties of the RBC suspensions used for perfusion. Suspending phase properties were not altered as the coated RBC were suspended in autologous plasma; plasma viscosity was not changed unlike the experiments using high-molecular-mass dextrans to induce aggregation. This method also allowed precise control of the level of aggregation, therefore enabling testing of the influence of gradual alterations of RBC aggregation on flow resistance (Yalcin et al. 2004).

The poloxamer coating method to alter RBC aggregability was also used to study the pressure dependence of flow resistance in an isolated-perfused guinea pig heart (Langendorf) preparation (Yalcin et al. 2005). The coronary vasculature was perfused via a catheter in the ascending aorta using a pressure-servo-controlled pump system (Figure 7.5) and flow resistance was calculated by dividing the perfusion pressure by the flow rate of RBC suspensions. Figure 7.18a demonstrates the results obtained in preparations with intact vascular smooth muscle tone. Flow resistance was not significantly influenced by perfusion pressures between 30 and 100 mmHg during perfusion with normally aggregating guinea pig blood (Figure 7.18a). However, when the coronary vasculature was perfused with poloxamer-coated RBC suspensions having enhanced aggregation (i.e., 100% increase of erythrocyte sedimentation rate), flow resistance was found to be dependent on perfusion pressure: flow resistance at perfusion pressures of 40 mmHg or below were significantly higher during perfusion with aggregating RBC suspensions (Figure 7.18a).

When the experiments were repeated after papaverin treatment (at 10^{-4} M concentration) of the isolated vasculature to inhibit smooth muscle tone, flow resistance was significantly reduced due to the maximal dilation of blood vessels during perfusion with both normal and hyperaggregating RBC suspensions (Figure 7.18b). However, the significant differences between flow resistances during perfusion with RBC suspensions could be detected in a wider range of perfusion pressures (60 mmHg or below). This finding also underlines the importance of vasomotor control mechanisms in compensating the hemodynamic effects of alterations in RBC aggregation.

7.3.3.1 Discrepancies between *in Vivo* Experiments

It should be obvious from the previous discussion that the results of *in vivo* experiments exploring the influence of RBC aggregation on blood flow resistance are not in total agreement. However, an analysis of the experimental details also indicates important methodological differences between these studies. These differences can be classified into four categories: (1) the organ or tissue under investigation, (2) physiological status of the organ or tissue under investigation, (3) method used to study pressure–flow relationships, and (4) method to modify RBC aggregation.

7.3.3.1.1 Organ/Tissue under Investigation

It is well known that properties of the vasculature of various organs or tissues differ significantly from each other (Shepro 2006). Perfusion studies comparing the influence of hemorheological alterations on flow resistance in different organs have shown that these differences are reflected in hemodynamic measurements (Chen et al. 1989; Fan et al. 1980; Simchon et al. 1987). These studies employed radio-labeled microspheres to measure regional blood flow, thus allowing simultaneous determination of flow resistance in a variety of organs under the influence of hemorheological factors,

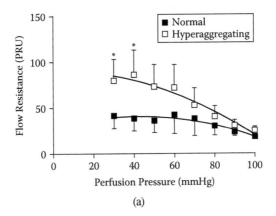

(a)

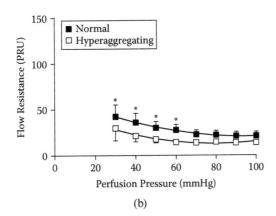

(b)

FIGURE 7.18 Blood flow resistance expressed in peripheral resistance units (PRU) measured in isolated-perfused guinea pig heart (Langendorf) preparations perfused with suspensions of normal and poloxamer F98-coated RBC. The poloxamer concentration during coating was 0.025 mg/mL, corresponding to a 100% enhancement of aggregation. Experiments were conducted in two series on preparations with intact vascular smooth muscle tone (a) and after inhibition of vascular smooth muscle tone with papaverin treatment (b). The asterisk (*) indicates significant difference from perfusion experiment using normal RBC suspensions. Redrawn using data from Yalcin, O., H. J. Meiselman, J. K. Armstrong, and O. K. Baskurt. 2005. "Effect of Enhanced Red Blood Cell Aggregation on Blood Flow Resistance in an Isolated-Perfused Guinea Pig Heart Preparation." *Biorheology* 42:511–520.

including alterations in hematocrit (Fan et al. 1980), plasma viscosity (Chen et al. 1989) and RBC deformability (Simchon et al. 1987). Comparison of the effects of these hemorheological factors indicated very different patterns of effect on regional hemodynamics.

Chen et al. (1989) studied regional blood flow in various organs of dogs following blood viscosity alterations induced by exchange transfusions of RBC suspended in plasma containing the strong proaggregant dextran 500 kDa. Their

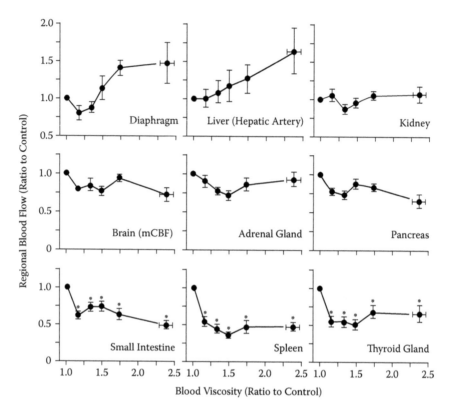

FIGURE 7.19 Regional blood flow in various organs of dog measured using radio-labeled microspheres for blood viscosity alterations induced by the strongly proaggregant dextran 500 kDa. RBC suspensions with dextran 500 in plasma were exchange transfused to dogs to eventually obtain blood viscosity up to 250% of control. The asterisk (*) indicates significantly different blood flow compared to control experiments with normal viscosity. Reproduced from Chen, R. Y., R. D. Carlin, S. Simchon, K. M. Jan, and S. Chien. 1989. "Effects of Dextran-Induced Hyperviscosity on Regional Blood Flow and Hemodynamics in Dogs." *American Journal of Physiology—Heart and Circulatory Physiology* 256:H898–H905 with permission.

study included a wide range of viscosity alterations: up to ~450% increment in plasma viscosity and ~250% increment in whole blood viscosity. Figure 7.19 shows changes in regional blood flow as a function of viscosity alterations in various organs and the different patterns of effect in various organs: almost no alteration in blood flow despite the 250% increment in blood viscosity in the kidney, significant reductions of blood flow under the same blood viscosity change in the small intestine and spleen, and increased liver and diaphragm blood flow. Although Chen et al. (1989) did not report RBC aggregation corresponding to the different levels of whole blood viscosity, it is obvious that the observed effects are influenced by the enhanced aggregation by dextran 500 kDa. No other studies of the effects of RBC aggregation on regional blood flow for this wide range of organs appear to exist.

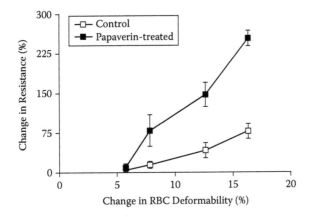

FIGURE 7.20 Change in blood flow resistance in an isolated-perfused rat hind limb preparation perfused with suspensions of RBC having progressively reduced deformability (5.5 to 17%) achieved by very low concentrations of glutaraldehyde. The experiments were done with normal vasculature and after inhibition of vascular tone with papaverin. The increase of flow resistance was ~70% at the greatest deformability decrease in preparations with intact vascular tone, whereas it was ~250% in preparations with vasomotor tone inhibited. Redrawn from Baskurt, O. K., O. Yalcin, and H. J. Meiselman. 2004a. "Hemorheology and Vascular Control Mechanisms." *Clinical Hemorheology and Microcirculation* 30:169–178.

7.3.3.1.2 Physiological Status of the Organ/Tissue under Investigation

Many physiological studies have clearly shown that the living vasculature has a remarkable ability to compensate for any effect that may interfere with adequate blood flow for homeostasis (Stainsby 1973). This ability is mainly related to the autoregulatory mechanisms of vascular tone in which vascular hindrance is adjusted to compensate for alteration of hemorheological parameters. Every organ and tissue has a specific range and degree of effectiveness of autoregulatory reserve (Baskurt et al. 1991), which can compensate for hemorheological challenges induced experimentally or during pathophysiological processes. Such different levels of effectiveness of vascular autoregulatory compensation may, at least in part, explain the observed differences in the hemodynamic changes in various organs.

The compensatory role of vasomotor control mechanisms has been demonstrated by experiments comparing vasculature with intact or papaverin-paralyzed vascular smooth muscles (Figures 7.17 and 7.18). These studies include those investigating the impact of enhanced RBC aggregation (Yalcin et al. 2004; Yalcin et al. 2005), but the similar role of vasomotor control mechanisms in compensating for the influence of gradual impairment of RBC deformability in the isolated-perfused rat hind limb was also demonstrated (Figure 7.20) (Baskurt et al. 2004a).

The experimental results indicating the dependence of the effect of RBC aggregation on perfusion pressure or flow rate (Yalcin et al. 2005) discussed previously (Figure 7.18) can also be interpreted in terms of the role of vasomotor reserve. Obviously, vasomotor reserve might be partly or completely depleted in order to maintain adequate metabolic conditions in perfused tissues if perfusion pressure is impaired or flow rate is decreased. Thus, in addition to the well-known dependence of RBC aggregation on perfusion pressure, the influence of vasomotor reserve should be considered in interpreting pressure–flow results in various vascular beds.

Gustafsson et al. (1981) also used papaverin to inhibit smooth muscle tone during *in vivo* viscometry experiments in isolated-perfused dog hind limb preparations, and reported no effect of enhanced RBC aggregation induced by high-molecular-mass dextrans. Bishop et al. (2001d) has suggested that this lack of effect might be due to the maximal dilation of blood vessels, thereby increasing shear rates in blood vessels, even with normal arterial pressures and hence preventing the development of RBC aggregates. This alternative interpretation of the effect of inhibiting vascular smooth muscle tone does not agree with experimental results presented in the preceding paragraphs. However, it should be kept in mind that the method to induce increased RBC aggregation used by Gustafsson et al. (i.e., dextran infusion) differed from the poloxamer-coating technique used by Baskurt and coworkers (Yalcin et al. 2004; Yalcin et al. 2005). In addition, RBC aggregation has recently been demonstrated to interfere with endothelial function (Baskurt et al. 2004a; Yalcin et al. 2008): enhanced RBC aggregation may down-regulate endothelial nitric oxide (NO) synthesizing mechanisms, which in turn, may influence vasomotor tone and flow resistance (see Section 7.4.4).

7.3.3.1.3   Method Used to Study the Pressure–Flow Relationship

It can be deduced from the previously mentioned *in vivo* experimental studies investigating the influence of RBC aggregation that conflicting results are mostly obtained by whole-organ perfusion studies, while experiments using intravital microscopy to study pressure–flow relationships always indicate increased flow resistance with enhanced RBC aggregation.

It is important to recall that intravital microscopy utilizes thin tissues spread on a special microscope stage, which are usually kept in a horizontal position; all blood vessels in which pressure–flow relationships are studied are horizontally oriented. It has been demonstrated that the *in vitro* influence of RBC aggregation depends on the orientation of the flow system (e.g., glass tubes): enhanced RBC aggregation results in increased flow resistance in horizontally oriented tubes while flow resistance is decreased in vertically oriented tubes (Alonso et al. 1989; Alonso et al. 1993; Cokelet and Goldsmith 1991). The increased flow resistance in horizontally oriented tubes with enhanced RBC aggregation is due to RBC sedimentation on the lower, dependent side of the tube (see Chapter 6, Section 6.5.5). Bishop et al. (2001b) demonstrated that this sedimentation effect also occurs in venules of the rat spinotrapezius muscle if RBC aggregation is enhanced by using a microscopic system that allows changeable orientation of the tissue with gravity. This dependence on the orientation of the flow system may, at least in part, explain the consistent findings of increased flow resistance with enhanced aggregation obtained by intravital microscopy. Alternatively, whole-organ preparations contain blood vessels of various sizes and different orientations with respect to gravity that collectively contribute to the changes of flow resistance with enhanced RBC aggregation. Additional hemodynamic mechanisms that are discussed in more detail in Section 7.4.4 may contribute to the overall response of a whole-organ preparation.

7.3.3.1.4   Method to Modify RBC Aggregation

Adding high-molecular-mass polymers or proteins (e.g., dextrans ≥ 70 kDa) to the suspending medium has two effects (i.e., increase aggregation, increase medium viscosity). Increased suspending phase viscosity is a side effect of the dextrans used for increasing RBC aggregation, and this viscosity increase may have *in vivo* effects on blood flow dynamics that cannot be accurately differentiated from aggregation effects. Furthermore, increased suspending phase (plasma) viscosity has been demonstrated to interfere with vasomotor control mechanisms (see Section 7.4.4) (Tsai et al. 2005) and thus has significant potential to interfere with experimental results. Modifying RBC aggregability by surface coating of RBC with copolymers has been used to modify RBC aggregation without altering suspending phase composition and properties (Yalcin et al. 2004; Yalcin et al. 2003; Yalcin et al. 2008). *In vivo* studies using this method are progressing and are expected to provide a better understanding of the mechanisms of hemodynamic alterations induced by enhanced RBC aggregation.

7.3.3.2   Importance of Vascular Control Mechanisms

It follows from the Poiseuille Equation that *in vivo* blood flow resistance has two components: (1) vascular hindrance (i.e., factors related to the geometry of blood vessels), and (2) blood viscosity (i.e., rheological behavior of blood as determined by such factors as hematocrit, plasma viscosity, RBC deformability and aggregation). A very important feature of vascular hindrance is its variability, which is the key factor of the physiological mechanisms to maintain adequate blood flow at the tissue level. With rare exceptions (e.g., kidney), blood flow is precisely regulated by modulating vascular tone in order to match the metabolic needs of the perfused tissue (Duling et al. 1987; Stainsby 1973). This regulatory mechanism can effectively counterbalance the hemodynamic impact of a hemorheological alteration (Baskurt et al. 2004a). The effectiveness of this ability to compensate depends on vasomotor reserve, which in turn is a function of vascular smooth muscle tone under basal physiological conditions (Feigl 1983; Gallagher et al. 1982). This autoregulatory reserve in some tissues allows maintenance of blood flow sufficient to match the metabolic needs of the tissue at rest, even if the vascular geometry is significantly challenged (Baskurt et al. 1991). These mechanisms may also compensate for serious hemorheological alterations if the vasomotor reserve has not been exhausted by other challenges (i.e., geometric alterations of blood vessels) (Baskurt et al. 1991).

The effectiveness of vasomotor control mechanisms should be considered as an important factor in evaluating the influence of hemorheological alterations, including changes in RBC aggregation, on *in vivo* blood flow dynamics (Figures 7.18 and 7.20). Furthermore, the role of these compensatory vasomotor mechanisms should also be considered under pathophysiological conditions encountered during various disease processes. Hemorheological alterations can be effectively counterbalanced by autoregulatory mechanisms if vasomotor reserve is sufficient. However, if vascular geometry is also challenged and vasomotor reserve is depleted, the affected organ or tissue may become much more sensitive to hemorheological alterations (Volger et al. 1986).

7.4 HEMODYNAMIC MECHANISMS INFLUENCED BY RED BLOOD CELL AGGREGATION

The discussion in Section 7.3 suggests that the role of RBC aggregation as a determinant of *in vivo* flow dynamics is quite complicated and thus unlike the more easily understandable influence of alterations in RBC deformability (Lipowsky et al. 1993; Simchon et al. 1987). A better understanding of this role requires further consideration of specific hemodynamic mechanisms that are influenced by alterations of the degree of RBC aggregation.

7.4.1 AXIAL MIGRATION OF RED BLOOD CELLS

Concentric fluid layers moving in a blood vessel during laminar flow have velocities changing from a maximum in the center of the vessel (i.e., central flow zone) to a minimum in the layers adjacent to the vessel wall (i.e., marginal flow zone) (Pries et al. 1996). The velocity gradient between laminae (shear rate) in the central flow zone is smaller than the velocity gradient in the marginal flow zone (Figure 7.21). RBC tend to move toward the central flow zone, as this zone is more stable in terms of shear forces compared to the region adjacent to the vessel wall (Figure 7.21). This phenomenon is known as axial migration and is discussed in more detail in Chapter 6.

Axial accumulation of RBC can be perceived as the result of two opposing effects: (1) hydrodynamic forces tending to move RBC toward the central flow zone, and (2) elastic compression of the RBC core tending to counter this movement (Bishop et al. 2001c; Goldsmith and Mason 1961). Axial accumulation of RBC is strongly influenced by aggregation (Kim et al. 2009), which tends to reduce the influence of elastic compression. Bishop et al. used fluorescent-labeled RBC and traced their trajectory in postcapillary venules of rat spinotrapezius muscle and observed that axial migration of RBC was slightly increased by aggregation achieved by dextran 500 kDa infusions (Bishop et al. 2001c). This effect of enhanced aggregation was only observed if shear forces were reduced by lowering perfusion pressures.

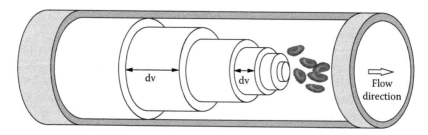

FIGURE 7.21 Red blood cells tend to migrate toward the central flow zone, moving at a higher speed in a cylindrical tube (e.g., blood vessels). The velocity gradient (dv) between the flow lamina (shear rate) in this central zone is smaller compared to the marginal flow zone near the vessel wall.

7.4.1.1 Formation of a Cell-Poor Layer in the Marginal Flow Zone

An RBC-poor (i.e., RBC-depleted) fluid layer is formed near the vessel wall in the marginal flow zone as a result of axial migration. This marginal fluid layer is often termed a *cell-free* layer, but it is unlikely that there is a total absence of RBC in this region; the term *cell-poor* will thus be used to indicate this region. The formation of cell-poor layers has been investigated using the microcirculation of various animal models (Bishop et al. 2001b; Kim, Kong, et al. 2007; Maeda et al. 1996; Soutani et al. 1995). Earlier studies were focused on blood vessels with different properties and employed manual analysis of video microscopic images. More recently, computerized image analysis methods have been used to analyze images obtained by high-speed video microscopy, thereby enabling a detailed evaluation of the temporal and spatial variations of the cell-poor layer thickness (Kim, Kong, et al. 2006; Kim, Kong, et al. 2007).

The thickness of the cell-poor layer is affected by factors related to both the vasculature and the blood, including vessel diameter, glycocalyx properties, flow rate, and RBC rheological properties (i.e., aggregation and deformability) (Kim et al. 2009). It has been demonstrated that the ratio of cell-poor layer thickness to vessel diameter is a function of vessel diameter, increasing as diameter decreases, approaching 12% in arterioles of 25 μm diameter (Kim, Kong, et al. 2007). This diameter-dependent behavior is consistent with previously reported data (Soutani et al. 1995; Tateishi et al. 1994). Soutani et al. (1995) clearly demonstrated the increment of cell-poor layer thickness in 25 to 35 μm-diameter vessels when RBC aggregation was enhanced by dextran 70 kDa: there was a ~100% change in cell-poor layer thickness from ~1.5 μm to ~3 μm with increased dextran 70 concentration (see Figure 7.12).

Formation of a cell-poor layer is a time-dependent process that develops over several seconds, at least in glass tubes having lengths on the order of >100 times the tube diameter. Bishop et al. (2001d) demonstrated that the formation of a cell-poor layer is attenuated with frequent branching in venous vessels since converging branches introduce RBC into the downstream marginal flow layer. Unbranched segment lengths in the venous vasculature are usually only 1 to 4 times the vessel diameter and thus the transit time of RBC or aggregates through these segments is relatively short. Such short transit times may only allow effective cell-poor layer formation during reduced flow rates (i.e., reduced perfusion pressures) (Bishop et al. 2001d).

The time dependence of cell-poor layer formation has been suggested to be an important factor in evaluating the influence of RBC aggregation. It has been proposed that the time course of RBC aggregation may not be fast enough to allow the development of aggregates that affect axial migration and cell-poor layer formation. That is, transit times corresponding to pressure gradients in the physiological range may be too short (Bishop et al. 2001c) for effective RBC aggregation. However, RBC aggregates are frequently observed in venous vessels with frequent branching, and the time course of their formation has been demonstrated to be significantly faster than that determined *ex vivo* (see Figures 7.10 and 7.11) (Kim et al. 2005). Therefore, the time course of aggregation may not be a limiting factor for cell-poor layer formation, even in venous blood vessels with frequent branching (Bishop et al. 2001c).

The hemodynamic consequences of the formation of a cell-poor layer are detailed in the following four subsections.

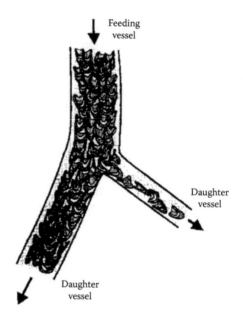

FIGURE 7.22   Plasma skimming at a bifurcation. (Modified from Pries, A. R., T. W. Secomb, and P. Gaehtgens. 1996. "Biophysical Aspects of Blood Flow in The Microvasculature." *Cardiovascular Research* 32:654–667. With permission)

7.4.1.1.1   Plasma Skimming

Plasma skimming is a process occurring at branching points of the arterial system as a consequence of phase separation. The blood stream is divided at a branch point and daughter vessels may not receive blood with the same composition. It has been experimentally observed that the daughter branch with a lower flow receives blood with a lower hematocrit (Figure 7.22) (Pries et al. 1989; Pries et al. 1996). This phenomenon is closely related to the axial migration of RBC and formation of the cell-poor layer (Goldsmith et al. 1989; Schmid-Schönbein 1988), with enhanced axial migration resulting in greater plasma skimming.

Plasma skimming leads to a reduced hematocrit in smaller blood vessels, and is one of the mechanisms resulting in lower microvascular and tissue hematocrit values (see Section 7.4.2).

7.4.1.1.2   Reduced Apparent Viscosity in Microvessels

The formation of a cell-poor layer adjacent to the vessel wall leads to reduced flow resistance and hence a lower apparent viscosity of blood. This has been investigated experimentally using glass tubes (Alonso et al. 1995; Cokelet and Goldsmith 1991; Reinke et al. 1986; Reinke et al. 1987). Interestingly, RBC aggregation and the formation of the cell-poor layer have been shown to reduce apparent viscosity in vertically oriented tubes, but not if RBC sedimentation occurs in horizontal tubes (Cokelet and Goldsmith 1991) (see Chapter 6).

The effect of a cell-poor layer on flow resistance can be explained by the movement of a RBC-rich core within the lubricating, lower-viscosity marginal zone. This effect is an important element of the mechanisms behind the Fåhraeus-Lindqvist effect, which is the reduction of apparent viscosity with decreasing tube diameter (see Chapter 6, Section 6.4). Kim et al. (2009) have indicated that the shape of the interface between the RBC-rich core and cell-poor layer is also an important determinant of apparent viscosity. Variations in cell-poor layer thickness significantly influence flow resistance in microvessels (Kim, Kong, et al. 2006). These variations of layer thickness are smaller if RBC aggregation is enhanced, thereby reducing their influence on flow resistance (Kim et al. 2009).

7.4.1.1.3 Reduced Wall Shear Stress in Microvasculature

Wall shear stress is the tangential force exerted on a vessel wall (i.e., endothelial cell layer) by flowing blood, and is determined by the local viscosity and velocity of the fluid in the marginal flow zone (Katritsis et al. 2007; Reneman et al. 2006). Axial migration of RBC, resulting in a cell-poor fluid zone near the vessel wall, reduces local viscosity and hence wall shear stress (Baskurt et al. 2004b; Yalcin et al. 2008). Note that wall shear stress is an important determinant of endothelial cell function as further discussed in Section 7.4.3.

7.4.1.1.4 Reduced Nitric Oxide Scavenging

RBC contribute to the delicate balance between generation and scavenging of NO that determines its local bioavailability (Imig et al. 1993; Liao et al. 1999). Hemoglobin encapsulated in RBC is a potent scavenger for NO originating from endothelium. A cell-poor layer near the vessel wall increases the distance between the source (i.e., endothelium) and sink (i.e., hemoglobin) and therefore acts as a barrier for NO scavenging (Kim et al. 2009). An increase of cell-poor layer thickness may therefore be expected to enhance the local NO concentration thereby decreasing vascular smooth muscle tone. However, it has recently been demonstrated that RBC have active NO synthesizing mechanisms and may also be a source of NO (Bor-Kucukatay et al. 2003; Kleinbongard et al. 2006; Ozuyaman et al. 2008). Furthermore, these mechanisms are activated by exposure of RBC to shear forces (Ulker et al. 2009; Ulker et al. 2011). Therefore, having the majority of RBC away from the vessel wall may reduce the endothelial effects of NO originating from RBC. The net effect of increased cell-poor layer thickness is still debatable.

7.4.1.2 Difference between Red Blood Cell and Plasma Velocity during Flow

Mean RBC velocity in a given vessel segment is higher than average blood velocity (Pries et al. 1996). This is due to the higher flow rate in the central flow zone compared to the marginal flow zone. Average transit time of RBC is reduced, resulting in a lower average hematocrit value within the cross section of the blood vessel, but the hematocrits of blood entering and leaving the vessel are equal (i.e., the Fåhraeus effect).

7.4.1.2.1 Fåhraeus Effect

The ratio of the hematocrit of blooding given vessel segment (i.e., tube hematrocrit) to the hematocrit of blood discharged from that segment decreases as the vessel diameter becomes smaller The Fåhraeus effect only becomes significant in tubes or vessels with an internal diameter below 300 μm (Goldsmith et al. 1989) and is discussed in more detail in Chapter 6 (Section 6.3). An obvious *in vivo* consequence of the Fåhraeus effect is lower hematocrit values in microcirculatory blood vessels compared to large vessel hematocrit (see Section 7.4.2).

7.4.1.3 Movement of White Blood Cells and Platelets to Marginal Flow Zone

Axial accumulation of RBC displaces other blood cells (i.e., white blood cells, platelets) to the marginal zone (Goldsmith et al. 1999). This margination of white blood cells and platelets has been called the *inverse Fåhraeus effect* (Uijttewaal et al. 1993) and produces results opposite to those for RBC. This effect is enhanced by RBC aggregation that promotes exclusion of cells other than RBC from the central core (Bishop et al. 2001d; Nash et al. 2008).

From a pathophysiological point of view, margination of white blood cells and platelets is essential for their function, especially for white blood cells in the case of inflammatory reactions. Enhanced RBC aggregation during inflammatory responses helps to support this function. However, it should also be noted that increased rolling and adhesion of white blood cells to the vessel wall due to margination would result

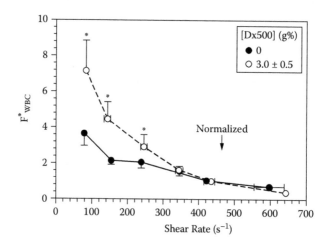

FIGURE 7.23   Rolling white blood cell flux (F_{WBC}) in cells/min normalized with respect to the value determined at a 450 s⁻¹ shear rate, in postcapillary venules of the rat mesentery. RBC aggregation was increased ~200% above control with 3 g/dl dextran 500 kDa. The increments in rolling flux at shear rates equal to or below 150 s⁻¹ were significant (marked by asterisks, *). Modified from Pearson, M. J., and H. H. Lipowsky. 2000. "Influence of Erythrocyte Aggregation on Leukocyte Margination in Postcapillary Venules of Rat Mesentery." *American Journal of Physiology—Heart and Circulatory Physiology* 279:H1460–H1471 with permission.

in increased flow resistance (Bishop et al. 2001d; Das et al. 2000). Enhanced RBC aggregation to ~200% above control via dextran 500 kDa has been shown to increase the fraction of rolling and adhering white cells, even under high shear rates (Pearson and Lipowsky 2000). Figure 7.23 demonstrates the influence of enhanced RBC aggregation on rolling white blood cell flux in postcapillary venules of rat mesentery.

7.4.2 Microvascular Hematocrit and Red Blood Cell Aggregation

It is well accepted that hematocrit is the major determinant of the rheological properties of blood and that blood viscosity increases rapidly with hematocrit. However, it is not well appreciated that hematocrit is not constant throughout the circulatory tree and that there is a very significant difference between the hematocrit of blood flowing in large blood vessels (i.e., arteries or veins) and in vessels of the microcirculation (Schmid-Schönbein 1988). Estimates of hematocrit values for blood flowing in microcirculatory blood vessels indicate that microvascular hematocrit values might be as low as only 20% of larger vessel hematocrit (Table 7.2). These low hematocrit values measured in microvessels indicate that the hematocrit measured in venous blood does not represent the average volume fraction of RBC in the blood filling the entire circulatory system.

The mean hematocrit value for *all* sizes of blood vessels in a given tissue can be determined using methods allowing the estimation of RBC and plasma volumes in that. These mean values are known as *tissue hematocrit*, and correspond to 75–90% of large vessel hematocrit (Table 7.3).

TABLE 7.2
Microvascular and Large Vessel Hematocrits Determined Using Various Methods

Species	Organ/Tissue	Microvascular/ Systemic Hematocrit	Method	Reference
Cat	Mesentery	0.08/0.33	Micro-occlusion	(Lipowsky et al. 1978)
Hamster	Cremaster Muscle	0.14/0.49	Optical	(Klitzman and Duling 1979)
Rat	Cremaster Muscle	0.18/0.37	Optical	(House and Lipowsky 1987)
Dog	Gastrocnemius- Plantaris Muscle	0.38/0.50	Isotope washout	(Frisbee and Barclay 1998)

Source: Data reproduced from Baskurt, O. K., O. Yalcin, F. Gungor, and H. J. Meiselman. 2006. "Hemorheological Parameters as Determinants of Myocardial Tissue Hematocrit Values." *Clinical Hemorheology and Microcirculation* 35:45–50 with permission.

TABLE 7.3

Tissue Hematocrit Values Determined Using Various Methods

Species	Organ/Tissue	Tissue/Systemic Hematocrit	Method	Reference
Rat	Heart (subepicardium)	0.32/0.39	Double isotope labeling	(Vicaut and Levy 1990)
Guinea pig	Heart (subepicardium)	0.34/0.38	Double isotope labeling	(Yalcin et al. 2006)
Human	Brain	0.38/0.43	PET	(Okazawa et al. 1996)
Rat	Brain	0.33/0.43	Double isotope labeling	(Todd et al. 1992)

Source: Data reproduced from Baskurt, O. K., O. Yalcin, F. Gungor, and H. J. Meiselman. 2006. "Hemorheological Parameters as Determinants of Myocardial Tissue Hematocrit Values." *Clinical Hemorheology and Microcirculation* 35:45–50 with permission.

7.4.2.1 Mechanisms of Hematocrit Reduction in Circulatory Networks

The two major mechanisms responsible for a reduced microvascular hematocrit are plasma skimming (see Section 7.4.1.1.1) and the Fåhraeus effect (see Section 7.4.1.2.1), both being a consequence of RBC axial migration. Flow partitioning at vascular bifurcations (Pries et al. 1986) and arteriovenous shunting (Klitzman and Duling 1979) should also be considered to be mechanisms contributing to the lower microvascular hematocrit values.

Microvascular and tissue hematocrit values are influenced by hemodynamic conditions including perfusion pressure and blood flow rate to the tissue (Brizel et al. 1993; Klitzman and Johnson 1982). It has been observed that microvascular hematocrit in skeletal muscle blood vessels is altered by muscle contraction (Frisbee and Barclay 1998; Klitzman and Duling 1979). A tissue hematocrit gradient in the left ventricular myocardium has been reported, being highest in the epicardial layer and decreasing toward the endocardium (Vicaut and Levy 1990).

Microvascular and/or tissue hematocrit values are also affected by the flow properties of blood. Reduction of systemic hematocrit increases the difference between microvascular and large vessel hematocrit, while the absolute values of microvascular hematocrit are also reduced (Lipowsky and Firrell 1986; Todd et al. 1992). Impaired RBC deformability has been reported to be associated with alterations in the microvascular hematocrit (Baskurt et al. 1995b; Lipowsky et al. 1993).

7.4.2.2 Effect of Red Blood Cell Aggregation on Microvascular Hematocrit

RBC aggregation promotes axial migration and related mechanisms and therefore influences microvascular hematocrit. Pearson and Lipowsky (2000) demonstrated a very significant decrease in microvascular hematocrit in postcapillary venules of rat mesentery when RBC aggregation was increased by 200% by infusion of dex-

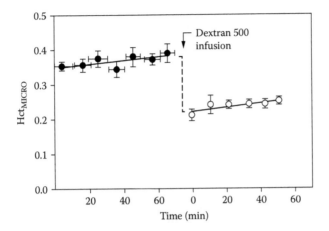

FIGURE 7.24 Microvascular hematocrit (Hct_{MICRO}) in postcapillary venules of rat mesentery before and after infusion of dextran 500 kDa. The arrow marks the time of dextran infusion. (Modified from Pearson, M. J., and H. H. Lipowsky. 2000. "Influence of Erythrocyte Aggregation on Leukocyte Margination in Postcapillary Venules of Rat Mesentery." *American Journal of Physiology—Heart and Circulatory Physiology* 279:H1460–H1471. With permission.)

tran 500 kDa to achieve a 3% final concentration in the perfusing RBC suspension (Figure 7.24).

The influence of RBC aggregation on myocardial tissue hematocrit and transmural hematocrit gradient was evaulated in a series of studies (Baskurt and Edremitlioglu 1995; Yalcin et al. 2006). The results indicated that enhanced RBC aggregation, achieved by either changing the macromolecular composition of the suspending phase or by modifying the surface properties of RBC, affects the myocardial hematocrit gradient, which was demonstrated to exist by Vicaut and Levy (1990). Use of suspensions of RBC coated with Pluronic with higher aggregability abolishes the gradient, while with dextran 500 kDa there is no gradient near the epicardial region, then a precipitous fall occurs toward the endocardial surface (Yalcin et al. 2006); the physiological significance of this gradient and its abolition are not yet clear.

7.4.2.3 Significance of Reduced Microvascular and Tissue Hematocrit

Obviously, reduced hematocrit has two physiological consequences: (1) lower mean hematocrit of blood perfusing microcirculation is an important factor for the lower *in vivo* apparent viscosity of blood. A reduction of hematocrit in microcirculation to 20% of large-vessel hematocrit very significantly influences the flow resistance at the microcirculatory level and underlies the discrepancies between *in vivo* and *in vitro* measurements of blood viscosity, (2) Reduced hematocrit results in lower oxygen carrying capacity. These two consequences of hematocrit alterations can be combined in the so-called *optimal hematocrit* concept. That is, any reduction in oxygen carrying capacity due to hematocrit decreases may be offset by reduced flow resistance and higher rate of blood flow, thereby maintaining adequate oxygen transfer to tissues (Birchard 1997; Fan et al. 1980). Derjardins and Duling postulate that oxygen transfer to tissues is not determined by microvascular hematocrit values, but rather by the discharge hematocrit of microvessels (Desjardins and Duling 1987).

The influence of enhanced RBC aggregation on microvascular and/or tissue hematocrit has important physiological implications. It follows from the previous discussion that the mechanisms causing reduced microvascular hematocrit (i.e., axial migration, phase separation, and plasma skimming) should be operative primarily on the arterial side of the circulatory network. Therefore, the relationship between RBC aggregation and microvascular hematocrit values suggests that RBC aggregates should exist on the arterial side of the microcirculation. This suggestion is at least relevant to the central flow region of arterioles where shear forces are lower compared to the marginal flow zone and RBC are accumulated due to axial migration. Note that the existence of RBC aggregates in the arterial part of the circulation requires that these aggregates must be dispersed (i.e., disaggregated) into individual RBC as they move toward the true capillaries with diameters only one-third of RBC diameter. This disaggregation process has an energy cost that contributes to overall blood flow resistance (Baskurt 2008).

7.4.3 RED BLOOD CELL AGGREGATION AND ENDOTHELIAL FUNCTION

The endothelium is a functionally important part of the blood vessel wall (Brevetti et al. 2008; Michiels 2003). In addition to its classically known permeability barrier function, it plays a critical role in the regulation of vascular smooth muscle tone and hence vascular hindrance, especially at the microcirculatory level (Baskurt and Meiselman 2010). It also contributes to hemostasis, inflammatory response, and angiogenesis and vessel remodeling. NO is an important part of the regulatory function of endothelium, not only because of its well-known role in vasomotor control (Furchott and Zawadski 1980; Ignarro et al. 1999), but also because of its role in maintaining the quiescent status of endothelium (Deanfield et al. 2007).

7.4.3.1 Wall Shear Stress and Endothelial Function

It has been demonstrated that wall shear stress (WSS) has significant effects on endothelial cell morphology and function, including the synthesis and secretion of a variety of endothelial cell–originated biomolecules (e.g., NO) influencing the regulation of vascular functions (Chien 2007; Dimmeler et al. 1999; Muller et al. 2004). The details related to this mechanotransduction are now well understood (Chien 2007; Muller et al. 2004).

WSS is a function of the flow velocity and viscosity of the fluid in touch with the vessel wall (Katritsis et al. 2007; Reneman et al. 2006). General or local reductions in blood flow in the arterial system are expected to reduce WSS, and disturbances of the flow regime (e.g., local changes of flow from laminar to turbulent) also alter effective WSS and endothelial function (Malek et al. 1999). The viscosity of the fluid in contact with the vessel wall is also an important determinant of WSS. As discussed previously in this chapter, the composition and flow properties of blood are not uniform throughout the vascular system (Goldsmith and Mason 1961). Further, the radial distribution of RBC in a blood vessel is not homogenous but rather there is an accumulation in the central flow zone and a cell-poor, plasma-rich layer near the vessel wall (see Section 7.4.1). This marginal, cell-poor layer represents the fluid that is in touch with the vessel wall, and therefore its properties should be considered in estimating WSS.

It has been demonstrated that NO-synthesizing mechanisms are up-regulated by acute or chronic physical exercise (Baskurt and Meiselman 2010; Laughlin et al. 2001; Miyauchi et al. 2003; Varin et al. 1999), indicating the influence of increased shear forces in the vascular network. This relationship has also been demonstrated in endothelial cell cultures and the molecular mechanisms have been elucidated (Fisslthaler et al. 2000; Fleming and Busse 2003).

Plasma viscosity has been accepted as a factor affecting the regulation of capillary perfusion (Tsai et al. 1998; Tsai et al. 2005). It has been shown that perivascular NO concentrations in the microcirculation are affected by plasma viscosity alterations by infusion of high-molecular-mass dextran solutions. Tsai et al. (2005) used specially designed microelectrodes to sense NO concentrations around arterioles and venules in a hamster dorsal skin chamber model; the electrodes were placed as close as possible to the walls of the blood vessel under investigation. They observed significantly higher perivascular NO concentrations in the animals with extreme hemodilution (i.e., hematocrit reduced to 0.11 l/l) if their plasma viscosity was above control (Figure 7.25). This effect of increased plasma viscosity may play a significant

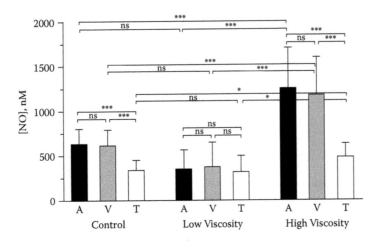

FIGURE 7.25 Perivascular nitric oxide (NO) concentrations in hamster dorsal skin chamber model, measured by microelectrodes with a tip diameter of 5 µm during extreme hemodilution. The properties of infused solutions during hemodilution resulted in plasma viscosity levels that were normal, low, or high. NO concentration near arterioles (A) and venules (V) were found to be very significantly increased (3 asterisks [***]: $p < 0.001$) if plasma viscosity was high (1.64 times control). Tissue (T) NO concentrations were also higher (one asterisk [*]: $p < 0.05$). Reproduced from Tsai, A. G., C. Acero, P. R. Nance, P. Cabrales, J. A. Frangos, D. G. Buerk, and M. Intaglietta. 2005. "Elevated Plasma Viscosity in Extreme Hemodilution Increases Perivascular Nitric Oxide Concentration and Microvascular Perfusion." *American Journal of Physiology—Heart and Circulatory Physiology* 288:H1730–H1739 with permission.

role in maintaining adequate tissue oxygenation during extreme hemodilution by regulating vascular geometry and functional capillary density (Tsai et al. 2005) and thus be beneficial for tissue perfusion (Salazar Vazquez et al. 2009). These findings also support the role of fluid viscosity in contact with the vessel wall as contributing to the WSS for maintaining normal endothelial function. The role of plasma viscosity would be expected to be especially important in blood vessels with enhanced axial migration of RBC.

7.4.3.2   Effect of Red Blood Cell Aggregation on Endothelial Function

It follows from the previous discussion that the formation of a cell-poor layer adjacent to vessel walls tends to reduce the shear forces acting on the endothelium since this layer has reduced viscosity. More effective axial migration and phase separation due to enhanced RBC aggregation would also be expected to result in an additional reduction in WSS. Reduced WSS due to enhanced RBC aggregation has been demonstrated indirectly by investigating the NO-synthesizing mechanisms in endothelial cells in a "chronically-enhanced RBC aggregation model" (Baskurt et al. 2004b). Poloxamer-coated RBC suspensions with enhanced aggregation were exchange transfused into rats, thereby enhancing RBC aggregation in the circulating blood during a 4-day period following exchange transfusions. NO-dependent vascular mechanisms were found to be significantly blunted in skeletal muscle arterioles of these rats on the fourth day. Both flow-mediated dilation (Figure 7.26) and acetylcholine-induced dilation responses (data not shown) were suppressed in the arteriolar segments. Furthermore, the expression of endothelial NO synthase (eNOS) in tissue samples was significantly down-regulated. This experiment was interpreted as reflecting the reduced wall shear stress due to enhanced aggregation for the several days before the isolation of blood vessels for analysis. Note that RBC aggregation in this experiment was increased by a surface coating of RBC with poloxamer F98

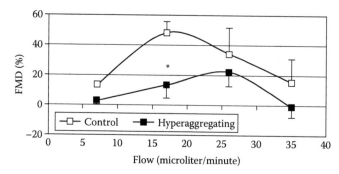

FIGURE 7.26   Flow-mediated dilation (FMD) in arterioles isolated from gracilis muscles of rats exchange transfused with normal and hyperaggregating RBC suspensions. The FMD curve shifted to the right in arteriolar segments of hyperaggregating animals thus indicating lower maximal dilation and higher flow rate is required to induce maximal dilation. Redrawn from Baskurt, O. K., O. Yalcin, S. Ozdem, J. K. Armstrong, and H. J. Meiselman. 2004b. "Modulation of Endothelial Nitric Oxide Synthase Expression by Red Blood Cell Aggregation." *American Journal of Physiology—Heart and Circulatory Physiology* 286:H222–H229.

and that these RBC were suspended in unmodified rat plasma: plasma viscosity was unaltered while RBC aggregation was enhanced.

The impact of increased suspending phase viscosity together with enhanced RBC aggregation on endothelial cells was investigated by Yalcin et al. (2008). Briefly, glass tubes of 1 mm diameter and 75 mm length were coated with human umbilical vein endothelial cells (HUVEC) on the inner surface and perfused with RBC suspensions having normal and enhanced aggregation due to either poloxamer surface coating (i.e., no change in suspending phase viscosity) or including dextran 500 kDa in the suspending medium (i.e., increased suspending phase viscosity and increased aggregation). As anticipated, the results indicated that the activated form of eNOS, phosphorylated at the serine 1177 position (Fleming and Busse 2003), was enhanced when the tubes were perfused with normally aggregating human blood. However, this enhancement was prevented if poloxamer-coated RBC suspensions were used at the same nominal wall shear stress; this effect of RBC aggregation was not seen with RBC suspensions in which RBC aggregation and suspending phase viscosity were increased by dextran 500 kDa. This *in vitro* experiment supports the previously mentioned *in vivo* findings and further indicates the important role of suspending phase viscosity.

It should be noted that, during pathophysiological processes, *in vivo* RBC aggregation might be increased by mechanisms resulting in blood rheology changes similar to those tested by the two modes of the *in vitro* experiment: (1) altered plasma composition and viscosity as in acute phase reactions, infections, and malignant diseases, and (2) altered RBC properties as in diabetes, ischemia–reperfusion injury, oxidant stress, and so on. It is thus likely that effects on endothelial function by these two modes of enhancing RBC aggregation may differ significantly from each other.

7.4.4 CONTRASTING EFFECTS OF AGGREGATION IN CIRCULATORY NETWORKS

It can be deduced from the discussion in this section (Section 7.4) that RBC aggregation can be related to various hemodynamic-vascular mechanisms with contrasting outcomes in terms of the influence on *in vivo* blood flow resistance. A summary of physiological factors with the potential to be influenced by RBC aggregation is provided below. It should be noted that the word *potential* does not necessarily mean an uncertainty about the relationship discussed, but rather that the impact of a given alteration in RBC aggregation may depend on the magnitude and the mode of alteration as well as the properties of the flow system (e.g., microcirculation) under investigation.

A. *Macroscopic viscosity:* RBC aggregation is the major determinant of low-shear-rate viscosity during flow in large geometry systems and large blood vessels, during bulk flow. *Increased RBC aggregation leads to increased blood viscosity and hence predicts increased flow resistance.*

B. *Disaggregation at microcirculatory entrance:* RBC aggregates within arterial blood vessels must disaggregate as they approach the microcirculation and enter capillary vessels. *Increased RBC aggregation leads to increased disaggregation energy and hence predicts increased flow resistance.*

C. *Frictional resistance in the marginal flow zone:* Axial migration of RBC promoted by aggregation leads to lower hematocrit, and hence a lower viscosity fluid layer in the marginal flow zone. *Increased RBC aggregation leads to decreased frictional resistance and hence predicts decreased flow resistance.*

D. *Microvascular hematocrit:* Another consequence of axial migration of RBC and phase separation is lower microvascular hematocrit, mainly due to plasma skimming and the Fåhraeus effect. *Increased RBC aggregation leads to decreased microvascular hematocrit and hence predicts decreased flow resistance.*

E. *Vasomotor control:* Formation of a cell-depleted flow zone near blood vessel walls results in lower wall shear stress exerted on endothelium, which in turn influences endothelial function, including NO-synthesizing mechanisms. *Increased RBC aggregation leads to down-regulation of NO synthesis in endothelium and increased vasomotor tone and hence predicts increased flow resistance.*

The net *in vivo* result of an alteration in RBC aggregation is determined by the relative contribution of each effect summarized above. The multifactorial nature of the interaction of aggregation with hemodynamic mechanisms is most likely the reason for the diverse *in vivo* findings of RBC aggregation (Baskurt 2008; Baskurt and Meiselman 2007; Zhang et al. 2009) (discussed in Section 7.3).

7.5   RED BLOOD CELL AGGREGATION: GOOD OR BAD FOR TISSUE PERFUSION?

The overall role of RBC aggregation in determining *in vivo* blood flow and tissue perfusion has been debated for decades. Knisely and his followers defended the opinion that RBC aggregation is a pathological phenomenon and is responsible for *sludged flow* in the microcirculation observed during disease processes (Knisely 1965; Knisely 1947). RBC aggregation and blood sludge flow were accepted as the major reason for increased flow resistance and even circulatory failure at the tissue level (Goldsmith et al. 1989). These ideas were formulated based on the observations of RBC aggregates in microcirculatory blood vessels in pathological conditions. It has been repeatedly demonstrated that flow resistance and several other parameters relevant to tissue perfusion (e.g., functional capillary density, microvascular hematocrit) are significantly influenced by RBC aggregation. Combining these findings with the fact that RBC aggregation is enhanced during many pathophysiological processes (see Chapter 8) may help to understand Knisely's point of view. Alternatively, concepts based on the original ideas of Robin Fåhraeus, discussed in Chapter 6 and Section 7.4 of this chapter, predict the opposite: RBC aggregation yields decreased flow resistance, at least under certain circumstances (Goldsmith et al. 1989). An important reason for this disagreement seems to be that the methodology used by different investigators leads to different observations and findings. Knisely based his opinions on intravital microscopy studies, yet his results may be unintentionally biased due to the orientation of the blood vessels being studied (see Section 7.3.3.1.3 of this chapter). The studies by Fåhraeus in glass tubes may also have been affected by orientation with respect to gravity (Cokelet and Goldsmith 1991).

It is now accepted that RBC aggregation is a physiological phenomenon that is an important factor in the maintenance of "normal" hemodynamics, at least in microcirculatory blood vessels. This influence includes venous flow dynamics and the establishment of microcirculatory hematocrit values, both of which are important determinants of capillary flow dynamics. More specifically, RBC aggregation may help minimize changes in capillary blood pressure during exercise and increased tissue blood flow: (1) capillary blood pressure is determined by the ratio of pre- to post–resistance (i.e., arteriolar and venous resistance); (2) increased tissue metabolism results in increased blood flow due to arteriolar vasodilation (i.e., lower preresistance), a decrease of the ratio, and would be expected to yield increased capillary blood pressure; (3) the viscosity of normally aggregating blood decreases with increased flow and shear rate, and hence venous resistance falls; and (4) the ratio is restored toward control. Thus capillary blood pressure becomes less affected by blood flow due to adjusting the venous resistance according to blood flow (Bishop et al. 2001d; Cabel et al. 1997).

Supporting this concept of RBC aggregation being an important homeostatic factor in maintaining normal microcirculatory flow dynamics, species with higher athletic capacity (e.g., horse, dog) are characterized by higher RBC aggregation compared to those with lower athletic capacity (e.g., cow, sheep) (Popel et al. 1994). Athletic capacity, as judged by maximal oxygen consumption, clearly indicates more efficient mechanisms for enhancing tissue metabolism and hence blood flow when needed. Of course, other factors should also be considered when comparing the effects of RBC aggregation among species, including microvascular flow dynamics and blood vessel sizes reflecting total body size (see Chapter 9).

It should also be noted that the characteristics of RBC aggregation may differ as a consequence of the pathophysiology or the experimental approach to altering aggregation. These characteristics include the extent and time course of aggregation together with aggregate strength (i.e., the force necessary for dispersion of aggregates). Different alterations of these parameters may, in turn, have different influences on flow dynamics. For example, increased aggregate strength is expected to interfere with disaggregation in arterioles as blood approaches the capillaries (see Section 7.4.2.3), tending to increase flow resistance. An altered time course of aggregation would mainly affect aggregate formation in postcapillary venules, thus influencing venous vascular resistance. Furthermore, the magnitude of RBC aggregation should also be considered as a factor determining whether tissue perfusion is promoted or impaired (Mchedlishvili 1998; Yalcin et al. 2004). Therefore, alterations of RBC aggregation properties due to pathological conditions may result in impaired tissue perfusion, while aggregation within physiological limits may promote tissue perfusion. The impact of various pathophysiological processes on RBC aggregation is discussed in more detail in Chapter 8.

Recent findings related to the influence of RBC aggregate formation on endothelial mechanisms indicate the need to consider this aspect of aggregation; the reduced WSS due to enhanced RBC aggregation is an important factor affecting endothelial cell function (see Section 7.4.3). Enhanced RBC aggregation leads to increased vascular hindrance (i.e., decreased vessel radius due to increased vascular smooth muscle tone) as a result of down-regulation of NO-synthesizing mechanisms. The resulting increased peripheral vascular resistance has been shown to lead to increased arterial pressure (Baskurt et al. 2004b). However, it is also important to note that NO generation by the endothelium is a key factor maintaining the quiescent, nonactivated status of endothelial cells. Disturbance of these mechanisms is known to enhance degenerative changes in the vessel wall and hence to promote atherosclerosis (Lloyd-Jones and Bloch 1996; Nerem et al. 1998). Therefore, chronically enhanced RBC aggregation may be related to the development of atherosclerotic lesions.

Finally, by promoting leukocyte margination, RBC aggregation and the resulting enhanced axial accumulation of RBC may play a role in pathophysiological mechanisms related to inflammation and leukocyte recruitment. However, margination may also result in increased flow resistance in microvessels due to rolling and adhesion of leukocytes to endothelial cells (Lipowsky 2006). In overview, RBC aggregation is a physiological process that plays significant roles in the maintenance of normal *in vivo* hemodynamic conditions, with the specific effects on tissue perfusion dependent on the magnitude and mode of alteration of aggregation.

LITERATURE CITED

Alonso, C., A. R. Pries, and P. Gaehtgens. 1989. "Time-Dependent Rheological Behavior of Blood Flow at Low Shear in Narrow Horizontal Tubes." *Biorheology* 26:229–246.

Alonso, C., A. R. Pries, and P. Gaehtgens. 1993. "Time-Dependent Rheological Behavior of Blood at Low Shear in Narrow Vertical Tubes." *American Journal of Physiology—Heart and Circulatory Physiology* 265:H553–H561.

Alonso, C., A. R. Pries, K. D. Lerche, and P. Gaehtgens. 1995. "Transient Rheological Behavior of Blood in Low-Shear Tube Flow: Velocity Profiles and Effective Viscosity." *American Journal of Physiology—Heart and Circulatory Physiology* 268:H25–H32.

Armstrong, J. K., H. J. Meiselman, R. B. Wenby, and T. C. Fisher. 2005. "Modulation of Red Blood Cell Aggregation and Blood Viscosity by the Covalent Attachment of Pluronic Copolymers." *Biorheology* 38:239–247.

Baskurt, O. K. 2008. "*In Vivo* Correlates of Altered Blood Rheology." *Biorheology* 45:629–638.

Baskurt, O. K., M. Bor-Kucukatay, and O. Yalcin. 1999. "The Effect of Red Blood Cell Aggregation on Blood Flow Resistance." *Biorheology* 36:447–452.

Baskurt, O. K., and M. Edremitlioglu. 1995. "Myocardial Tissue Hematocrit: Existence of a Transmural Gradient and Alterations after Fibrinogen Infusions." *Clinical Hemorheology* 15:97–105.

Baskurt, O. K., M. Edremitlioglu, and A. Temiz. 1995b. "Effect of Erythrocyte Deformability on Myocardial Hematocrit Gradient." *American Journal of Physiology—Heart and Circulatory Physiology* 268:H260–H264.

Baskurt, O. K., R. A. Farley, and H. J. Meiselman. 1997. "Erythrocyte Aggregation Tendency and Cellular Properties in Horse, Human, and Rat: A Comparative Study." *American Journal of Physiology—Heart and Circulatory Physiology* 273:H2604–H2612.

Baskurt, O. K., E. Levi, S. Caglayan, N. Dikmenoglu, O. Ucer, R. Guner, and S. Yorukan. 1991. "The Role of Hemorheological Factors in the Coronary Circulation." *Clinical Hemorheology* 11:121–127.

Baskurt, O. K., and H. J. Meiselman. 2007. "Hemodynamic Effects of Red Blood Cell Aggregation." *Indian Journal of Experimental Biology* 45:25–31.

Baskurt, O. K., and H. J. Meiselman. 2010. "Endothelial Function and Physical Activity." In *Exercise Physiology: From a Cellular to an Integrative Approach*, ed. P. Connes, O. Hue, and S. Perrey, 230–244. Amsterdam, Berlin, Oxford, Tokyo, Washington, DC: IOS Press.

Baskurt, O. K., O. Yalcin, and H. J. Meiselman. 2004a. "Hemorheology and Vascular Control Mechanisms." *Clinical Hemorheology and Microcirculation* 30:169–178.

Baskurt, O. K., O. Yalcin, S. Ozdem, J. K. Armstrong, and H. J. Meiselman. 2004b. "Modulation of Endothelial Nitric Oxide Synthase Expression by Red Blood Cell Aggregation." *American Journal of Physiology—Heart and Circulatory Physiology* 286:H222–H229.

Baskurt, O. K., O. Yalcin, F. Gungor, and H. J. Meiselman. 2006. "Hemorheological Parameters as Determinants of Myocardial Tissue Hematocrit Values." *Clinical Hemorheology and Microcirculation* 35:45–50.

Benis, A. M., S. Usami, and S. Chien. 1970. "Effect of Hematocrit and Inertial Losses on Pressure-Flow Relations in the Isolated Hindpaw of the Dog." *Circulation Research* 27:1047–1068.

Berne, R. M., and M. N. Levy. 2001. *Cardiovascular Physiology*. St. Louis, London, Philadelphia, Sydney, Toronto: Mosby Inc.

Birchard, G. F. 1997. "Optimal Hematocrit: Theory, Regulation and Implications." *American Zoologist* 37:65–72.

Bishop, J. J., P. R. Nance, A. S. Popel, M. Intaglietta, and P. C. Johnson. 2001a. "Effect of Erythrocyte Aggregation on Velocity Profiles in Venules." *American Journal of Physiology—Heart and Circulatory Physiology* 280:H222–H236.

Bishop, J. J., P. R. Nance, A. S. Popel, M. Intaglietta, and P. C. Johnson. 2001b. "Erythrocyte Margination and Sedimentation in Skeletal Muscle Venules." *American Journal of Physiology—Heart and Circulatory Physiology* 281:H951–H958.

Bishop, J. J., A. S. Popel, M. Intaglietta, and P. C. Johnson. 2001c. "Effects of Erythrocyte Aggregation and Venous Network Geometry on Red Blood Cell Axial Migration." *American Journal of Physiology—Heart and Circulatory Physiology* 281:H939–H950.

Bishop, J. J., A. S. Popel, M. Intaglietta, and P. C. Johnson. 2001d. "Rheological Effects of Red Blood Cell Aggregation in the Venous Network: A Review of Recent Studies." *Biorheology* 38:263–274.

Bishop, J. J., A. S. Popel, M. Intaglietta, and P. C. Johnson. 2002. "Effect of Aggregation and Shear Rate on the Dispersion of Red Blood Cells Flowing in Venules." *American Journal of Physiology—Heart and Circulatory Physiology* 283:H1985–H1996.

Bishop, J. J., P. R. Nance, A. S. Popel, M. Intaglietta, and P. C. Johnson. 2004. "Relationship between Erythrocyte Aggregate Size and Flow Rate in Skeletal Muscle Venules." *American Journal of Physiology—Heart and Circulatory Physiology* 286:H113–H120.

Bor-Kucukatay, M., R. B. Wenby, H. J. Meiselman, and O. K. Baskurt. 2003. "Effects of Nitric Oxide on Red Blood Cell Deformability." *American Journal of Physiology—Heart and Circulatory Physiology* 284:H1577–H1584.

Brevetti, G., V. Schiano, and M. Chiariello. 2008. "Endothelial Dysfunction: A Key to the Pathophysiology and Natural History of Peripheral Arterial Disease?" *Atherosclerosis* 197:1–11.

Brizel, D. M., B. Klitzman, J. M. Cook, J. Edwards, G. Rosner, and M. W. Dewhirst. 1993. "A Comparison of Tumor and Normal Tissue Microvascular Hematocrits and Red Cell Fluxes in a Rat Window Chamber Model." *International Journal of Radiation Oncology Biology and Physiology* 25:269–276.

Cabel, M., H. J. Meiselman, A. S. Popel, and P. C. Johnson. 1997. "Contribution of Red Blood Cell Aggregation to Venous Vascular Resistance in Skeletal Muscle." *American Journal of Physiology—Heart and Circulatory Physiology* 272:H1020–H1032.

Charansonney, O., S. Mouren, J. Dufaux, M. Duvelleroy, and E. Vicaut. 1993. "Red Blood Cell Aggregation and Blood Viscosity in an Isolated Heart Preparation." *Biorheology* 30:75–84.

Chen, R. Y., R. D. Carlin, S. Simchon, K. M. Jan, and S. Chien. 1989. "Effects of Dextran-Induced Hyperviscosity on Regional Blood Flow and Hemodynamics in Dogs." *American Journal of Physiology—Heart and Circulatory Physiology* 256:H898–H905.

Chien, S. 2007. "Mechanotransduction and Endothelial Cell Homeostasis: The Wisdom of the Cell." *American Journal of Physiology—Heart and Circulatory Physiology* 292:H1209–H1224.

Cokelet, G. R., and H. L. Goldsmith. 1991. "Decreased Hydrodynamic Resistance in the Two-Phase Flow of Blood through Small Vertical Tubes at Low Flow Rates." *Circulation Research* 68:1–17.

Das, B., P. C. Johnson, and A. S. Popel. 2000. "Computational Fluid Dynamic Studies of Leukocyte Adhesion Effects on Non-Newtonian Blood Flow through Microvessels." *Biorheology* 37:239–258.

Deanfield, J. E., J. P. Halcox, and T. J. Rabelink. 2007. "Endothelial Function and Dysfunction: Testing and Clinical Relevance." *Circulation* 115:1285–1295.

Desjardins, C., and B. R. Duling. 1987. "Microvessel Hematocrit: Measurement and Implications for Capillary Oxygen Transport." *American Journal of Physiology—Heart and Circulatory Physiology* 252:H494–H503.

Dimmeler, S., I. Fleming, B. Fisslthaler, C. Hermann, R. Busse, and A. M. Zeiher. 1999. "Activation of Nitric Oxide Synthase in Endothelial Cells by Akt-Dependent Phosphorylation." *Nature* 399, no. 6736:601.

Djojosugito, A. M., B. Folkow, B. Oberg, and S. White. 1970. "A Comparison of Blood Viscosity Measured *in Vitro* and in a Vascular Bed." *Acta Physiologica Scandinavica* 78:70–84.

Duling, B. R., R. D. Hogan, B. L. Langille, P. Lelkes, S. S. Segal, S. Vatner, H. Weigelt, and M. A. Young. 1987. "Vasomotor Control: Functional Hyperemia and Beyond." *Federation Proceedings* 46:251–263.

Durussel, J. J., M. F. Berthault, G. Guiffant, and J. Dufaux. 1998. "Effects of Red Blood Cell Hyperaggregation on the Rat Microcirculation Blood Flow." *Acta Physiologica Scandinavica* 163, no. 1:25–32.

Eliassen, E., B. Folkow, and B. Oberg. 1973. "Are There Significant Inertial Losses in the Vascular Bed?" *Acta Physiologica Scandinavica* 87:567–569.

Fan, F., R. Y. Z. Chen, G. B. Schuessler, and S. Chien. 1980. "Effects of Hematocrit Variations on Regional Hemodynamics and Oxygen Transport in the Dog." *American Journal of Physiology—Heart and Circulatory Physiology* 238:H545–H552.

Feigl, E. O. 1983. "Coronary Physiology." *Physiological Reviews* 63:1–205.

Fisslthaler, B., S. Dimmeler, C. Hermann, R. Busse, and I. Fleming. 2000. "Phosphorylation and Activation of the Endothelial Nitric Oxide Synthase by Fluid Shear Stress." *Acta Physiologica Scandinavica* 168, no. 1:81–88.

Fleming, I., and R. Busse. 2003. "Molecular Mechanisms Involved in the Regulation of the Endothelial Nitric Oxide Synthase." *American Journal of Physiology-Regulatory, Integrative and Comparative Physiology* 284, no. 1:R1–12.

Frisbee, J. C., and J. K. Barclay. 1998. "Microvascular Hematocrit and Permeability: Surface Area Product in Contracting Canine Skeletal Muscle in Situ." *Microvacscular Research* 55:153–164.

Fukumura, D. 2006. "Window Models." In *Microvascular Research*, ed. D. Shepro, 151–163. Burlington, San Diego, London: Elsevier Academic Press.

Furchott, R. F., and J. V. Zawadzki. 1980. "The Obligatory Role of Endothelial Cells in the Relaxation of Arterial Smooth Muscle by Acetylcholine." *Nature* 288:373–376.

Gallagher, K. P., G. Osakada, M. Matsuzaki, W. S. Kemper, and J. Ross. 1982. "Myocardial Blood Flow and Function with Critical Coronary Stenosis in Exercising Dogs." *American Journal of Physiology—Heart and Circulatory Physiology* 243:H698–H707.

Goldsmith, H. L., D. N. Bell, S. Spain, and F. A. McIntosh. 1999. "Effect of Red Blood Cells and Their Aggregates on Platelets and White Cells in Flowing Blood." *Biorheology* 36:461–468.

Goldsmith, H. L., G. R. Cokelet, and P. Gaehtgens. 1989. "Robin Fåhraeus: Evolution of His Concepts in Cardiovascular Physiology." *American Journal of Physiology—Heart and Circulatory Physiology* 257:H1005–H1015.

Goldsmith, H. L., and S. G. Mason. 1961. "Axial Migration of Particles in Poiseuille Flow." *Nature* 190:1095–1096.

Greve, J. M., A. S. Les, B. T. Tang, M. T. D. Blomme, N. M. Wilson, R. L. Dalman, N. J. Pelc, and C. A. Taylor. 2006. "Allometric Scaling of Wall Shear Stress from Mice to Humans: Quantification Using Cine Phase-Contrast MRI and Computational Fluid Dynamics." *American Journal of Physiology—Heart and Circulatory Physiology* 291:H1700–H1708.

Gustafsson, L., L. Appelgren, and H. E. Myrvold. 1981. "Effects of Increased Plasma Viscosity and Red Blood Cell Aggregation on Blood Viscosity *in Vivo*." *American Journal of Physiology—Heart and Circulatory Physiology* 241:H513–H518.

Hester, R. 2006. "The Cremaster Muscle for Microcirculatory Studies." In *Microvascular Research*, ed. D. Shepro, 159–163. Burlington, San Diego, London: Elsevier Academic Press.

House, S. D., and P. C. Johnson. 1986. "Microvascular Pressure in Venules of Skeletal Muscle during Arterial Pressure Reduction." *American Journal of Physiology—Heart and Circulatory Physiology* 250:H838–H845.

House, S. D., and H. H. Lipowsky. 1987. "Microvascular Hematocrit and Red Cell Flux in Rat Cremaster Muscle." *American Journal of Physiology—Heart and Circulatory Physiology* 252:H211–H222.

Ignarro, L. J., G. Cirino, A. Casini, and C. Napoli. 1999. "Nitric Oxide as a Signaling Molecule in the Vascular System: An Overview." *Journal of Cardiovascular Pharmacology* 34:879–886.

Imig, J. D., D. Gebremedhin, D. R. Harder, and R. J. Roman. 1993. "Modulation of Vascular Tone in Renal Microcirculation by Erythrocytes: Role of EDRF." *American Journal of Physiology—Heart and Circulatory Physiology* 264:H190–H195.

Katritsis, D., L. Kaiktsis, A. Chaniotis, J. Pantos, P. Efstathopoulos, and V. Marmarelis. 2007. "Wall Shear Stress: Theoretical Considerations and Methods of Measurements." *Progress in Cardiovascular Diseases* 49:307–329.

Kim, S., R. L. Kong, A. S. Popel, M. Intaglietta, and P.C. Johnson. 2007. "Temporal and Spatial Variations of Cell-Free Layer Width in Arterioles." *American Journal of Physiology—Heart and Circulatory Physiology* 293:H1526–H1535.

Kim, S., R. L. Kong, A. S. Popel, M. Intaglietta, and P. C. Johnson. 2006. A computer-based method for determination of the cell-free layer width in microcirculation. *Microcirculation* 13:199–207.

Kim, S., P. K. Ong, O. Yalcin, M. Intaglietta, and P. C. Johnson. 2009. "The Cell-Free Layer in Microvascular Blood Flow." *Biorheology* 46:181–189.

Kim, S., A. S. Popel, M. Intaglietta, and P. C. Johnson. 2005. "Aggregate Formation of Erythrocytes in Postcapillary Venules." *American Journal of Physiology—Heart and Circulatory Physiology* 288:H584–H590.

Kim, S., A. S. Popel, M. Intaglietta, and P. C. Johnson. 2006. "Effect of Erythrocyte Aggregation at Normal Human Levels on Functional Capillary Density in Rat Spinotrapezius Muscle." *American Journal of Physiology—Heart and Circulatory Physiology* 290:H941–H947.

Kim, S., J. Zhen, A. S. Popel, M. Intaglietta, and P. C. Johnson. 2007. "Contributions of Collision Rate and Collision Efficiency to Erythrocyte Aggregation in Postcapillary Venules at Low Flow Rates." *American Journal of Physiology—Heart and Circulatory Physiology* 293:H1947–H1954.

Kleinbongard, P., R. Schutz, T. Rassaf, T. Lauer, A. Dejam, T. Jax, I. Kumara, et al. 2006. "Red Blood Cells Express a Functional Endothelial Nitric Oxide Synthase." *Blood* 107:2943–2951.

Klitzman, B., and B. R. Duling. 1979. "Microvascular Hematocrit and Red Cell Flow in Resting and Contracting Striated Muscle." *American Journal of Physiology—Heart and Circulatory Physiology* 237:H481–H490.

Klitzman, B., and P. C. Johnson. 1982. "Capillary Network Geometry and Red Cell Distribution in Hamster Cremaster Muscle." *American Journal of Physiology—Heart and Circulatory Physiology* 242:H211–H219.

Knisely, M. H. 1947. "Sludged Blood." *Science* 106:431–440.

Knisely, M. H. 1965. "Intravascular Erythrocyte Aggregation (Blood Sludge)." In *Handbook of Physiology*, 2249–2292. Bethesda, MD: American Physiological Society.

Laughlin, M. H., J. S. Pollock, J. F. Amann, M. L. Hollis, C. R. Woodman, and E. M. Price. 2001. "Training Induces Nonuniform Increases in eNOS Content along the Coronary Arterial Tree." *Journal of Applied Physiology* 90:501–510.

Lawall, H., and B. Angelkort. 1999. "Correlation between Rheological Parameters and Erythrocyte Velocity in Nailfold Capillaries in Patients with Diabetes Mellitus." *Clinical Hemorheology and Microcirculation* 20:41–47.

Liao, J. C., T. W. Hein, M. W. Vaughn, K. T. Huang, and L. Kuo. 1999. "Intravascular Flow Decreases Erythrocyte Consumption of Nitric Oxide." *Proceedings of the National Academy of Sciences of the United States of America* 96:8757–8761.

Lipowsky, H. H. 1995. Shear Stress in the Circulation. In *Flow Dependent Regulation of Vascular Function*. J. Bevan and G. Kaley, eds. New York: Oxford University Press. 28-45.

Lipowsky, H. H. 2006. "Rheology of Blood Flow in the Microcirculation." In *Microvascular Research*, ed. D. Shepro, 233–238. Amsterdam: Elsevier.

Lipowsky, H. H., L. E. Cram, W. Justice, and M. J. Eppihimer. 1993. "Effect of Erythrocyte Deformability on *in Vivo* Red Cell Transit Time and Hematocrit and Their Correlation with *in Vitro* Filterability." *Microvascular Research* 46, no. 1:43–64.

Lipowsky, H. H., and J. C. Firrell. 1986. "Microvascular Hemodynamics during Systemic Hemodilution and Hemoconcentration." *American Journal of Physiology—Heart and Circulatory Physiology* 250:H908–H922.

Lipowsky, H. H., S. Kovalcheck, and B. W. Zweifach. 1978. "The Distribution of Blood Rheological Parameters in the Microvasculature of Cat Mesentery." *Circulation Research* 43:738–749.

Lloyd-Jones, D. M., and K. D. Bloch. 1996. "The Vascular Biology of Nitric Oxide and Its Role in Atherogenesis." *Annual Review of Medicine* 47:365–375.

Maeda, N., Y. Suzuki, J. Tanaka, and N. Tateishi. 1996. "Erythrocyte Flow and Elasticity of Microvessels Evaluated by Marginal Cell-Free Layer and Flow Resistance." *American Journal of Physiology—Heart and Circulatory Physiology* 271:H2454–H2461.

Malek, A., S. L. Alper, and S. Izumo. 1999. "Hemodynamic Shear Stress and Its Role in Atherosclerosis." *JAMA* 282:2035–2042.

Maricq, H. R. 2006. "The Significance of the Nail Fold in Clinical and Experimental Studies." In *Microvascular Research*, ed. D. Shepro, 173–180. Burlington, San Diego, London: Elsevier Academic Press.

McDonald, D.A. 1974. *Blood Flow in Arteries*. Baltimore: The Williams and Wilkins Co. 356-378.

Mchedlishvili, G. 1998. "Disturbed Blood Flow Structuring as Critical Factor of Hemorheological Disorders in Microcirculation." *Clinical Hemorheology and Microcirculation* 19:315–325.

Mchedlishvili, G., L. Gobejishvili, and N. Beritashvili. 1993. "Effect of Intensified Red Blood Cell Aggregability on Arterial Pressure and Mesenteric Microcirculation." *Microvascular Research* 45:233–242.

Michiels, C. 2003. "Endothelial Cell Functions." *Journal of Cellular Physiology* 196:430–443.

Miyauchi, T., S. Maeda, M. Iemitsu, T. Koboyashi, Y. Kumagai, I. Yamaguchi, and M. Matsuda. 2003. "Exercise Causes a Tissue-Specific Change of NO Production in the Kidney and Lung." *Journal of Applied Physiology* 94:60–68.

Muller, S., V. Labrador, N. Da Isla, D. Dumas, R. Sun, X. Wang, L. Wei, et al. 2004. "From Hemorheology to Vascular Mechanobiology: An Overview." *Clinical Hemorheology and Microcirculation* 30:185–200.

Nash, G. B., T. Watts, C. Thornton, and M. Barigou. 2008. "Red Cell Aggregation as a Factor Influencing Margination and Adhesion of Leukocytes and Platelets." *Clinical Hemorheology and Microcirculation* 39:303–310.

Nerem, R. M., R. W. Alexander, D. C. Chappell, R. M. Medford, S. E. Vagner, and W. R. Taylor. 1998. "The Study of the Influence of Flow on Vascular Endothelial Biology." *American Journal of Medical Science* 316:169–175.

Okazawa, H., Y. Yonekure, Y. Fujibayashi, H. Yamauchi, K. Ishizu, S. Nishizawa, Y. Magata, et al. 1996. "Measurement of Regional Cerebral Plasma Pool and Hematocrit with Copper-62-Labeled HSA-DTS." *Journal of Nuclear Medicine* 37:1080–1085.

Ozuyaman, B., M. Grau, M. Kelm, M. W. Merx, and P. Kleinbongard. 2008. "RBC NOS: Regulatory Mechanisms and Therapeutic Aspects." In *Trends in Molecular Medicine* 14:314–322.

Pearson, M. J., and H. H. Lipowsky. 2000. "Influence of Erythrocyte Aggregation on Leukocyte Margination in Postcapillary Venules of Rat Mesentery." *American Journal of Physiology—Heart and Circulatory Physiology* 279:H1460–H1471.

Popel, A. S., P. C. Johnson, M. V. Kameneva, and M. A. Wild. 1994. "Capacity for Red Blood Cell Aggregation is Higher in Athletic Mammalian Species Than in Sedentary Species." *Journal of Applied Physiology* 77:1790–1794.

Pries, A. R., K. Ley, and P. Gaehtgens. 1986. "Generalization of the Fåhraeus Principle for Microvessel Networks." *American Journal of Physiology—Heart and Circulatory Physiology* 251:H1324–H1332.

Pries, A. R., K. Ley, M. Claaben, and P. Gaehtgens. 1989. "Red Cell Distribution at Microvascular Bifurcations." *Microvascular Research* 38:81–101.

Pries, A. R., T. W. Secomb, T. Gebner, M. B. Sperandio, J. F. Gross, and P. Gaehtgens. 1994. "Resistance to Blood Flow in Microvessels *in Vivo*." *Circulation Research* 75:904–915.

Pries, A. R., T. W. Secomb, and P. Gaehtgens. 1996. "Biophysical Aspects of Blood Flow in The Microvasculature." *Cardiovascular Research* 32:654–667.

Reinke, W., P. Gaehtgens, and P. C. Johnson. 1987. "Blood Viscosity in Small Tubes: Effect of Shear Rate, Aggregation, and Sedimentation." *American Journal of Physiology—Heart and Circulatory Physiology* 253:H540–H547.

Reinke, W., P. C. Johnson, and P. Gaehtgens. 1986. "Effect of Shear Rate Variation on Apparent Viscosity of Human Blood in Tubes of 29 to 94 Microns Diameter." *Circulation Research* 59:124–132.

Reneman, R. S., T. Arts, and Arnold P. G. Hoeks. 2006. "Wall Shear Stress—An Important Determinant of Endothelial Cell Function and Structure in the Arterial System *in Vivo*." *Journal of Vascular Research* 43:251–269.

Rogausch, H. 1987. "The Apparent Viscosity of Aggregating and Non-aggregating Erythrocyte Suspensions in the Isolated Perfused Liver." *Biorheology* 24:163–171.

Salazar Vazquez, B. Y., J. Martini, A. Chavez Negrete, P. Cabrales, A. G. Tsai, and M. Intaglietta. 2009. "Microvascular Benefits of Increasing Plasma Viscosity and Maintaining Blood Viscosity: Counterintuitive Experimental Findings." *Biorheology* 46:167–179.

Sapuppo, F., M. Bucolo, M. Intaglietta, P. C. Johnson, L. Fortuna, and P. Arena. 2007. "An Improved Instrument for Real-Time Measurement of Blood Flow Velocity in Microvessels." *IEEE Transactions on Instrumentation and Measurement* 56, no. 6:2663–2671.

Schmid-Schönbein, H. 1988. "Fluid Dynamics and Hemorheology *in Vivo*: The Interactions of Hemodynamic Parameters and Hemorheological 'Properties' in Determining the Flow Behavior of Blood in Microvascular Networks." In *Clinical Blood Rheology*, ed. G. D. Lowe, 129–219. Boca Raton, FL: CRC Press.

Schmid-Schönbein, H., G. Grunau, H. Brauer. 1980. "Exempla hämorheologica 'Das strömende Organ Blut.'" Wiesbaden: Albert-Roussel Pharma GmbH.

Secomb, T. W., and A. R. Pries. 2007. "Basic Principles of Hemodynamics." In *Handbook of Hemorheology and Hemodynamics*, eds. O. K. Baskurt, M. R. Hardeman, M. W. Rampling, and H. J. Meiselman, 289–306. Amsterdam, Berlin, Oxford, Tokyo, Washington, DC: IOS Press.

Seki, J. 1990. "Fiberoptic Laser-Doppler Anemometer Microscope Developed for the Measurement of Microvascular Red Cells Velocity." *Microvascular Research* 40:302–316.

Shepro, D. 2006. "Organ Microvascular Adaptations." In *Microvascular Research*, ed. D. Shepro, 361–509. San Diego, CA: Elsevier Academic Press.

Simchon, S., K. M. Jan, and S. Chien. 1987. "Influence of Reduced Red Cell Deformability on Regional Blood Flow." *American Journal of Physiology—Heart and Circulatory Physiology* 253:H898–H903.

Soutani, M., Y. Suzuki, N. Tateishi, and N. Maeda. 1995. "Quantificative Evaluation of Flow Dynamics of Erythrocytes in Microvessels: Influence of Erythrocyte Aggregation." *American Journal of Physiology—Heart and Circulatory Physiology* 268:H1959–H1965.

Stainsby, W. N. 1973. "Local Control of Regional Blood Flow." *Annual Review of Physiology* 35:151–168.

Svensjo, E. 2006. "The Hamster Cheek Pouch as a Research Model of Inflammation." In *Microvascular Research*, ed. D. Shepro, 195–200. Burlington, San Diego, London: Elsevier Academic Press.

Tateishi, N., Y. Suzuki, M. Soutani, and N. Maeda. 1994. "Flow Dynamics of Erythrocyte in Microvessels of Isolated Rabbit Mesentery: Cell-Free Layer and Flow Resistance." *Journal of Biomechanics* 27:1119–1125.

Thulesius, O., and P. C. Johnson. 1966. "Pre- and Postcapillary Resistance in Skeletal Muscle." *American Journal of Physiology* 210:869–872.

Todd, M. M., J. B. Weeks, and D. S. Warner. 1992. "Cerebral Blood-Flow, Blood Volume, and Brain-Tissue Hematocrit during Isovolemic Hemodilution with Hetastarch in Rats." *American Journal of Physiology—Heart and Circulatory Physiology* 263:H75–H82.

Tsai, A. G., C. Acero, P. R. Nance, P. Cabrales, J. A. Frangos, D. G. Buerk, and M. Intaglietta. 2005. "Elevated Plasma Viscosity in Extreme Hemodilution Increases Perivascular Nitric Oxide Concentration and Microvascular Perfusion." *American Journal of Physiology—Heart and Circulatory Physiology* 288:H1730–H1739.

Tsai, A. G., B. Friesenecker, M. McCarthy, H. Sakai, and M. Intaglietta. 1998. "Plasma Viscosity Regulates Capillary Perfusion during Extreme Hemodilution in Hamster Skinfold Model." *American Journal of Physiology—Heart and Circulatory Physiology* 275:H2170–H2180.

Uijttewaal, W. S. J., E. J. Nijhof, P. J. H. Bronkhorst, E. Denhartog, and R. M. Heethaar. 1993. "Near-Wall Excess of Platelets Induced by Lateral Migration of Erythrocytes in Flowing Blood." *American Journal of Physiology—Heart and Circulatory Physiology* 264:H1239–H1244.

Ulker, P., L. Sati, C. Celik-Ozenci, H. J. Meiselman, and O. K. Baskurt. 2009. "Mechanical Stimulation of Nitric Oxide Synthesizing Mechanisms in Erythrocytes." *Biorheology* 46:121–132.

Ulker, P., N. Yaras, O. Yalcin, C. Celik-Ozenci, P.C. Johnson, H. J. Meiselman, and O.K. Baskurt. 2011. "Shear Stress Activation of Nitric Oxide Synthase and Increased NO Levels in Human Red Blood Cells." *Nitric Oxide.* 24: 177-184.

Varin, R., P. Mulder, V. Richard, F. Tamion, C. Devaux, J. Henry, F. Lallemand, G. Larebours, and C. Thuillez. 1999. "Exercise Improves Flow-Mediated Vasodilation of Skeletal Muscle Arteries in Rats with Chronic Heart Failure: Role of Nitric Oxide, Prostanoids and Oxidant Stress." *Circulation* 99:2951–2957.

Verkeste, C. M., P. F. Boekkooi, P. R. Saxena, and L. L. Peeters. 1992. "Increased Red Cell Aggregation Does Not Reduce Uteroplacental Blood Flow in the Awake, Hemoconcentrated, Late-Pregnant Guinea Pig." *Pediatric Research* 31:91–93.

Vicaut, E. 1995. "Opposite Effects of Red Blood Cell Aggregation on Resistance to Blood Flow." *Journal of Cardiovascular Surgery* 36:361–368.

Vicaut, E., X. Hou, L. Decuypere, A. Taccoen, and M. Duvelleroy. 1994. "Red Blood Cell Aggregation and Microcirculation in Rat Cremaster Muscle." *International Journal of Microcirculation Clinical and Experimental* 14:14–21.

Vicaut, E., and B. I. Levy. 1990. "Transmural Hematocrit Gradient in Left Ventricular Myocardia of Rats." *American Journal of Physiology—Heart and Circulatory Physiology* 259:H403–H408.

Volger, E., C. Pfafferott, R. M. Bauersachs, U. Busch, F. Gaim, and M. Stoiber. 1986. "Haemorheological Aspects of Myocardial Ischaemia." *Clinical Hemorheology* 6:229–243.

Wayland, H., and P. C. Johnson. 1967. "Erythrocyte Velocity Measurement in Microvessels by a Two-Slit Photometric Method." *Journal of Applied Physiology* 22:333–337.

Whittaker, S. R. F., and F. R. Winton. 1933. "The Apparent Viscosity of Blood Flowing in the Isolated Hindlimb of the Dog and Its Variation with Corpuscular Concentration." *Journal of Physiology London* 78:339–368.

Yalcin, O., F. Aydin, P. Ulker, M. Uyuklu, F. Gungor, J. K. Armstrong, H. J. Meiselman, and O. K. Baskurt. 2006. "Effects of Red Blood Cell Aggregation on Myocardial Hematocrit Gradient Using Two Approaches to Increase Aggregation." *American Journal of Physiology—Heart and Circulatory Physiology* 290:H765–H771.

Yalcin, O., A. Erman, S. Muratli, M. Bor-Kucukatay, and O. K. Baskurt. 2003. "Time Course of Hemorheological Alterations after Heavy Anaerobic Exercise in Untrained Human Subjects." *Journal of Applied Physiology* 94:997–1002.

Yalcin, O., H. J. Meiselman, J. K. Armstrong, and O. K. Baskurt. 2005. "Effect of Enhanced Red Blood Cell Aggregation on Blood Flow Resistance in an Isolated-Perfused Guinea Pig Heart Preparation." *Biorheology* 42:511–520.

Yalcin, O., P. Ulker, U. Yavuzer, H. J. Meiselman, and O. K. Baskurt. 2008. "Nitric Oxide Generation of Endothelial Cells Exposed to Shear Stress in Glass Tubes Perfused with Red Blood Cell Suspensions: Role of Aggregation." *American Journal of Physiology— Heart and Circulatory Physiology* 294:H2098–H2105.

Yalcin, O., M. Uyuklu, J. K. Armstrong, H. J. Meiselman, and O. K. Baskurt. 2004. "Graded Alterations of RBC Aggregation Influence *in Vivo* Blood Flow Resistance." *American Journal of Physiology—Heart and Circulatory Physiology* 287:H2644–H2650.

Zhang, J. F., P. C. Johnson, and A. S. Popel. 2009. "Effects of Erythrocyte Deformability and Aggregation on the Cell-Free Layer and Apparent Viscosity of Microscopic Blood Flows." *Microvascular Research* 77:265–272.

8 Alterations in Red Blood Cell Aggregation

Red blood cell (RBC) aggregation is a physiological phenomenon with important hemodynamic consequences (see Chapter 7). The degree of RBC aggregation is determined by both cellular and suspending phase properties (see Chapter 2) and changes in these properties result in various degrees of aggregation in RBC suspensions (e.g., blood). Alterations of RBC aggregation due to physiological and pathophysiological challenges have been reported extensively; this chapter is devoted to a discussion of such variations in RBC aggregation. Alterations of RBC aggregation during the course of a variety of disease or clinical states are also summarized briefly in Section 8.3.

8.1 "NORMAL" RANGES OF RED BLOOD CELL AGGREGATION

The difficulty in defining the "normal" range of RBC aggregation has been discussed in Chapter 4, Section 4.4. This difficulty is due to the nature of the measured aggregation parameters as determined by various approaches and instruments. The widely used dimensionless aggregation indexes strongly depend on the technical specifications of the measuring system (e.g., geometry of aggregation chamber, calibration of electronics), and thus the results from instruments manufactured by different companies or laboratories are not directly comparable. Therefore, all data related to alterations of RBC aggregation during a specific physiological or pathophysiological process should be reported together with control data obtained using the same instrument under the same measurement conditions. Additionally, RBC aggregation is influenced by biological factors such as gender and age as well as by physiological processes such as pregnancy and menopause. Therefore, in the material below, the discussion on the alterations of RBC aggregation in almost all cases are based on "comparisons" within separate studies.

8.1.1 Gender Difference in Red Blood Cell Aggregation

It is well known that the erythrocyte sedimentation rate (ESR), which is a direct reflection of the extent of RBC aggregation, is higher in females compared to males (Piva et al. 2001; Rampling 1988). Higher RBC aggregation has also been reported for samples obtained from healthy female subjects (Pignon et al. 1994). However, findings of gender differences for RBC aggregation, as measured by different methods, do not always agree with this pattern. Korotaeva et al. (2007) reported lower aggregation (i.e., smaller aggregate size) for healthy females compared to healthy male subjects in blood samples with hematocrit adjusted to 0.4 1/l. They measured aggregation in a Couette-type shearing device and quantitated the extent of aggregation (i.e., aggregate size) by monitoring light backscattering (Firsov et al. 1998; Korotaeva et al. 2007). It should be noted that ESR values are usually measured in whole blood diluted 1/5 with a citrate solution or undiluted at native hematocrit. In general, hematocrit values are lower in female subjects; ESR is known to increase as hematocrit decreases (Poole and Summers 1952) (see Chapter 4, Section 4.1.1.4), thus possibly explaining the higher ESR values for female populations. Supporting this explanation, Kamaneva et al. (1999) has reported higher ESR for males versus females when blood samples are adjusted to a hematocrit of 0.4 l/l.

It is possible to argue that differences in sex hormones may explain the gender differences of RBC aggregation. In support of this possibility, Pignon et al. (1994) have reported smaller differences in aggregation between male and female elderly subjects. In contrast with their findings, RBC aggregation has been reported to be enhanced in female subjects following menopause (Demiroglu et al. 1997). Further, hormone replacement therapy has been shown to be effective in reducing RBC aggregation in postmenopausal women (Spengler et al. 2003).

8.1.2 ALTERATIONS IN RED BLOOD CELL AGGREGATION WITH SUBJECT AGE

Subject age has been reported to be a significant factor influencing RBC aggregation. Hammi et al. (1994) investigated RBC aggregation using ultrasound interferometry and backscattering in two groups of subjects with age ranges of 20–30 years and 66–89 years: RBC aggregation was enhanced in the older group (Hammi et al. 1994). Comparisons of various aggregation parameters measured by the Myrenne cone-plate aggregometer and the Sefam aggregometer in young (mean age ~25 years) and middle-aged (mean age ~52 years) healthy male subjects indicate higher RBC aggregation in the older group, with the difference for disaggregation shear rate being significant (Manetta et al. 2006). Christy et al. (2010) has also reported increased M and M1 Myrenne aggregometer indexes with age in a group of healthy volunteers between 20 and 59 years. Aggregation indexes measured in a standard suspending medium (i.e., 3% dextran 70 kDa solution) that reflect RBC aggregability (see Chapter 4, Section 4.3) were not different between younger and older subjects (Christy et al. 2010). Feher et al. (2006) studied RBC aggregation in blood samples of 6,236 patients with vascular problems: Myrenne aggregometer indexes were significantly lower in young subjects with a mean age of ~36 years compared to middle-aged and aged groups (mean ages of ~57 and ~71 years, respectively). However, when they repeated the comparison in a subgroup of patients selected to match risk factors, age was no longer found to be a determinant of hemorheological parameters (Feher et al. 2007). Cavestri et al. (1992) investigated the relationship between RBC aggregation and cerebral blood flow in subjects aged from 20 to 74 years and found a significant negative correlation between the two parameters for subjects over 45 years of age but no correlation for subjects below 45 years of age.

In most cases, the increased RBC aggregation in elderly subjects can be explained by altered concentrations of acute phase reactants (e.g., fibrinogen) (Ajmani and Rifkind 1998; Christy et al. 2010; Hammi et al. 1994). Therefore, age-related changes of aggregation are in agreement with reports indicating increased acute phase reactants in elderly human subjects (Ballou et al. 1996; Hager et al. 1994).

8.1.3 ALTERATIONS IN RED BLOOD CELL AGGREGATION DURING PREGNANCY, LABOR, AND DELIVERY

RBC aggregation, determined using different measurement methods, has been reported to be gradually increased during normal pregnancy (Huisman et al. 1988; Ozanne et al. 1983b). Huisman et al. (1988) studied the aggregation time course based on syllectometry (see Chapter 4, Section 4.1.6) in women with uncomplicated pregnancy. They found a decrease in aggregation half-time from 5.2 s at the 14th week of gestation to 3.3 s at the 37th week, indicating a significant acceleration of RBC aggregation rate; aggregation half-time returned to normal 8 weeks after delivery. Bollini et al. (2005) used a custom-built photometric system for evaluating RBC aggregation in the second trimester of normal pregnancy and reported increased aggregate size and aggregation rate. ESR is also enhanced during pregnancy, being significantly higher during the second half of pregnancy compared to the first half (van den Broek and Letsky 2001). These alterations in RBC aggregation paralleled plasma fibrinogen concentrations (Bollini et al. 2005; Huisman et al. 1988).

Labor and delivery have been found to further influence RBC aggregation. Aggregation indexes measured by a photometric instrument (Sefam erythroaggregometer) begin to increase with the start of labor and continue to increase until delivery, returning to prelabor values following delivery of the placenta (Brun et al. 1994). In addition, a gradual decrease of disaggregation shear rate (see Chapter 4, Section 4.1.6.3) during labor was observed, indicating that RBC aggregates are more easily dispersed by shear forces. The same investigators later reported unchanged M and M1 indexes measured by a Myrenne aggregometer during most of the labor process, with significant decreases at the time of delivery (Brun et al. 1995). It is clear from both studies that RBC aggregation decreases with delivery; Brun et al. (1995) attributed this change to a reduction of plasma fibrinogen during delivery, thereby causing the return of aggregation to nonpregnant levels.

8.1.4 RED BLOOD CELL AGGREGATION IN FETAL BLOOD

Human fetal blood obtained by cord venipuncture at various gestational ages is characterized by very low RBC aggregation. Table 8.1 presents RBC aggregation values measured by a Myrenne aggregometer at gestational ages of less than 22 to more than 36 weeks of normal pregnancy. The aggregation index M was below one for all gestational ages, while the M index in maternal blood samples was ~12 (El Bouhmadi et al. 2000). Large differences between fetal and maternal blood samples also exist for M1 indexes that are measured under low shear. Aggregation indexes (i.e., M and M1) increased toward the end of gestation starting after 32 weeks. Plasma fibrinogen concentrations were also low during earlier gestational weeks and increased gradually with gestational age. El Bouhmadi et al. (2000) studied the roles of plasma and cellular factors in determining the difference between fetal and

TABLE 8.1

Red Blood Cell Aggregation Indexes (M) and (M1) Measured in Fetal Blood at Gestational Ages of 22 to 36 Weeks

Gestational Age (Weeks)	M index[a]	M1 index[a]
<22	0.01 ± 0.01	3.30 ± 0.72
22–23	0.04 ± 0.02	3.04 ± 0.57
24–25	0.02 ± 0.02	2.67 ± 0.64
26–27	0.06 ± 0.03	3.14 ± 0.80
28–29	0.04 ± 0.03	2.82 ± 0.52
30–31	0.05 ± 0.02	1.22 ± 0.36
32–33	0.16 ± 0.12	3.05 ± 1.06
34–35	0.35 ± 0.07	3.55 ± 0.85
>36	0.74 ± 0.15	4.84 ± 0.89
Maternal	11.73 ± 0.45	18.05 ± 0.58

Source: Data from El Bouhmadi, A. P. Boulot, F. Laffargue, and J. F. Brun. 2000. "Rheological Properties of Fetal Red Cells with Special Reference to Aggregability and Disaggregability Analyzed by Light Transmission and Laser Backscattering Techniques." *Clinical Hemorheology and Microcirculation* 22:79–90.

[a] M index at stasis, M1 at very low shear.

maternal blood. When they suspended maternal RBC in fetal plasma the aggregation index was lower than that measured in maternal blood. Fetal RBC suspended in maternal plasma were also characterized by lower aggregation, indicating that both plasma and cellular factors play a role in the significantly lower RBC aggregation of fetal blood.

In contrast with findings in human subjects, RBC aggregation higher than maternal blood has been reported for blood obtained from fetal lambs in the third quarter of gestation (Windberger et al. 1998). However, adult sheep blood exhibits almost no aggregation (see Chapter 9) and hence the results are not directly comparable with those obtained in human subjects.

Hematocrit increases with gestational age and can reach ~0.5 l/l at 40 weeks for humans (Linderkamp 1996). However, it has been shown that the flow properties of fetal blood in glass tubes are quite different than adult blood: fetal blood has a lower viscosity at similar hematocrit values (Linderkamp 1996). The lower aggregation of human fetal blood may contribute to this difference in rheological behavior and may provide an explanation for the maintenance of adequate blood flow in the fetus toward the end of gestation despite the increased hematocrit.

8.1.5   Red Blood Cell Aggregation in Neonatal Blood

RBC aggregation remains low in neonatal blood compared to adults. Linderkamp et al. (1984) used a counter-rotating cone plate rheoscope system to study RBC aggregation in preterm and term neonates, with aggregation parameters calculated based on light transmission intensity. Their results indicated significantly lower RBC aggregation in placental blood from term neonates, with even lower values for pre-term neonates (Linderkamp et al. 1984). Adult RBC suspended in neonatal plasma exhibited aggregation characteristics very similar to fetal blood. This pattern indicates that plasma factors are of primary importance in the observed differences; consistent with this finding, Linderkamp et al. (1984) also reported differences in fibrinogen concentration in neonatal and adult plasma.

Neonatal RBC suspended in 1% dextran 500 kDa have aggregation characteristics similar to those of adult RBC (Linderkamp et al. 1984), suggesting no important contributions of cellular factors to the difference in aggregation (Rampling et al. 2004). However, a later report (Whittingstall and Meiselman 1991a) indicated a similar aggregation pattern for adult and newborn RBC suspended in both autologous plasma and 3% dextran 70 kDa: aggregation was higher for adult RBC in both media. Whittingstall and Meiselman (1991a) also confirmed similar aggregation for adult and neonatal RBC aggregation in dextran 500 solutions at concentrations of 1% or less, yet lower neonatal aggregation (i.e., reduced aggregability) at higher dextran 500 concentrations. Thus, testing adult versus neonatal RBC only in 1% dextran 500 led to correct but not fully informative results; the different influence of various molecular mass polymers has been confirmed by other reports (Baskurt et al. 2000; Rampling et al. 2004).

Similar differences in RBC aggregation between adults and newborns have also been demonstrated for other species. Castellini et al. (2006) reported very low aggregation in the blood of Weddell seal pups obtained on the first day after birth. A sharp increase in aggregation, almost reaching adult levels, was observed in samples obtained on 7th day (Castellini et al. 2006); similar differences were found for RBC suspended in standard polymer solutions (e.g., 3% dextran 70), thus emphasizing the role of cellular factors.

Meiselman (1993) has suggested that since neonatal RBC have a shorter lifespan, the difference in aggregability between neonatal and adult RBC can be explained by the longer *in vivo* conditioning for adult RBC compared to neonatal RBC. This suggestion is consistent with findings that older RBC that have spent more time in the circulation have higher aggregability compared to younger RBC (see Section 8.1.6). Therefore, differences in RBC properties and the lower concentrations of proaggregant factors (e.g., fibrinogen) in plasma both contribute to the markedly low aggregation in neonatal blood.

8.1.6   Changes in Aggregation with *In Vivo* Aging of Red Blood Cells

In vivo aging of red blood cells increases cellular density, and this property of cells with different ages can be used to obtain fractions with various average *in vivo* ages by high-speed centrifugation (Murphy 1973). It has been demonstrated that RBC

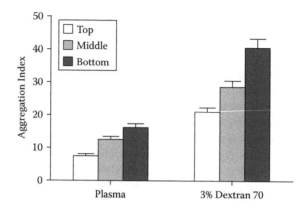

FIGURE 8.1 Aggregation indexes for density-separated human red blood cells resuspended in autologous plasma or 3% dextran 70 kDa. "Top" corresponds to the 10% youngest RBC population, whereas "Bottom" corresponds to the 10% oldest and thus cells with the longest time in the circulation; "Middle" is the remaining 80% of the RBC population. Redrawn from Meiselman, H. J. 1993. "Red Blood Cell Role in RBC Aggregation: 1963–1993 and Beyond." *Clinical Hemorheology* 13:575–592 with permission.

with different *in vivo* ages have different aggregation patterns (Bauersachs et al. 1989b; Nash et al. 1987; Nordt 1983; Sowemimo-Coker et al. 1989). Figure 8.1 presents aggregation indexes for density-separated RBC as measured by the Myrenne aggregometer. The RBC fraction with lowest density (the youngest population, "Top" in Figure 8.1) exhibits the lowest aggregation, while the fraction with highest density (the oldest population, "Bottom" in Figure 8.1) has the highest aggregation. This pattern was observed for RBC suspensions in autologous plasma and in 3% dextran 70 kDa (Meiselman 1993). These findings confirm the earlier findings of Nordt (1983) who indicated ~100% difference in aggregation between the 10% least dense and the 10% most dense RBC after separation by centrifugation then resuspension in autologous plasma. The difference in aggregation behavior between more and less dense RBC suspended in solutions of other polymers, including 360 kDa polyvinylpyrrolidone, 61.2 kDa poly-l-glutamic acid, and sodium heparin, have also been reported (Whittingstall et al. 1994). The difference between younger and older still existed after removing sialic acid and immunoglobulin G from the RBC surface (Whittingstall and Meiselman 1991b). Hadengue et al. (1998) have reported increased disaggregation shear rate and decreased aggregation time for denser, older RBC as measured by the SEFAM erythroaggregometer. These differences of aggregation were associated with age-related decreased levels of membrane sialic acid, which is the main factor contributing to the RBC surface charge (Hadengue et al. 1998).

Denser, older RBC have decreased deformability (Linderkamp et al. 1982; Tugral et al. 2002), which may contribute to alterations in RBC aggregation. However, decreased deformability is expected to result in decreased RBC aggregation, especially if the method used to quantitate RBC aggregation is sensitive to the extent of aggregation. Other important cellular factors relevant to RBC aggregation include their surface properties, which may change throughout the *in vivo* life span of RBC

(Meiselman 2009). Surface properties of RBC may be altered by various micro-environmental factors, including interaction with activated leukocytes and free radicals from various sources (Baskurt et al. 1998; Baskurt and Meiselman 1998). The extent of such alterations increase with time of exposure in the circulation and hence with *in vivo* age, resulting in marked differences of the oldest RBC compared to the youngest RBC. Tugral et al. (2002) reported smaller differences between young and old RBC populations in aged humans (60–85 years of age) compared to subjects below 45 years of age, a finding that may be related to a more aggressive environment for the cells and hence accelerated RBC aging in older people.

8.1.7 OTHER PHYSIOLOGICAL INFLUENCES ON RED BLOOD CELL AGGREGATION

Measured RBC aggregation may depend on the site of blood sampling, since comparisons of RBC aggregation in arterial and venous blood show slightly higher values in venous blood (Mokken et al. 1996). In contrast with Mokken et al., Hever et al. (2010) reported higher aggregation indexes in arterial blood samples versus venous samples from rats. In addition, hemorheological parameters, including RBC aggregation, may depend on the vein that is sampled since the blood from that vein has been exposed to the local conditions in the tissue or organ drained by that particular vessel (Baskurt et al. 1991). Sampling conditions should thus be standardized as discussed in detail in Chapter 4, Section 4.2.1.

RBC aggregation can be modified due to alterations of the hydration status of the donor, most probably due to a contraction of plasma volume and a concomitant increase in fibrinogen concentration (Tikhomirova et al. 2002). Blood viscosity has a diurnal variation, with an early morning minimum and a peak value later in the afternoon (Acciavatti et al. 1993). Interestingly, plasma fibrinogen concentrations did not exhibit a diurnal variation; RBC aggregation was not studied by Acciavatti et al.

Vaya et al. (2004) reported increased RBC aggregation indexes in postmenopausal women compared to a premenopausal control group. Aggregation was found to be decreased with hormone replacement therapy in the postmenopausal group (Vaya et al. 2004). In contrast, Puniyani and Sonar (1991) did not find any significant differences in RBC aggregation between pre- and postmenopausal women. Estrogens are known to affect plasma fibrinogen concentrations, and therefore RBC aggregation might be expected to be altered during the menstrual cycle (KaBer et al. 1987).

Gaudard et al. (2004) investigated the relationship between nutritional status and RBC aggregation in a group of athletes. They reported a negative correlation between RBC disaggregation shear rate and daily calorie, protein, and lipid intake. Their study did not include an investigation of cellular properties affecting aggregation (i.e., RBC aggregability), but the dependence of fibrinogen concentration on nutritional parameters would seem to explain the relationship between RBC aggregation and nutrition (Gaudard et al. 2004). Cicha et al. reported that postprandial increases in plasma triglyceride concentration may lead to increased RBC aggregation (Cicha et al. 2004).

8.2 ALTERATIONS OF RED BLOOD CELL AGGREGATION WITH EXTREME CONDITIONS

8.2.1 PHYSICAL ACTIVITY

The influence of physical activity on hemorheological parameters including RBC aggregation has been extensively investigated. Alterations of these parameters are determined by factors related to physical activity (i.e., intensity, duration, type) as well as to subject properties (e.g., trained or sedentary individuals) and sampling time relative to the exercise activity. Section 8.2.1 presents a brief summary of the influence of physical activity on RBC aggregation; a more complete discussion of the hemorheological consequences of physical exercise can be found elsewhere (e.g., Brun et al. 2010; Connes et al. 2010; El Sayed et al. 2005).

8.2.1.1 Acute Effects of Exercise on Red Blood Cell Aggregation

RBC aggregation in sedentary male human subjects has been found to be significantly decreased following an exhausting, cycling exercise episode (i.e., Wingate protocol) (Figure 8.2). The decrease was first detected in samples obtained 30 minutes after the end of the exercise episode, and remained significantly lower compared to the preexercise value for 12 hours (Yalcin et al. 2003). Aggregation indexes were also measured for RBC suspended in 1% dextran 500 kDa and were found to follow a similar pattern, indicating that cellular properties play an important role in the decreased RBC aggregation following the exhausting exercise episode. Yalcin et al. (2003) also studied leukocyte counts and activation following the exercise episode and reported significant increases in total leukocyte and granulocyte counts with a

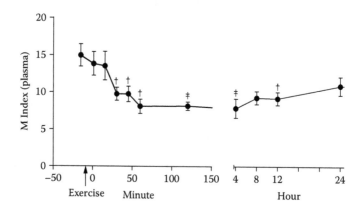

FIGURE 8.2 Time course of red blood cell aggregation measured in autologous plasma following an exhausting exercise episode in sedentary human subjects. Exercise protocol was started at the point marked by the arrow. The first sample was obtained right after the end of an exercise episode (time = 0) and repeated at 15-minute intervals during the first hour. †,‡ indicate significant differences from pre-exercise value. Reproduced from Yalcin, O., A. Erman, S. Muratli, M. Bor-Kucukatay, and O. K. Baskurt. 2003. "Time Course of Hemorheological Alterations after Heavy Anaerobic Exercise in Untrained Human Subjects." *Journal of Applied Physiology* 94, no. 3:997–1002 with permission.

timing similar to the alterations of RBC aggregability; leukocyte respiratory burst was also enhanced at one hour following exercise. Blood lactate concentrations followed a different pattern, with the highest value at five minutes following exercise, returning to the preexercise value within one hour. RBC deformability was found to be significantly decreased one minute after the exercise episode, with a time course similar to that of the increment in lactate; RBC deformability remained decreased for 12 hours. It can be argued that RBC aggregation may be decreased due to impaired RBC deformability following exercise (Rampling et al. 2004) and also that RBC surface properties may be altered as a consequence of leukocyte activation (Baskurt and Meiselman 1998; Camus et al. 1994) and/or enhanced free radical attack (Baskurt et al. 1998; Ji and Leichtweis 1997; Sen et al. 1994).

Reports regarding the acute effects of strenuous exercise on RBC aggregation are not always in concordance. Ajmani et al. (2003) found no alterations of the Myrenne M and M1 aggregation indexes for healthy male and female sedentary subjects aged between 33 and 88 years following a treadmill exercise that was continued until exhaustion. Nageswari et al. (2000) also reported unaltered RBC aggregation indexes based on low-shear viscometry for sedentary, healthy male volunteers following 15 minutes of heavy cycling exercise. In contrast, Senturk et al. (2005) reported enhanced RBC aggregation measured in sedentary, young male subjects following cycle ergometry performed at maximal oxygen consumption for 8–12 minutes with the period of exercise determined by the exhaustion time of the subjects.

The response to strenuous exercise in terms of RBC aggregation has been shown to differ between sedentary and trained subjects. In the study by Senturk et al. (2005), no alterations of aggregation were observed in exercise-trained subjects following exhausting cycling exercise, while age-matched sedentary subjects performing the same exercise protocol had enhanced aggregation. Using elite athletes, Varlet-Marie et al. (2003b) studied the effect of 25 minutes of cycling, with the last 15 minutes performed at 85% of theoretical maximal heart rate, on RBC aggregation measured by two different methods. They found no alterations of Myrenne M and M1 indexes, whereas aggregation times and disaggregation shear rates measured by the Sefam erythroaggregometer were altered, indicating slightly faster aggregation and increased aggregate strength. There was a strong correlation between resting plasma fibrinogen concentration and the aggregation index at the end of the exercise episode (Varlet-Marie et al. 2003b). Conversely, Connes et al. reported unaltered RBC aggregation during and following heavy cycling exercise protocols in trained subjects (Connes et al. 2007).

RBC aggregation has also been found to be altered during prolonged exercise. Ernst et al. (1991) observed enhanced RBC aggregation during the first hour of a three-hour cycling exercise in a mixed group of subjects including sedentary ($n = 5$) and endurance-trained ($n = 10$) individuals. RBC aggregation returned to resting levels by the end of the three-hour exercise episode (Ernst et al. 1991). Wood et al. reported enhanced RBC aggregation above control for well-trained runners following a mountain run of 48 km (Wood et al. 1991). Conversely, Neuhaus et al. (1992) observed no significant alterations of RBC aggregation in recreational runners following a marathon run. Cakir-Atabek et al. (2009) studied RBC aggregation before and after resistance training in healthy young male subjects, and found that

RBC aggregation measured by the Laser-Assisted Optical Rotational Cell Analyzer (LORCA) light backscattering method was enhanced following resistance training compared to values measured prior to the training.

Animal experiments have been performed to study the influence of exercise on RBC aggregation. Yalcin et al. (2000) studied RBC aggregation in rats following a 60-minute swimming exercise. Aggregation indexes measured by a plate-on-plate photometric rheoscope were found to be significantly enhanced at 24 hours after swimming compared to control, nonswimming rats. Six weeks of training partially prevented the increase of aggregation induced by the 60-minute period of swimming. RBC aggregation was also measured in pooled, standard plasma obtained from healthy, nonexercised rats in order to assess alterations of cellular factors affecting RBC aggregability. These measurements indicated that RBC properties were affected by the swimming exercise for untrained rats, and that this effect of cellular factors diminished significantly in trained animals (Yalcin et al. 2000).

The previously mentioned discrepancies among published reports dealing with exercise are most likely due to differences in study design (e.g., exercise protocol, blood sampling), properties of the subjects (i.e., sedentary or trained), and the methods used to assess RBC aggregation. These factors are of importance since both plasma factors (e.g., fibrinogen concentration) (Varlet-Marie et al. 2003b) and cellular properties might be altered acutely by exercise episodes (Yalcin et al. 2003; Yalcin et al. 2000). It seems possible that endurance training may prevent the acute alterations of RBC aggregation induced by short-term episodes of exhaustive exercise, and that enhanced aggregation may be a part of the physiological adaptation response during long-term exercise protocols (e.g., marathon run) (Wood et al. 1991). The nutritional and antioxidant status of the subjects may also affect the pattern of changes inasmuch as Senturk et al. (2005) have demonstrated that treatment with antioxidant vitamins prior to exhaustive exercise testing may prevent acute alterations of RBC aggregation.

8.2.1.2 Alteration of Red Blood Cell Aggregation by Training

Exercise training is known to influence blood rheology, mostly due to the alterations of plasma volume and hematocrit (Brun et al. 2010; El-Sayed et al. 2005; Neuhaus and Gaehtgens 1994). However, the effect of training on RBC aggregation is controversial and contrasting findings have been reported, most probably reflecting differences in exercise protocol and the characteristics of the subjects.

Dintenfass and Lake (1976) reported an inverse relationship between fitness level and RBC aggregation in patient groups having anxiety, osteoarthritis, hypertension, and coronary artery disease. However, this effect of fitness was not confirmed by other studies on healthy individuals (Brun et al. 1989; Ernst et al. 1985). Reports of cross-sectional studies on sportsmen have also not been conclusive. Significantly lower RBC aggregation has been reported in elite athletes (i.e., long-distance runners, rowers, cyclists, and canoeists) compared to nonexercising controls (Neuhaus and Gaehtgens 1994). However, Ernst et al. (1991) have reported that compared to sedentary subjects, there is markedly higher RBC aggregation in endurance-trained subjects both at rest and during performance of cycle ergometry.

Interestingly, there is a general consensus that the impact of an acute exercise episode on hemorheological parameters, including RBC aggregation, is significantly blunted in exercise-trained individuals and that this consensus is valid for various types of exercise models. For example, the significant enhancement in RBC aggregation induced by a single episode of resistance training was found to be diminished following a six-week period of training with three resistance training sessions per week (Cakir-Atabek et al. 2009). Other examples are presented above in Section 8.2.1.1. It has been suggested that enhanced RBC aggregation may play a role in the "overtraining syndrome" in which the athlete's performance deteriorates despite extensive training (Brun et al. 2010). Varlet-Marie et al. (2003a) reported an association between RBC aggregation parameters, as measured by both the Myrenne aggregometer and the Sefam Erythroaggregometer, with the feeling of "heavy legs," a common sign of overtraining.

There is no current consensus on the influence of RBC aggregation on tissue perfusion and oxygenation (see Chapter 7). Controversy also exists when considering the relationship between RBC aggregation and exercise performance. A negative correlation has been reported between postexercise RBC aggregation and lactate-clearing rate in a mixed population including athletes and sedentary subjects (Brun et al. 2002), thus obviously supporting a negative effect of RBC aggregation on performance. However, there tend to be higher levels of "normal" RBC aggregation in athletic species (e.g., dog, antelope, horse) compared to sedentary species (e.g., sheep, cow), with active and sedentary classification based on maximal oxygen consumption (Popel et al. 1994); this observation thus counters the belief that enhanced aggregation negatively affects athletic performance (see Chapter 9, Section 9.4.1).

8.2.2 RED BLOOD CELL AGGREGATION UNDER EXTREME ENVIRONMENTAL CONDITIONS

RBC aggregation has been found to be enhanced in subjects exposed to high altitude (i.e., 4,559 m) during the first 18 hours (Reinhart et al. 1991). This increase was not related to any pathological condition (e.g., acute mountain sickness, high-altitude pulmonary edema) but rather to the acute phase response. Prolonged exposure to high altitude has been found to significantly increase fibrinogen levels (i.e., 53% after three months and 61% after 13 months at 4,100–4,500 m) compared to values at sea level prior to going to high altitude (Vij 2009). Vij (2009) did not measure RBC aggregation, but the greater than 50% increase of fibrinogen is certainly expected to result in increased levels of aggregation.

Exposure of rats to a simulated high altitude of 6,000 m did not induce any changes in RBC aggregation as estimated based on blood viscometry, but did significantly enhance the influence of local cold exposure, resulting in markedly increased aggregation (Yang et al. 1999). Yang et al. (2003) studied the effect of local cold exposure in rats acclimatized to cold (i.e., raised at 4°C) and reported that RBC aggregation indexes were significantly lower in cold-acclimatized rats. RBC aggregation was increased in control animals raised at 20°C and in cold-acclimatized rats with local cold exposure (i.e., −20°C), although the change of RBC aggregation indexes were smaller in the cold-acclimatized group (Yang et al. 2003).

Taylor et al. (1998) studied RBC aggregation under hyperbaric conditions by simulating underwater diving conditions at 66 feet and 300 feet. They observed very prominent increases of RBC aggregation and an almost threefold increase of aggregate size following a two-hour exposure to pressure corresponding to 66 feet of water. Further increases of aggregate size were detected after increasing the pressure to 9.85 atmospheres corresponding 300 feet (i.e., 9.85 atmospheres) on the second day (Taylor et al. 1998). Two important points should be mentioned: First, the observed effects on RBC aggregation were related to hydrostatic pressure since inert gases were used to increase pressure, and oxygen partial pressure was maintained at a selected level. It has been previously shown that *in vitro* exposure of RBC to hydrostatic pressures between ~4 and 16 atmospheres also results in a significant enhancement of aggregate size (Chen et al. 1994). Secondly, aggregation was assessed in suspensions prepared in standardized suspending medium containing 0.5% dextran 500 (Taylor et al. 1998), and therefore reflect changes of RBC cellular properties and not of the suspending medium.

Dintenfass examined RBC under microgravity conditions during space shuttle flights in 1985 and 1988 (Dintenfass 1989; Dintenfass et al. 1985). These experiments were conducted on blood samples obtained on Earth using macro and microphotographic methods. The results of both experimental series indicated altered RBC aggregate morphology under zero gravity: aggregate size was significantly diminished in native plasma and in 2% dextran 200 kDa. The mechanisms responsible for the altered aggregation are not clear and need further evaluation.

8.3 ALTERATIONS OF RED BLOOD CELL AGGREGATION IN PATHOPHYSIOLOGICAL PROCESSES

8.3.1 ACUTE PHASE REACTION

The acute phase reaction is the extensive, nonspecific systemic response of the organism to local or systemic disturbances in homeostasis (Gordon and Koy 1985; Gruys et al. 2005). This reaction is an important component of inflammation caused by infections, tissue injuries, trauma, neoplastic growth, and immunological disorders. The acute phase reaction is initiated by cytokines (e.g., interleukins and tumor necrosis factor-α) secreted by inflammatory cells that affect the liver so as change the rate of synthesis of a group of plasma factors (Gruys et al. 2005). These factors include *positive acute phase proteins* (e.g., fibrinogen, C-reactive protein, amyloid, ceruloplasmin, ferritin, haptoglobin, α2-macroglobulin), which have increased plasma concentrations, and *negative acute phase proteins* (e.g., albumin, transferrin, transcortin) with decreased plasma concentrations. Positive acute phase proteins, especially fibrinogen and C-reactive protein (CRP), are generally accepted markers of inflammation.

Increased plasma fibrinogen concentrations due to the acute phase reaction are directly related to the observed increases of RBC aggregation and erythrocyte sedimentation rate (ESR). Other acute phase reactants have also been demonstrated to correlate with RBC aggregation (Weng et al. 1996). ESR is probably the most widely used test in clinical pathology laboratories as a nonspecific indicator of inflammation and the acute phase reaction. However, RBC aggregation is at least equally effective in monitoring the acute phase reaction (Stuart 1991) and has significant methodological advantages (see Chapter 4, Sections 4.1.1 and 4.1.6). A method based on photometric measurement of RBC aggregation has been reported to correlate better than the Westergren ESR with plasma factors of acute phase response in patients with infections, autoimmune disorders, and malignancies (Cha et al. 2009).

8.3.2 INFLAMMATORY CONDITIONS

These relationships between acute phase reactions, increased concentrations of aggregating macromolecules (e.g., fibrinogen) in plasma, and RBC aggregation underlie many findings of altered aggregation in the disease processes discussed in the following text. There are a large number of published examples of enhanced RBC aggregation due to inflammation, including the response to trauma or surgery, allergic disorders, and rheumatoid diseases. Other disease processes such as infections, circulatory disorders, obesity, diabetes, and a variety of other pathologic conditions also affect aggregation.

If an inflammatory process is not complicated by extensive tissue or organ damage (e.g., infections or ischemia), the alterations of RBC aggregation may be due to plasmatic factors without changes of RBC aggregability. It has been experimentally shown that a simple laparotomy in rats results in significantly enhanced RBC aggregation within 18 hours when measured in autologous plasma, but no changes are observed when measured in a standard suspending medium (Baskurt et al. 1997); see Figure 8.3 and the related text in Section 8.3.3.1. Alterations of RBC aggregation due to the acute phase response have also been observed in humans following elective

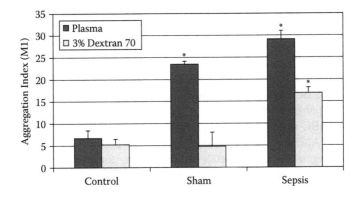

FIGURE 8.3 Red blood cell aggregation index (M1) in blood samples of control, sham-operated, and septic rats measured in autologous plasma and in 3% dextran 70 kDa using a Myrenne aggregometer. The asterisk (*) indicates difference from corresponding Control; $p < 0.05$. Data from Baskurt, O. K., A. Temiz, and H. J. Meiselman. 1997. "Red Blood Cell Aggregation in Experimental Sepsis." *Journal of Laboratory and Clinical Medicine* 130:183–190.

abdominal surgery (Caswell et al. 1992). Obviously, such a response is expected to
be systemic since altered synthesis of acute phase reactants (e.g., fibrinogen) should
play an important role. In contrast with this explanation, Kumsishvili et al. (2004)
reported increased RBC aggregation in venous blood samples draining inflamed
ears of rabbits while there were no alterations in blood sampled from the contralat-
eral ear. The aseptic inflammation in their model was induced by the local application of turpentine (Kumsishvili et al. 2004). Unfortunately, they did not study RBC
aggregability and only report aggregation indexes estimated based on image analysis
of RBC in autologous plasma. Alterations in RBC aggregation in samples returning
from tissues affected by circulatory abnormalities have also been reported (Baskurt
et al. 1991; Endre et al. 2010; Kayar et al. 2001), and may reflect the influence of local
damage to RBC by activated leukocytes or factors released from damaged tissues
(e.g., oxygen free radicals) (Baskurt et al. 1998; Baskurt and Meiselman 1998).

Rheumatic diseases are among the most common examples of clinical conditions with enhanced RBC aggregation due to chronic inflammation. RBC aggregation, as indexed by ESR values, is among the markers included in the calculation of
Disease Activity Score that are used for clinical evaluation of rheumatoid arthritis
(Crowson et al. 2009; Paulus et al. 1999). ESR is also among the variables used
for planning and monitoring treatment protocols for other rheumatic diseases (de
Vries et al. 2009). Hemorheological abnormalities, including increased plasma viscosity and enhanced RBC aggregation, have been demonstrated in patients with
rheumatic diseases (Luquita et al. 2009). Highly significant correlations between
RBC aggregation indexes and acute phase reactants (i.e., CRP and fibrinogen) in
rheumatoid arthritis patients were also reported (Luquita et al. 2009). Lee and Kim
(2009) have suggested that RBC aggregation measured by a photometric instrument
is a more useful parameter than ESR in monitoring disease activity in rheumatoid
arthritis. Although significant increments of plasma viscosity have been detected in
patients with allergic skin disorders, RBC aggregation was not altered (Puniyani,
Annapurna, et al. 1988).

8.3.3 INFECTIONS

Given the previous discussion regarding the acute phase response, it is not surprising
that enhanced RBC aggregation is found in a variety of infectious diseases. There
are numerous publications indicating relationships between inflammatory reaction
markers and RBC aggregation parameters in various acute and chronic infectious
diseases (e.g., Almog et al. 2005; Zilberman et al. 2005). Ben Ami et al. (2001)
studied RBC aggregation parameters obtained using an image analysis technique
(see Chapter 4, Section 4.1.5) in patients with severe infectious diseases, including
pneumonia, urinary and soft tissue infections, and septicemia. In addition to a significantly increased aggregate size (~100% greater average number of RBC per aggregate) in blood samples obtained from these patients compared to healthy controls,
the percentage of very large aggregates (i.e., aggregates with 33 or more RBC) was
63.5% whereas it was only 4.8% in control subjects (Ben Ami et al. 2001). Ben Ami
et al. (2001) also reported significant positive correlations between the acute phase
reactants CRP and fibrinogen and the percentage of large aggregates in patients

with infectious diseases. These investigators studied RBC aggregation parameters in 0.5% dextran 500 kDa in order to differentiate between cellular and plasmatic factors, and concluded that ~65% of the impact of infections on RBC aggregation is related to plasmatic factors (Ben Ami et al. 2001). Of course the other 35% represents aggregability, thus underlining the involvement of cellular factors in the altered RBC aggregation seen in infectious diseases.

Acute phase reactions caused by infections are usually more severe compared to noninfectious inflammation, and proper quantification of the degree of the reaction might be useful in differentiating between related clinical situations (Urbach et al. 2005). The simple *slide test* (see Chapter 4, Section 4.1.5.1), which is based upon determining RBC aggregation, has been shown to be an effective diagnostic method for differential diagnosis of pediatric bacterial infections and to be easy to employ even at primary care locations. RBC aggregation alterations differ between bacterial and viral infections. Using the slide test, Goldin et al. (2007) studied RBC aggregation in patients with acute bacterial and viral infections and also measured ESR and plasma fibrinogen concentration. Compared to controls, RBC aggregation, ESR, and fibrinogen levels were significantly elevated in patients with bacterial infections. Conversely, RBC aggregation did not differ from controls in patients with acute viral infections and ESR and fibrinogen values were only slightly increased (Goldin et al. 2007).

Human immunodeficiency virus (HIV) infection is of current global interest and RBC aggregation in this disease has been extensively studied. RBC aggregation assessed by a computerized photometric rheoscope has been shown to be significantly higher in HIV-infected individuals compared to HIV-negative controls (Kim et al. 2006); the degree of altered aggregation was not associated with the severity of immunodeficiency. Monsuez et al. (2000) also reported enhanced RBC aggregation (i.e., shortened aggregation time) and increased disaggregation shear rate (indicating aggregate strength was higher) in asymptomatic HIV-infected patients versus controls. Other chronic infections have also been reported to be characterized by higher RBC aggregation (Puniyani, Agashe, et al. 1988). Aggregation indexes measured using a photometric rheoscope were significantly higher in pediatric patients with chronic respiratory tract infections and tuberculosis, with no differences between infected subjects with and without tuberculosis (Puniyani, Agashe, et al. 1988).

8.3.3.1 Sepsis and Septic Shock

Sepsis is the systemic, severe inflammatory response to infection (Matot and Sprung 2001) that may lead to multiorgan dysfunction with high mortality rates (Vincent et al. 1998) and to arterial hypotension and perfusion anomalies. Enhanced RBC aggregation is a frequent finding in sepsis (Alt et al. 2004; Berliner et al. 2000; Kirshenbaum et al. 2000). This increase is partly due to increased concentrations of acute phase reactants, most importantly fibrinogen, in response to severe inflammation (Alt et al. 2004). However, it has been repeatedly shown that RBC properties are also altered during sepsis, especially if organ or tissue perfusion problems exist as in septic shock (Piagnerelli et al. 2003).

Figure 8.3 shows the results of an experimental study that demonstrated the differential contribution of plasmatic and cellular factors to alterations of RBC aggregation (Baskurt et al. 1997). Sepsis was induced by cecal ligation-puncture (CLP) in rats and blood samples were obtained at 18 hours. Samples were also obtained from a group of sham-operated animals who underwent only laparotomy and handling of the cecum but not CLP. RBC aggregation indexes in autologous plasma measured by a Myrenne aggregometer were found to be increased significantly in both the sham-operated and sepsis groups with larger changes in the septic group. RBC aggregation measured in 3% dextran 70 kDa was also significantly enhanced in septic rats, indicating that aggregability (i.e., the intrinsic properties of RBC affecting aggregation) was altered in sepsis; there were no alterations of aggregability in sham-operated rats. These data clearly indicate that the enhanced RBC aggregation in sham-operated animals was simply due to the acute phase reaction and increased proaggregating plasma factors, while in sepsis both plasma and RBC properties contribute to the enhanced RBC aggregation. Fibrinogen concentrations measured in both groups were higher compared to control animals, with a significantly higher level in septic animals than sham-operated rats, again indicating a more intense acute phase reaction in sepsis (Baskurt et al. 1997). Further studies using a cell electrophoresis technique confirmed alterations of RBC surface properties (i.e., surface charge) in experimental sepsis, with the decreased surface charge contributing to the enhanced aggregation in both autologous plasma and the standard aggregation medium (Baskurt et al. 2002). Experimental studies have also indicated that parameters reflecting both the extent (i.e., aggregate size) and time course of RBC aggregation as measured by a plate–plate photometric rheoscope are altered in CLP-induced sepsis (Baskurt and Mat 2000).

8.3.4 Cardiovascular Diseases

8.3.4.1 Hypertension

Hypertension is one of the cardiovascular disorders where hemorheological properties have been extensively studied. Hematological, hemorheological, and coagulation abnormalities in arterial hypertension have been reported (Ajmani 1997; Meiselman 1999), and there are studies indicating enhanced RBC aggregation in patients with hypertension (Caimi et al. 1993; Cicco et al. 1999; Cicco and Pirrelli 1999; Foresto et al. 2005; Lebensohn et al. 2009; Leschke et al. 1987; Longhini et al. 1986; Petralito et al. 1985; Puniyani et al. 1987; Vaya et al. 1992). RBC aggregation has also been found to be enhanced in animal models of hypertension (Bor-Kucukatay et al. 2000; Hacioglu et al. 2002; Lominadze et al. 1998).

Both clinical and experimental studies of hypertension have indicated that various aggregation parameters are altered in hypertension, including those related to the time course and extent of aggregation as well as to aggregate strength. Foresto et al. (2005) reported that aggregate shapes in blood samples from hypertensive patients differed from healthy controls, being more spherical and having an increased aggregate shape parameter (see Chapter 4, Section 4.1.5.1). The disaggregation shear stress, reflecting aggregate strength, has also been reported to be significantly increased in samples from hypertensive patients (Razavian et al. 1991). The degree of alteration of aggregation parameters was related to the level of arterial pressure and was determined by the type of hypertension (Ajmani 1997): aggregation was changed more in renal hypertension compared to essential hypertension (Leschke et al. 1987; Longhini et al. 1986).

Since elevated fibrinogen concentration is a common finding in hypertensive patients, increased RBC aggregation in this disease has frequently been related to alterations of plasma factors (Bogar 2002; Lip 1995; Longhini et al. 1986; Vaya et al. 1992) and RBC aggregability has been reported to be unaltered in hypertensive patients (Vaya et al. 1992). However, RBC aggregation indexes measured in standard aggregating media are altered in various animal models of hypertension, thus indicating a contribution of cellular factors to the enhanced aggregation (Hacioglu et al. 2002). It seems possible that the increase of RBC aggregation in hypertensives can be attributed to microvascular damage during the course of the disease and the resulting acute phase reaction (Ajmani 1997; Pirrelli 1999).

The close association between hemorheology and hypertension has been discussed for many years based on clinical observations and experimental studies. However, the exact nature of this relationship cannot yet be clearly described (Bogar 2002; Meiselman 1999). It has been widely accepted that arterial hypertension is the result of increased total peripheral vascular resistance (TPVR) since cardiac output remains essentially unaltered in most cases of hypertension (Meiselman 1999). In turn, TPVR is determined by vascular hindrance and blood rheological behavior, yet the relative contributions of vascular geometry and rheology are still uncertain. Based on fundamental hemodynamic considerations, hemorheological alterations including enhanced RBC aggregation might be among the factors that cause arterial hypertension; relationships between RBC aggregation and vascular resistance have been demonstrated in hypertensive patients (Mchedlishvili et al. 1997). Conversely, hemorheological alterations may result from microvascular damage caused by increased blood pressure. Such microvascular damage and related mechanisms (e.g., acute phase reaction, leukocyte activation) may, in turn, lead to abnormalities of blood rheologic behavior. Note that increased concentrations of acute phase reactants (e.g., fibrinogen) have been shown to affect endothelial function as well as RBC aggregation, thereby contributing to the vascular component of TPVR (Lominadze et al. 2010).

Obviously, the two-way interaction between hemorheology and hypertension can be viewed as a vicious cycle that may contribute to the progression of the clinical course of hypertension. Attempting to unravel the relations between hemorheology and hypertension appears to be a "Which came first ... the chicken or egg?" situation (Meiselman 1999): Is altered hemorheology (e.g., elevated RBC aggregation) the result or the cause of hypertension? Are rheological abnormalities *makers* or *markers* of the disease? If hemorheological alterations were the result of hypertension, they would be expected to be reversed by treatment of hypertension. Hemorheological factors, including RBC aggregation, have been found to be improved following normalization of blood pressure in many but not in all hypertensive subjects (Bogar 2002; Meiselman 1999). Thus the hypothesis that RBC aggregation may be a result of enhanced arterial pressure and the ensuing microvascular damage appears to have merit.

Recent experimental reports support a linkage between vascular and rheological components of TPVR and lend support to increased RBC aggregation as a cause of hypertension. It has been observed that blood pressure in rats with experimentally enhanced RBC aggregation increases gradually over a four-day follow-up period (Figure 8.4a) (Baskurt et al. 2004). In these studies aggregation was enhanced by coating RBC with a polymer (Pluronic F98) that increases the intensity of aggregation when these cells are suspended in unmodified, autologous plasma.

Aggregation was maximal immediately after the exchange transfusions and decreased gradually over the four-day period (Figure 8.4b); aggregation was reduced by about 40% on the fourth day when highest arterial pressures were observed (Figure 8.4a). This inverse relationship between the level of RBC aggregation and mean arterial pressure suggests that enhanced aggregation, per se, may not be the

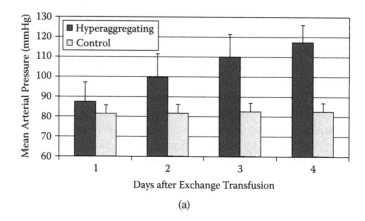

(a)

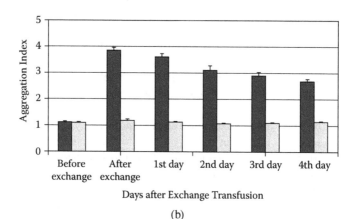

(b)

FIGURE 8.4 (a) Mean arterial pressure measured by a noninvasive method in rats during four days following an exchange transfusion with poylmer-coated, hyperaggregating red blood cell suspensions. (b) Aggregation indexes for blood samples before and after exchange transfusion. See the text for explanations. Data from Baskurt, O. K., O. Yalcin, S. Ozdem, J. K. Armstrong, and H. J. Meiselman. 2004. "Modulation of Endothelial Nitric Oxide Synthase Expression by Red Blood Cell Aggregation." *American Journal of Physiology—Heart and Circulatory Physiology* 286:H222–H229.

direct cause of increased blood pressure, but rather may have triggered mechanisms that gradually increase arterial pressure. It has been shown that nitric oxide
(NO)-related vasomotor mechanisms are impaired, such as decreased endothelial
NO synthase eNOS) expression in resistance arteries in rats with greatly elevated
RBC aggregation (see Chapter 7, Sections 7.1 and 7.4.3.2). Down-regulation of eNOS
expression and activity can be explained by an enhanced axial migration of RBC
induced by RBC aggregation (see Chapter 6, Section 6.5), thereby decreasing wall
shear stress and hence reducing eNOS synthesis and NO levels in endothelial cells
(Baskurt et al. 2004; Yalcin et al. 2008). Such findings indicate that RBC aggregation may contribute to alterations of vascular geometry and hence arterial pressure.

8.3.4.2 Atherosclerosis

Atherosclerosis is widely accepted as an inflammatory disease (Ross 1999), with
acute phase reactants, including fibrinogen, frequently cited as early indicators of
the atherosclerotic process in asymptomatic individuals (Ben Assayag et al. 2008;
Tribouilloy et al. 1998). Fibrinogen has always been among the major cardiovascular
risk factors (Lowe et al. 1993; Sweetnam et al. 1996; Vaya et al. 1996; Yarnell et
al. 1991) and RBC aggregation has been reported to correlate with asymptomatic
atherosclerotic processes (Ben Assayag et al. 2008). Berliner et al. (2005) proposed
RBC aggregation as an early marker of the inflammatory process related to the
development of atherosclerotic disease.

The "chicken and egg" problem also characterizes the relationship between RBC
aggregation and atherosclerosis (see Section 8.3.4.1). RBC aggregation may simply
reflect the inflammatory process related to the atherosclerosis or, alternatively, RBC
aggregation may play a role in the development of endothelial dysfunction and atherosclerosis. Wall shear stress is an important factor in maintaining the quiescent
status of endothelium and any disturbance of hemodynamic forces such as changes
of magnitude or flow pattern will affect the endothelium, resulting in activation of
endothelial cells (Baskurt and Meiselman 2010; Chien 2006; Chien 2007; Malek
et al. 1999). RBC aggregation has been demonstrated to influence effective hemodynamic forces (i.e., wall shear stress) and interfere with endothelial function (see
Section 8.3.4.1 and Chapter 7, Section 7.4.3.2). Therefore, enhanced aggregation may
be expected to interfere with endothelial function and trigger atherosclerotic development in the vasculature (Boisseau 2006; Forconi and Gori 2009; Morariu et al. 2004).

8.3.4.3 Myocardial Ischemia and Infarction

RBC aggregation has been shown to be significantly enhanced in patients with
ischemic heart diseases (Ben Ami et al. 2001; Caimi et al. 2003; Justo et al. 2009;
Lee et al. 2008a; Lee et al. 2008b; Marton et al. 2003; Neumann et al. 1989;
Pfafferott et al. 1999; Shapira et al. 2001; Zorio et al. 2008). The degree of increase
depends, in general, on the severity of the ischemic event, being highest in acute
myocardial infarction and markedly less in stable angina (Ben Ami et al. 2001;
Lee et al. 2008a). Lee et al (2008a) used a Myrenne aggregometer to measure
RBC aggregation indexes in patients with stable and unstable angina and acute
myocardial infarction (Figure 8.5). Both the M index at stasis and the M1 index
at low shear were significantly higher in patients with stable and unstable angina

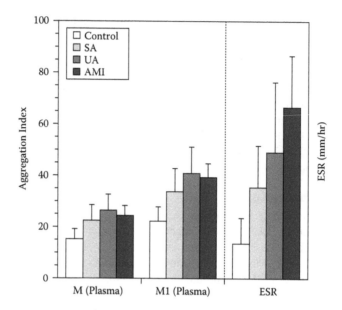

FIGURE 8.5 Red blood cell aggregation indexes (M, M1) and erythrocyte sedimentation rate (ESR) for blood samples of patients with stable angina (SA), unstable angina (UA), and acute myocardial infarction (AMI). ESR values can be read from the aggregation index axis. Reproduced from Lee et al. (2008a). With permission.

compared to control. Both aggregation indexes for patients with acute myocardial infarction were similar to those for subjects with unstable angina, although ESR measurements showed significantly higher values for patients with acute myocardial infarction. The authors suggested that the Myrenne aggregometer might have failed to indicate the strong aggregation in infarct patients due to insufficient disaggregation forces (see Chapter 4, Section 4.1.6.8.2). Ben Ami et al. (2001), using cell flow analysis, have reported higher aggregation indexes for patients with acute myocardial infarction versus those with unstable angina.

Contributions of plasmatic and cellular factors to RBC aggregation have been reported to differ between unstable angina and acute myocardial infarction, with both factors contributing to the alterations in ischemic heart diseases (Ben Ami et al. 2001; Lee et al. 2008a). However, plasma factors contribute more prominently to the significantly higher RBC aggregation in acute myocardial infarction (Ben Ami et al. 2001), with this finding supported by significantly increased acute phase reactants (Lee et al. 2008a). Alterations in RBC aggregability may be expected to result from local alterations in myocardial tissue caused by ischemia. In an animal model of myocardial ischemia induced by a critical coronary artery stenosis plus increased hematocrit, RBC mechanical properties were significantly altered in blood from the coronary sinus draining the ischemic myocardium (Baskurt et al. 1991). During the acute phase of myocardial infarction, the extensive tissue injury triggers an intense acute phase reaction that further contributes to the alterations of RBC aggregation.

Increased RBC aggregation, together with increased inflammatory markers, plasma viscosity, and hematocrit-corrected whole blood viscosity, have been reported in cardiac syndrome X and suggested to contribute to the pathophysiology of this cardiac syndrome (Lee et al. 2008b). An important role has been attributed to RBC aggregation in determining the prognosis of unstable angina (Neumann et al. 1991). There was a significant difference between the event-free survival rates of patients with high and low RBC aggregation measured when the patients were admitted to the hospital (Figure 8.6); the prognostic value of hemorheological parameters including RBC aggregation has also been supported by other studies (Resch et al. 1991).

8.3.4.4 Cerebral Ischemia and Stroke

Cerebrovascular disorders form a special group of circulatory problems with hemorheological findings similar to those related to ischemia or infarct of other tissues. Szapary et al. (2004) studied RBC aggregation together with other hemorheological parameters in patients with carotid artery stenosis and reported a strong correlation between RBC aggregation and severity of the stenosis. RBC aggregation has been found to be significantly enhanced in patients with cerebral infarction (Beamer et al. 1997; Bolokadze et al. 2006; Tanahashi et al. 1996; Zeltser et al. 2001) and is higher during the acute phase of stroke compared to the chronic phase (Tanahashi et al. 1996).

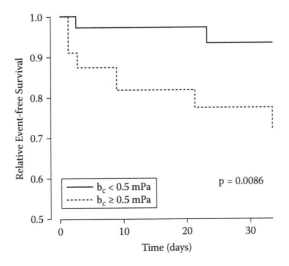

FIGURE 8.6 Comparison of relative event-free survival (Kaplan-Meier survival distribution) in unstable angina patients with high (dotted line) or low (solid line) RBC aggregation indexes. RBC aggregation was measured in samples obtained at admission to the hospital using a photometric rheoscope. The parameter b_c is the rate constant for aggregation obtained by multiplying the measured rate by plasma viscosity. Reproduced from Neumann, F-J., H. A. Katus, E. Hoberg, P. Roebruck, M. Braun, H. M. Haupt, H. Tillmanns, and W. Kubler. 1991. "Increased Plasma Viscosity and Erythrocyte Aggregation: Indicators of an Unfavourable Clinical Outcome in Patients with Unstable Angina Pectoris." *British Heart Journal* 66:425–430 with permission.

The degree of alterations of RBC aggregation has usually been reported to be correlated with the level of acute phase reactants and hence the severity of cerebral damage and related complications (Tanahashi et al. 1996). Lechner et al. (1994) suggested that plasma fibrinogen levels predict carotid atherosclerosis in neurologically asymptomatic patients more successfully than other factors included in their study. Both RBC aggregation and other hemorheological factors, together with parameters reflecting tissue damage, were found to be more affected if measured in blood samples draining the affected area (e.g., jugular vein of the affected brain hemisphere) (Bolokadze et al. 2006). However, alterations in hemorheological factors from these affected areas can modify blood in the systemic circulation and may further influence perfusion and function in previously unaffected areas of the brain (Mchedlishvili et al. 2004). Higher RBC aggregation indexes have been reported for patients with hemorrhagic stroke compared to those with ischemic stroke (Bolokadze et al. 2006).

Several investigators have reported that RBC aggregation alterations are more pronounced if cerebrovascular diseases were combined with or have an etiological relationship with pathologic conditions such as hypertension, hyperlipidemia, or diabetes (Beamer et al. 1997; Szapary et al. 2004). Beamer et al. (1997) suggested that better control of blood glucose levels or blood pressure may reduce RBC aggregation in cerebrovascular patients and may also affect their clinical course. RBC aggregation has been demonstrated to have a prognostic significance in stroke patients: the level of RBC aggregation during the acute phase of a cerebrovascular event was found to be associated with the probability of having a second stroke (Resch et al. 1991; Szapary et al. 2004).

Beridze et al. (2004) reported a negative correlation between blood NO concentrations and RBC aggregation assessed by an image analysis in patients with ischemic stress. Patients grouped as "mild stroke" according to widely accepted scales were characterized with lower aggregation indexes and higher NO concentrations. Beridze et al. (2004) concluded that NO may prevent extensive cerebral damage by various mechanisms including reducing RBC aggregation as demonstrated by Bor-Kucukatay et al. (2000).

8.3.4.5 Peripheral Vascular Diseases

Most circulatory disorders discussed in Section 8.3.4 are primarily related to the arterial side of the vasculature (e.g., ischemic heart and cerebrovascular diseases). It is now well known that endothelial dysfunction and consequent atherosclerotic processes are important pathophysiological mechanisms affecting arterial function. Deterioration of arterial function is not limited to coronary or cerebral circulations but rather affects all of the vasculature, including other organs and the peripheral circulation: peripheral occlusive arterial diseases (POAD) are relatively common and can cause serious health problems and major lifestyle changes.

The hemorheological alterations discussed previously for other circulatory disorders have also been demonstrated in patients suffering from peripheral arterial diseases. RBC aggregation as measured by the Sefam erythroaggregometer was found to be significantly elevated in POAD patients compared to control subjects (Koscielny et al. 2004). There was also an increase of plasma fibrinogen concentration in the patient group, indicating the contribution of plasmatic factors to the enhanced RBC aggregation. Interestingly, Koscielny et al. (2004) reported no difference in RBC aggregation between patient groups with POAD, coronary heart disease, and cerebrovascular insufficiency; the presence of cardiovascular risk factors did not predict differences of RBC aggregation. Le Devehat et al. (2001) have reported significantly altered aggregation parameters (i.e., extent and time course of RBC aggregation, disaggregation shear rate) in patients with lower limb ischemia as a complication of diabetes, thus indicating enhanced RBC aggregation.

Venous insufficiency also has hemorheological consequences, although the mechanisms related to the development of the disease process differ from those affecting the arterial side. Venous hypertension, a consequence of the upright position of humans, is the major cause of venous insufficiency, with the final severity influenced by age and genetic risk factors (Boisseau 2006; Boisseau et al. 2004). The time course of RBC aggregation was found to be accelerated in patients with vascular insufficiency, with the degree of alteration correlated with disease severity and the presence of complications (e.g., varicose veins, local skin disorders) (Khodabandehlou et al. 2004). Le Devehat et al. (1989b) have reported increased RBC aggregation indexes in blood obtained from an antecubital vein and from the affected veins of patients with lower limb venous insufficiency: RBC aggregation indexes were increased after the application of venous stasis for 30 minutes in patients but not in the control group. This finding confirms the effect of local tissue changes on the properties of blood obtained from such regions (Khodabandehlou et al. 2004). Plasma fibrinogen concentrations were also significantly higher in the patient group and were further increased following the 30-minute period of stasis (Le Devehat et al 1989b). These observations are consistent with RBC aggregation being a very important factor in venous blood flow since shear forces are in the range to allow prominent aggregate formation. Therefore, RBC aggregation may also play role in the development of complications (e.g., deep vein thrombosis) secondary to venous insufficiency (Le Devehat et al. 1989a; Mira et al. 1998).

Accelerated RBC aggregation has been reported in patients with Reynauld's phenomena, a vasomotor disorder affecting the peripheral circulation (Spengler et al. 2004): RBC aggregation rate was increased while aggregate size was not affected. These alterations and the associated increases of fibrinogen concentrations were also accompanied by impaired microcirculatory flow as studied by nailfold capillaroscopy.

8.3.4.6 Ischemia–Reperfusion Injury

The pathophysiological mechanisms that are involved in the hemorheological consequences of ischemia–reperfusion injury include increased free-radical generation and white blood cell (WBC) activation (Gori et al. 2006; Kayar et al. 2001). *In vitro* studies demonstrated that RBC aggregability can be affected by both factors: disaggregation shear rate indicating aggregate strength was found to be significantly increased in RBC suspensions exposed to free radicals (Baskurt et al. 1998) or to activated WBCs (Baskurt and Meiselman 1998). Myrenne aggregation indexes were lower as a consequence of free radical exposure (Baskurt et al. 1998), while the same indexes were higher for suspensions of RBC incubated with activated WBCs (Baskurt and Meiselman 1998). Further, ischemia-reperfusion injury may trigger the acute phase response, one of the major causes of increased RBC aggregation (Nemeth et al. 2004; Peto et al. 2007). Therefore, the effect of ischemia–reperfusion is expected to reflect the combination of these factors.

In a study using a rat hind limb preparation, Kayar et al. (2001) reported decreased RBC aggregability following 10 minutes of nonperfusion ischemia and 15 minutes of reperfusion in blood samples from the affected limb. These changes were correlated with changes in RBC surface properties (e.g., surface charge density) assessed using particle electrophoresis. Brath et al. (2010) studied the effect of intestinal ischemia–reperfusion in rats and reported decreased aggregation indexes in samples obtained from the portal vein following a 30-minute occlusion of the superior mesenteric artery plus 60 minutes of reperfusion. Tamas et al. (2010) investigated the time course of changes in RBC aggregation indexes in the skeletal muscle of dogs after ischemia–reperfusion. They observed increased aggregation indexes in the samples obtained 1 and 5 minutes after the start of reperfusion, following which the indexes returned to resting values. These results suggest that alterations in RBC aggregation status may be complex and time dependent, especially under *in vivo* conditions, most probably due to the contribution of various factors discussed in the previous paragraph.

8.3.4.7 Circulatory Shock

Circulatory shock refers to a condition of general hypoperfusion due to low arterial pressure, which may lead to significant tissue and organ damage. The pathophysiology of shock is complex and may involve a variety of factors including WBC activation and hemorheological alterations (Zhao 2005). RBC aggregation becomes an increasingly important factor with reduced perfusion pressure and slower blood flow (i.e., reduced shear rates), thereby contributing to a vicious cycle leading to irreversible shock (Ince 2005; Koppensteiner 1996). Circulatory shock may frequently be a consequence of trauma, which may involve severe hemorrhage. Experimental studies have indicated that traumatic shock may cause significantly higher alterations of RBC aggregation than hemorrhagic shock without severe tissue or organ injury (Tatarishvili et al. 2006).

8.3.5 Metabolic Disorders

8.3.5.1 Diabetes

Diabetes is a complex clinical condition characterized by a variety of pathophysiological problems (Dintenfass 1991). The major abnormality in diabetes is related to glucose utilization and hyperglycemia that can be due to the lack of insulin (type 1 diabetes) or tissues being unable to respond to insulin (insulin resistance, type 2 diabetes). A discussion of the complete pathophysiology of diabetes is beyond the scope of this section, but it should be noted that an important aspect of diabetes pathophysiology is related to vascular problems leading to important complications such as kidney failure, vision impairment. and peripheral artery diseases (McMillan 1992).

Diabetes has frequently been described as a vascular disease since most morbidity and mortality associated with this clinical entity relate to angiopathy developed in the course of disease (Toth and Kesmarky 2007).

RBC aggregation is elevated in blood from diabetic patients as would be expected in a pathophysiological condition characterized with angiopathy. Bauersachs et al. (1989a) studied RBC aggregation in type 2 (noninsulin-dependent) diabetic subjects using a Myrenne aggregometer modified to measure aggregation half time and disaggregation shear rate. The patients were poorly controlled with serum glucose levels over 21 mmol/l and a mean glycated hemoglobin (HbA1c) percentage of 15.7 ± 2.1% versus 4.9–7.5% for healthy controls. Aggregation half time was decreased and disaggregation shear rate was significantly increased; plasma fibrinogen levels were also significantly elevated (Bauersachs et al. 1989a). Conversely, Bertram et al. (1992) reported no alteration of RBC aggregation indexes in type 1 (insulin-dependent) diabetes, although plasma viscosity was found to be higher compared to control subjects. They also reported no effects of glycemic control as judged by percent HbA1c on RBC aggregation (Bertram et al. 1992). However, their subjects had lower HbA1c levels, even in their "bad glycemic control" group, compared to the patient group included in the study by Bauersachs et al (1989a). A report by Elishkevitz et al. (2002) supports the findings of Bertram et al. in that they report no relationships among glycemic control, RBC aggregation, and acute phase reactants.

In contrast to the studies by Bertram et al. (1992) and Elishkevitz et al. (2002), significant correlations between HbA1c as an indicator of glycemic control and RBC aggregation have been reported for type 1 but not type 2 diabetes (Zilberman-Kravits et al. 2006). Babu and Singh (2005) reported increased RBC aggregation in a group of patients with diabetes, but did not indicate the nature of the disease: RBC aggregation was more pronounced in the samples from patients with higher blood glucose concentrations. However, their findings may not be reflecting the direct effect of glucose on RBC aggregation since the severity of diabetes-related vascular problems may contribute to such differences. Khodabandehlou et al. (1998) reported decreased RBC aggregation obtained during an acute hyperglycemic period in type 1 diabetes compared to normoglycemic control samples from the same patients. Such discrepancies among data reported by different groups apparently reflect the diversity among patient groups, the variety of protocols, and the complex nature of diabetes (Dintenfass 1991).

A report by Chong-Martinez et al. (2003) is an example of longitudinal studies comparing RBC aggregation in diabetic patients before and after significantly improved diabetic control. RBC aggregation indexes (M and M1) measured by Myrenne aggregometer, as well as disaggregation shear rate, decreased significantly following 14 weeks of appropriate treatment for diabetic control. They also demonstrated that RBC aggregability studied in 3% dextran 70 kDa was diminished following the treatment (Chong-Martinez et al. 2003). Since 14 weeks (i.e., 98 days) is a large proportion of the 120-day *in vivo* RBC lifespan, their results indicate that "newer" cells generated during improved glycemic control have normalized properties affecting aggregation.

Reports of RBC aggregation in diabetic patients with vascular complications are fairly consistent. Momstselidze et al. (2006) reported increased aggregation indexes in diabetic patients with cerebral infarcts; the existence of diabetes in cerebral infarct patients predicted higher aggregation indexes compared to nondiabetic patients with similar cerebrovascular events. Mantskava et al. (2006) reported similar increases in RBC aggregation indexes in type 1 and type 2 diabetes patients with lower extremity gangrene. Vekasi et al. (2001) indicated enhanced RBC aggregation in diabetic retinopathy, with increases of aggregation index and disaggregation shear rate and decreased aggregation half time as measured with the LORCA aggregometer; aggregation parameters did not differ among patients with various stages of retinopathy.

Increased plasma fibrinogen is a common finding in patients with diabetes, especially if associated with vascular pathologies, and the fibrinogen molecule may be modified in diabetic patients (Khodabandehlou et al. 1996; Schpitz-Droz et al. 1991). In addition, alterations of RBC membrane properties (e.g., decreased sialic acid and glycophorin A content, modified phospholipid composition) have been found in diabetes and may contribute to the enhanced aggregation (Martinez et al. 1998).

8.3.5.2 Metabolic Syndrome and Obesity

Metabolic syndrome includes alterations in plasma lipid composition, obesity, impaired glucose tolerance, and hypertension, and leads to a highly increased risk for diabetes and cardiovascular disease (e.g., atherosclerosis) (Aloulou et al. 2006b). Hemorheological alterations including enhanced RBC aggregation have been demonstrated in patients with metabolic syndrome (Aloulou et al. 2006a; Aloulou et al. 2006b; Toker et al. 2005). Toker et al. (2005) reported a significant positive correlation between RBC aggregation and the number of existing components of metabolic syndrome in a given patient, while Aloulou et al. (2006a) found significant correlations between RBC aggregation and plasma cholesterol, but not with the clinical score of metabolic syndrome. This latter observation appears to mainly reflect the low-grade inflammation in metabolic syndrome (Toker et al. 2005) and the associated increase of acute phase reactants including fibrinogen.

RBC properties such as membrane lipid composition can be affected by alterations of plasma lipid profiles (Uyuklu et al. 2007), with these changes possibly associated with alterations of RBC aggregation (Vaya et al. 1997; Vaya et al. 1993). Obesity without other components of metabolic syndrome has been reported to be associated with enhanced RBC aggregation (Sola et al. 2004). These investigators report increased M and M1 indexes measured by Myrenne aggregometer in obese individuals (BMI: 44.9 ± 6.7) compared to the control group (BMI: 23.5 ± 4.8); weight loss was found to be effective in normalizing RBC aggregation indexes (Sola et al. 2007).

8.3.6 Hematological Disorders

Since RBC aggregation is determined by both cellular and plasmatic factors, any alterations of RBC properties are expected to influence the aggregation behavior of blood. Such alterations may be due to genetic disorders affecting hemoglobin structure and properties (i.e., hemoglobinopathies), alterations of membrane proteins leading to shape changes (e.g., hereditary spherocytosis, hereditary elliptocytosis) or defects in RBC metabolism (e.g., glucose-6-phosphate dehydrogenase deficiency) (Baskurt 2007). RBC properties may also be altered in acquired diseases such as anemia of various causes and hematological malignancies (Toth and Kesmarky 2007).

Thalassemia is a relatively common genetic disorder that has been reported to be manifested by altered aggregation. This disease is characterized by a defective synthesis of globin chains, which leads to inefficient hemoglobin formation and accumulation of unpaired globin chains in RBC (Schrier 1994). The unpaired globin chains interfere with normal RBC development and cell mechanics, thereby reducing RBC deformability. Oxidative stress is also enhanced in RBC with defective hemoglobin formation due to unpaired globins (Schrier and Mohandas 1992; Scott et al. 1993), and it is possible that this enhanced oxidative stress may be a mechanism causing altered aggregation in thalassemic patients.

Chen et al. (1996) reported significantly enhanced RBC aggregation in β-thalassemia major patients (i.e., homozygous patients with defective β-globin gene) using a cell-flow image analysis system (Figure 8.7). RBC aggregate size in blood from these subjects was increased more than twofold immediately prior to a periodic blood transfusion and decreased to control levels following the transfusion (Chen et al. 1996). RBC aggregability was also reported to be significantly enhanced indicating that RBC damage is involved in the observed alterations, suggesting that enhanced RBC aggregation may lead to circulatory problems that are frequently encountered in thalassemia major patients.

Reports on heterozygous thalassemia patients are not always in agreement. Vaya et al. (2003) have shown increased RBC aggregation in thalassemia carriers using a Myrenne aggregometer (Vaya et al. 2003), whereas the same group reported decreased aggregation indexes and disaggregation shear rate and increased aggregation time in similar patients when using a Sefam erythroaggregometer (Falco et al. 2003). It is not yet possible to resolve this apparent disagreement but it may relate to the specific design of the two aggregometers (see Chapter 4).

Plasma composition and the properties of plasma proteins are important determinants of RBC aggregation. Alterations in fibrinogen structure are known to induce prominent alterations of RBC aggregation and serious circulatory problems (Kwaan et al. 1997). Greatly enhanced RBC aggregation in plasma cell dyscrasias is a well-known finding and is due to the markedly increased gamma globulin concentration in plasma (Kwaan and Bongu 1999; Mehta and Singhal 2003). Zhao et al. (1999) reported very high aggregation in blood samples from patients with multiple myeloma that was so intense that it was not measurable by some instruments due to insufficient disaggregation shear stress.

8.3.7 Other Pathophysiological Conditions

Hemorheological alterations have been repeatedly demonstrated in a variety of malignant diseases (von Tempelhoff et al. 2003). These alterations (e.g., increased extent of RBC aggregation) are mainly due to major acute phase reactions induced by the process of malignancy. However, increased RBC aggregation may be both an indicator of this reaction and also a contributing factor promoting metastasis : slower

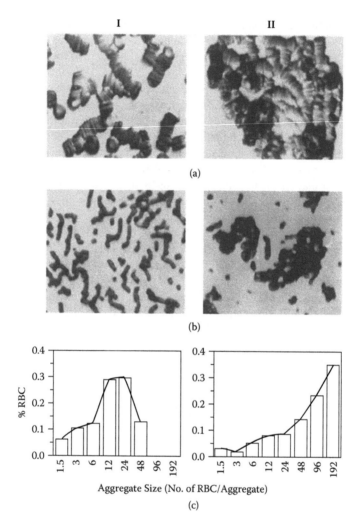

FIGURE 8.7   Red blood cell aggregation in blood of a control subject (I) and a β-thalassemia major patient (II). The lower panels show the distribution of aggregate size, as RBC per aggregate, in these samples; size is shifted to higher values in thalassemic patients. Reproduced from Chen, S., A. Eldor, G. Barshtein, S. Zhang, A. Goldfarb, E. Rachmilewitz, and S. Yedgar. 1996. "Enhanced Aggregability of Red Blood Cells of Beta-Thalassemia Major Patients." *American Journal of Physiology Heart and Circulatory Physiology* 270:H1951–H1956 with permission.

blood flow due to enhanced aggregation may favor entrapment of malignant cells (Dintenfass 1982).

Enhanced RBC aggregation has been reported in cases of nephrotic syndrome (Ozanne et al. 1983a) and chronic renal failure (Martinez et al. 1999), whereas decreased aggregation occurs in liver cirrhosis (Liu et al. 2006), with the latter finding probably due to the diminished synthesis of plasma factors and the abnormal

RBC shape in severe liver failure. Significantly enhanced RBC aggregation has been reported in a patient with primary sclerosing cholangitis complicated by biliary stricture due to elevated immunoglobulin concentrations (Anwar et al. 2004). Increased immunoglobulin (e.g., IgG, IgM) concentrations are known to increase RBC aggregation, especially if the levels are substantially higher than the normal physiological range (Rampling 1988).

Exposure of blood to a nonphysiological environment (e.g., extracorporeal circulation) may result in alterations of RBC aggregation (Kameneva and Antaki 2007). RBC aggregation is significantly enhanced after the implantation of a heart-assist device in cardiac patients and following implanting a ventricular-assist device in calves. These effects are, in part, due to the mechanical trauma to RBC encountered in these nonphysiological environments.

There are hundreds of articles in the literature reporting alterations of RBC aggregation in a wide variety of clinical situations, and there are several reports attempting to link the pathophysiology to mechanisms responsible for the altered aggregation. A complete review of this literature is beyond the scope of this chapter. The reader is referred to the relevant journals (e.g., *Clinical Hemorheology and Microcirculation*, IOS Press, Amsterdam) and other sources that discuss the clinical aspects of hemorheological alterations in more detail (Baskurt et al. 2007; Chien et al. 1987; Lowe 1988; Stoltz et al. 1999).

8.4 THERAPEUTIC APPROACH TO RED BLOOD CELL AGGREGATION

Given the significant alterations of RBC aggregation in various disease processes and the expected influence of these alterations on *in vivo* blood flow (see Chapter 7), several therapeutic strategies have been proposed. These include pharmacological treatments as well as hemapheresis to reduce plasma factors that affect RBC aggregation. However, none of these therapeutic methods have yet achieved well-documented clinical success and most remain experimental protocols. The treatment of the primary (i.e., the specific disease process) or secondary (e.g., acute phase reaction) causes of RBC aggregation can result in normalization of hemorheological parameters. Typical examples are reduced RBC aggregation following thrombolytic therapy (Ben-Ami et al. 2002) or improved glycemic control in diabetics (Chong-Martinez et al. 2003). However, such treatments are focused on therapy for the disease and not as therapies targeted at RBC aggregation.

8.4.1 PHARMACOLOGICAL AGENTS AFFECTING RED BLOOD CELL AGGREGATION

Enzymes that reduce fibrinogen levels (e.g., Ancrod) have been widely explored as treatments for severe circulatory problems including peripheral vascular diseases, coronary artery diseases, and ischemic stroke (Ernst 1987; Levy et al. 2009; Tanahashi et al. 1995). Ancrod is an enzyme isolated from snake venom that can be used parenterally, reducing plasma fibrinogen levels within hours; RBC aggregation, plasma and whole blood viscosities are reduced with Ancrod treatment. Several drugs designed for other purposes (e.g., clofibrate, stanozolol, ticlopidine, sulocti-dil, hydroxychroloquine) have fibrinogen reducing effects that may lead to reduced RBC aggregation. A more detailed description of such drugs can be found elsewhere (Ehrly 1991; Ernst, 1987).

Troxerutine is a pharmacological agent proposed to be useful for reducing RBC aggregation (Boisseau et al. 1989). It is a flavonoid, and marketed as a veno-protec-tive drug generally prescribed for venous disorders such as varicose veins and hem-orrhoids (Vin et al. 1994). Le Devehat (1989b) studied the hemorheological effects of Troxerutine in patients with venous insufficiency of the lower limbs. RBC aggrega-tion following 30 minutes of venous stasis was found to be significantly enhanced in the samples obtained from the veins of affected limbs. This increase of aggregation due to stasis was prevented by Troxerutine treatment for 15 days prior to the venous stasis studies (Le Devehat et al. 1989b).

Troxerutine has a powerful anti-inflammatory activity, which may explain some of the findings related to the antiaggregating activity of the drug. However, Troxerutine has also been shown to be effective in reducing aggregation in the presence of high-molecular-mass dextrans (Durussel et al. 1998). These findings suggested a direct action of Troxerutine on RBC properties affecting aggregability, possibly by attach-ment to the RBC glycocalyx and interference with fibrinogen interacting with the cell surface (Durussel et al. 1998). Rutosides are other flavonoids suggested to be effective in reducing RBC aggregation (Boisseau and Toth 2007; Wadworth and Faulds 1992); they have an anti-inflammatory activity and are effective in preventing white blood cell and endothelial cell activation, thus reducing the influence of such processes on RBC aggregation.

Buflomedil is another pharmacological agent with antiaggregating activity under *in vitro* conditions (Ben-Ami et al. 2002). It is also a vasoactive agent suggested to be especially effective in claudication but less effective in cerebral-vascular disease. Buflomedil has not yet been extensively studied clinically with respect to its effect on RBC aggregation.

8.4.2 SELECTIVE REMOVAL OF BLOOD COMPONENTS: HEMAPHERESIS

Hemapheresis is the selective removal of blood elements via extracorporeal devices and can be applied to various cell types and to plasma components (Fadul et al. 1997; Schooneman et al. 1997). Details of this approach can be found elsewhere (MacPherson and Kasprisin 2010). Briefly, it can be used to selectively remove high-molecular-mass plasma proteins (e.g., fibrinogen, $\alpha 2$-macroglobulin), which are important promoters of RBC aggregation. Hemapheresis therapy has been demon-strated to be effective in reducing RBC aggregation in various clinical disorders (Fadul et al. 1997; Klingel et al. 2002) and is especially useful in controlling the greatly enhanced RBC aggregation that may influence microcirculatory function in plasma cell dyscrasias (Schooneman et al. 1997).

8.4.3 OTHER TREATMENT APPROACHES

There is accumulating evidence for hemorheological alterations due to dietary changes. For example, a vegetarian diet has been reported to improve the hemorheological properties of blood (Ernst 1987; McCarty 2002). However, conclusive studies on the influence of various diets on RBC aggregation are still lacking. Exercise training is another factor that has been shown to improve hemorheology (see Section 8.2.1.2) and there is general agreement regarding the effect of physical fitness on acute changes of RBC aggregation induced by episodes of strenuous exercise (Senturk et al. 2005).

LITERATURE CITED

Acciavatti, A., D. Pieragalli, T. Provvedi, G. L. Messa, C. Frigerio, M. Saletti, C. Galigani, et al. 1993. "Circadian Variation of Platelet-Aggregation and Blood Rheology." *Clinical Hemorheology* 13:177–186.

Ajmani, R. S. 1997. "Hypertension and Hemorheology." *Clinical Hemorheology and Microcirculation* 17:397–420.

Ajmani, R. S., J. L. Fleg, A. A. Demehin, J. G. Wright, F. O'Connor, J. M. Heim, E. Tarien, and J. M. Rifkind. 2003. "Oxidative Stress and Hemorheological Changes Induced by Acute Treadmill Exercise." *Clinical Hemorheology and Microcirculation* 28:29–40.

Ajmani, R. S., and J. M. Rifkind. 1998. "Hemorheological Changes during Human Aging." *Gerontology* 44:111–120.

Almog, B., R. Gamzu, R. Almog, J. B. Lessing, I. Shapira, S. Berliner, D. Pauzner, S. Maslovitz, and I. Levin. 2005. "Enhanced Erythrocyte Aggregation in Clinically Diagnosed Pelvic Inflammatory Disease." *Sexually Transmitted Diseases* 32:484–486.

Aloulou, I., E. Varlet-Marie, J. Mercier, and J–F. Brun. 2006a. "Hemorheological Disturbances Correlate with the Lipid Profile but Not with the NCEP-ATPIII Score of the Metabolic Syndrome." *Clinical Hemorheology and Microcirculation* 35:207–212.

Aloulou, I., E. Varlet-Marie, J. Mercier, and J–F. Brun. 2006b. "The Hemorheological Aspects of the Metabolic Syndrome Are a Combination of Separate Effects of Insulin Resistance, Hyperinsulinemia and Adiposity." *Clinical Hemorheology and Microcirculation* 35:113–119.

Alt, E., B. R. Amann-Vesti, C. Madl, G. Funk, and R. Koppensteiner. 2004. "Platelet Aggregation and Blood Rheology in Severe Sepsis/Septic Shock." *Clinical Hemorheology and Microcirculation* 30:107–115.

Anwar, M. A., and M. W. Rampling. 2004. "Erythrocyte Hyper-Aggregation in a Patient Undergoing Orthotopic Transplantation for Primary Sclerosing Cholangitis Complicated by Biliary Stricture." *Clinical Hemorheology and Microcirculation* 31:169–172.

Babu, N., and M. Singh. 2005. "Analysis of Aggregation Parameters of Erythrocytes in Diabetes Mellitus." *Clinical Hemorheology and Microcirculation* 32:269–277.

Ballou, S. P., F. B. Lozanski, S. Hodder, D. L. Rzewnicki, L. C. Mion, J. D. Sipe, A. B. Ford, and I. Kushner. 1996. "Quantitative and Qualitative Alterations of Acute-Phase Proteins in Healthy Elderly Persons." *Age and Aging* 25: 224–230.

Baskurt, O. K. 2007. "Mechanisms of Blood Rheology Alterations." In *Handbook of Hemorheology and Hemodynamics*, ed. O. K. Baskurt, M. R. Hardeman, M. W. Rampling, and H. J. Meiselman, 170–190. Amsterdam, Berlin, Oxford, Tokyo, Washington, DC: IOS Press.

Baskurt, O. K., M. Bor-Kucukatay, O. Yalcin, H. J. Meiselman, and J. K. Armstrong. 2000. "Standard Aggregating Media to Test the 'Aggregability' of Rat Red Blood Cells." *Clinical Hemorheology and Microcirculation* 22:161–166.

Baskurt, O. K., M. R. Hardeman, M. W. Rampling, H. J. Meiselman, eds. 2007. *Handbook of Hemorheology and Hemodynamics*. Amsterdam, Berlin, Oxford, Tokyo, Washington, DC: IOS Press.

Baskurt, O. K., E. Levi, S. Caglayan, N. Dikmenoglu, O. Ucer, R. Guner, and S. Yorukan. 1991. "The Role of Hemorheological Factors in the Coronary Circulation." *Clinical Hemorheology* 11:121–127.

Baskurt, O. K., and F. Mat. 2000. "Importance of Measurement Temperature in Detecting the Alterations of Red Blood Cell Aggregation and Deformability Studied by Ektacytometry: A Study on Experimental Sepsis in Rats." *Clinical Hemorheology and Microcirculation* 23:43–49.

Baskurt, O. K., and H. J. Meiselman. 1998. "Activated Polymorphonuclear Leukocytes Affect Red Blood Cell Aggregability." *Journal of Leukocyte Biology* 63:89–93.

Baskurt, O. K., and H. J. Meiselman. 2010. "Endothelial Function and Physical Activity." In *Exercise Physiology: From a Cellular to an Integrative Approach*, ed. P. Connes, O. Hue, and S. Perrey, 230–244. Amsterdam, Berlin, Oxford, Tokyo, Washington, DC: IOS Press.

Baskurt, O. K., A. Temiz, and H. J. Meiselman. 1997. "Red Blood Cell Aggregation in Experimental Sepsis." *Journal of Laboratory and Clinical Medicine* 130:183–190.

Baskurt, O. K., A. Temiz, and H. J. Meiselman. 1998. "Effect of Superoxide Anions on Red Blood Cell Rheologic Properties." *Free Radicals in Biology and Medicine* 24:102–110.

Baskurt, O. K., E. Tugral, B. Neu, and H. J. Meiselman. 2002. "Particle Electrophoresis as a Tool to Understand the Aggregation Behavior of Red Blood Cells." *Electrophoresis* 23:2103–2109.

Baskurt, O. K., O. Yalcin, S. Ozdem, J. K. Armstrong, and H. J. Meiselman. 2004. "Modulation of Endothelial Nitric Oxide Synthase Expression by Red Blood Cell Aggregation." *American Journal of Physiology—Heart and Circulatory Physiology* 286:H222–H229.

Bauersachs, R. M., S. J. Shaw, A. Zeidler, and H. J. Meiselman. 1989a. "Red Blood Cell Aggregation and Blood Viscoelasticity in Poorly Controlled Type 2 Diabetes Mellitus." *Clinical Hemorheology* 9:935–952.

Bauersachs, R. M., R. B. Wenby, and H. J. Meiselman. 1989b. "Determination of Specific Red Blood Cell Aggregation Indices via an Automated System." *Clinical Hemorheology* 9:1–25.

Beamer, N., G. Giraud, W. Clark, M. Wynn, and B. Coull. 1997. "Diabetes, Hypertension and Erythrocyte Aggregation in Acute Stroke." *Cerebrovascular Diseases* 7:144–149.

Ben Ami, R., G. Barshtein, D. Zeltser, Y. Goldberg, I. Shapira, A. Roth, G. Keren, et al. 2001. "Parameters of Red Blood Cell Aggregation as Correlates of the Inflammatory State." *American Journal of Physiology—Heart and Circulatory Physiology* 280:H1982–H1988.

Ben Ami, R., G. Sheinman, S. Yedgar, A. Eldor, A. Roth, A. S. Berliner, and G. Barshtein. 2002. "Thrombolytic Therapy Reduces Red Blood Cell Aggregation in Plasma without Affecting Intrinsic Aggregability." *Thrombosis Research* 105:487–492.

Ben Assayag, E., I. Bova, A. Kesler, S. Berliner, I. Shapira, and N. M. Bornstein. 2008. "Erythrocyte Aggregation as an Early Biomarker in Patients with Asymptomatic Carotid Stenosis." *Disease Markers* 24:33–39.

Beridze, M., N. Momtselidze, R. Shakarishvili, and G. Mchedlishvili. 2004. "Effect of Nitric Oxide Initial Blood Levels on Erythrocyte Aggregability during 12 Hours from Ischemic Stroke Onset." *Clinical Hemorheology and Microcirculation* 30:403–406.

Berliner, A. S., I. Shapira, O. Rogowski, N. Sadees, R. Rotstein, R. Fusman, D. Avitzour, et al. 2000. "Combined Leukocyte and Erythrocyte Aggregation in the Peripheral Venous Blood during Sepsis. An indication of commonly shared adhesive protein(s)." *International Journal of Clinical & Laboratory Research* 30:27–31.

Berliner, S., O. Rogowski, S. Aharonov, T. Mardi, T. Tolshinsky, M. Rozenblat, D. Justo, et al. 2005. "Erythrocyte Adhesiveness/Aggregation: A Novel Biomarker for the Detection of Low-Grade Internal Inflammation in Individuals with Atherothrombotic Risk Factors and Proven Vascular Disease." *American Heart Journal* 149:260–267.

Bertram, B., S. Wolf, O. Arend, K. Schulte, T. W. Pesch, F. Jung, H. Kiesewetter, and M. Reim. 1992. "Blood Rheology and Retinopathy in Adult Type I Diabetes Mellitus." *Clinical Hemorheology* 12:437–448.

Bogar, L. 2002. "Hemorheology and Hypertension: Not 'Chicken or Egg' but Two Chickens from Similar Eggs." *Clinical Hemorheology and Microcirculation* 26:81–83.

Boisseau, M. R. 2006. "Hemorheology and Vascular Diseases: Red Cell Should Rub Up to the Wall, Leucocytes Should Cope with It." *Clinical Hemorheology and Microcirculation* 35:11–16.

Boisseau, M. R., and B. de La Giclais. 2004. "Chronic Venous Diseases: Roles of Various Pathophysiological Factors." *Clinical Hemorheology and Microcirculation* 31:67–74.

Boisseau, M. R., G. Freyburger, M. Busquet, and C. Beylot. 1989. "Pharmacological Aspects of Erythrocyte Aggregation, Effect of High-Doses of Troxerutine." *Clinical Hemorheology* 9:871–876.

Boisseau, M. R., and K. Toth. 2007. "Treatment in Clinical Hemorheology: A Current Overview." In *Handbook of Hemorheology and Hemodynamics*, ed. O. K. Baskurt, M. R. Hardeman, M. W. Rampling, and H. J. Meiselman, 433–444. Amsterdam, Berlin, Oxford, Tokyo, Washington, DC: IOS Press.

Bollini, A., G. Hernandez, M. B. Luna, L. Cinara, and M. Rasia. 2005. "Study of Intrinsic Flow Properties at the Normal Pregnancy Second Trimester." *Clinical Hemorheology and Microcirculation* 33, no. 2:155–161.

Bolokadze, N., I. Lobjanidze, N. Momtselidze, R. Shakarishvili, and G. Mchedlishvili. 2006. "Comparison of Erythrocyte Aggregability Changes during Ischemic and Hemorrhagic Stroke." *Clinical Hemorheology and Microcirculation* 35:265–267.

Bor-Kucukatay, M., O. Yalcin, O. Gokalp, D. Kipmen-Korgun, A. Yesilkaya, A. Baykal, M. Ispir, et al. 2000. "Red Blood Cell Rheological Alterations in Hypertension Induced by Chronic Inhibition of Nitric Oxide Synthesis in Rats." *Clinical Hemorheology and Microcirculation* 22:267–275.

Brath, E., N. Nemeth, F. Kiss, E. Sajtos, T. Hever, L. Matyas, L. Toth, I. Miko, and I. Furka. 2010. "Changes of Local and Systemic Hemorheological Properties in Intestinal Ischemia–Reperfusion Injury in the Rat Model." *Microsurgery* 30:321–326.

Brun, J-F., M. Sekkat, C. Lagoueyte, C. Fedou, and A. Orsetti. 1989. "Relationship between fitness and blood viscosity in untrained normal short children." *Clinical Hemorheology* 9:953–963.

Brun, J-F., E. Varlet-Marie, P. Connes, and I. Aloulou. 2010. "Hemorheological Alterations Related to Training and Overtraining." *Biorheology* 47:95–115.

Brun, J. F., H. Belhabas, M. Ch. Granat, C. Sagnes, G. Thoni, J. P. Micallef, and J. Mercier. 2002. "Postexercise Red Cell Aggregation Is Negatively Correlated with Blood Lactate Rate of Disappearance." *Clinical Hemorheology and Microcirculation* 26:231–239.

Brun, J. F., P. Boulot, J. P. Micallef, J. L. Viala, and A. Orsetti. 1995. "Physiological Modifications of Blood-Viscosity and Red-Blood-Cell Aggregation during Labor and Delivery." *Clinical Hemorheology* 15:13–24.

Brun, J. F., P. Boulot, O. Rousseau, A. Elbouhmadi, F. Laffargue, J. L. Viala, and A. Orsetti. 1994. "Modifications of Erythrocyte Aggregation during Labor and Delivery." *Clinical Hemorheology* 14:643–649.

Caimi, G., A. Contorno, A. Serra, A. Catania, R. L. Presti, A. Sarno, and G. Cerasola. 1993. "Red Cell Metabolic Parameters and Rheological Determinants in Essential Hypertension." *Clinical Hemorheology* 13:35–44.

Caimi, G., E. Hoffmann, M. Montana, B. Canino, F. Dispensa, A. Catania, and R. Lo Presti. 2003. "Haemorheological Pattern in Young Adults with Acute Myocardial Infarction." *Clinical Hemorheology and Microcirculation* 29:11–18.

Cakir-Atabek, H., P. Atsak, N. Gunduz, and M. Bor-Kucukatay. 2009. "Effects of Resistance Training Intensity on Deformability and Aggregation of Red Blood Cells." *Clinical Hemorheology and Microcirculation* 41:251–261.

Camus, G., G. Deby-Dupont, J. Duchateau, C. Deby, J. Pincemail, and M. Lamy. 1994. Are Similar Inflammatory Factors Involved in Strenuous Exercise and Sepsis? *Intensive Care Med.* 20:602–610.

Castellini, M., R. Elsner, O. K. Baskurt, R. B. Wenby, and H. J. Meiselman. 2006. "Blood Rheology of Weddell Seals and Bowhead Whales." *Biorheology* 43:57–69.

Caswell, M., M. P. Corlett, J. Stuart, and B. S. Bull. 1992. "Tests for Monitoring the Acute-Phase Response to Surgery." *Clinical Hemorheology* 12:407–413.

Cavestri, R., L. Radice, F. Ferrarini, M. Longhini, and E. Longhini. 1992. "Influence of Erythrocyte Aggregability and Plasma-Fibrinogen Concentration on Cbf with Aging." *Acta Neurologica Scandinavica* 85:292–298.

Cha, C. H., C. J. Park, Y. J. Cha, H. K. Kim, D. H. Kim, Honghoon, J. H. Bae, et al. 2009. "Erythrocyte Sedimentation Rate Measurements by TEST 1 Better Reflect Inflammation Than Do Those by the Westergren Method in Patients with Malignancy, Autoimmune Disease, or Infection." *American Journal of Clinical Pathology* 131:189–194.

Chen, S., A. Eldor, G. Barshtein, S. Zhang, A. Goldfarb, E. Rachmilewitz, and S. Yedgar. 1996. "Enhanced Aggregability of Red Blood Cells of Beta-Thalassemia Major Patients." *American Journal of Physiology Heart and Circulatory Physiology* 270:H1951–H1956.

Chen, S., B. Gavish, G. Barshtein, Y. Mahler, and S. Yedgar. 1994. "Red Blood Cell Aggregability Is Enhanced by Physiological Levels of Hydrostatic Pressure." *Biochimica Biophysica Acta* 1192:247–252.

Chien, S. 2006. "Molecular Basis of Rheological Modulation of Endothelial Functions: Importance of Stress Direction." *Biorheology* 43:95–116.

Chien, S. 2007. "Mechanotransduction and Endothelial Cell Homeostasis: The Wisdom of the Cell." *American Journal of Physiology Heart and Circulatory Physiology* 292:H1209–H1224.

Chien, S., J. Dormandy, E. Ernst, and A. Matrai. 1987. *Clinical Hemorheology.* Dordrecht, Boston, Lancaster: Martinus Nijhoff Publ.

Chong-Martinez, B., T. A. Buchanan, R. B. Wenby, and H. J. Meiselman. 2003. "Decreased Red Blood Cell Aggregation Subsequent to Improved Glycaemic Control in Type 2 Diabetes Mellitus." *Diabetic Medicine* 20:301–306.

Christy, R. M., O. K. Baskurt, G. C. Gass, A. B. Gray, and S. Marshall-Gradisnik. 2010. "Erythrocyte Aggregation and Neutrophil Function in an Aging Population." *Gerontology* 56:175–180.

Cicco, G., E. Dolce, P. Vicenti, G. D. Stingi, M. S. Tarallo, and A. Pirrelli. 1999. "Hemorheological Aspects in Hypertensive Menopausal Smoker Women Treated with Female Hormones." *Clinical Hemorheology and Microcirculation* 21:343–347.

Cicco, G., and Pirrelli, A. 1999. "Red Blood Cell (RBC) Deformability, RBC Aggregability and Tissue Oxygenation in Hypertension." *Clinical Hemorheology and Microcirculation* 21:169–177.

Cicha, I., Y. Suzuki, N. Tateishi, and N. Maeda. 2004. "Effects of Dietary Triglycerides on Rheological Properties of Human Red Blood Cells (Abstract)." *Clinical Hemorheology and Microcirculation* 30:301–305.

Connes, P., J-F. Brun, and O. K. Baskurt. 2010. "Blood Rheology and Exercise." In *Exercise Physiology: From a Cellular to an Integrative Approach*, ed. P. Connes, O. Hue, and S. Perrey, 213–229. Amsterdam, Berlin, Oxford, Tokyo, Washington, DC: IOS Press.

Connes, P., C. Caillaud, Py. Guillaume, J. Mercier, O. Hue, and J. F. Brun. 2007. "Maximal Exercise and Lactate Do Not Change Red Blood Cell Aggregation in Well Trained Athletes." *Clinical Hemorheology and Microcirculation* 36:319–326.

Crowson, C. S., M. U. Rahman, and E. L. Matteson. 2009. "Which Measure of Inflammation to Use? A Comparison of Erythrocyte Sedimentation Rate and C-Reactive Protein Measurements from Randomized Clinical Trials of Golimumab in Rheumatoid Arthritis." *Journal of Rheumatology* 36:1606–1610.

de Vries, M. K., I. C. van Eijk, I. E. van der Horst-Bruinsma, M. J. L. Peters, M. T. Nurmohamed, B. A. C. Dijkmans, and B. P. C. Hazenberg. 2009. "Erythrocyte Sedimentation Rate, C-Reactive Protein Level and Serum Amyloid: A Protein for Patient Selection and Monitoring of Anti-Tumor Necrosis Factor Treatment in Ankylosing Spondylitis." *Arthritis and Rheumatology* 61:1484–1490.

Demiroglu, H., I. Barista, and S. Dundar. 1997. "The Effects of Age and Menopause on Erythrocyte Aggregation." *Thrombosis and Haemostasis* 77:404.

Dintenfass, L. 1982. "Hemorheology of Cancer Metastases: An Example of Malignant Melanoma. Survival Times and Abnormality of Blood Viscosity Factors." *Clinical Hemorheology* 2:259–271.

Dintenfass, L. 1989. Second experiment on aggregation of red cells and blood viscosity under zero gravity: on STS 26, Sept/Oct 1988. *Clinical Hemorheology* 9:677–679.

Dintenfass, L. 1991. "Complex Haemorheology of Diabetes: Some Interfaces with Genetics and Immunology." *Clinical Hemorheology* 11:155–165.

Dintenfass, L., and B. Lake. 1976. "Exercise Fitness, Cardiac Work and Blood Viscosity Factors in Patients and Normals." *European Surgical Research* 8:174–184.

Dintenfass, L., P. Osman, and H. J. Jedrzejczyk. 1985. "First Hemorheological Experiment on NASA Space Shuttle *Discovery* STS 51-C: Aggregation of Red Cells." *Clinical Hemorheology* 5:917–936.

Durussel, J. J., M. F. Berthault, G. Guiffant, and J. Dufaux. 1998. "Effects of Red Blood Cell Hyperaggregation on the Rat Microcirculation Blood Flow." *Acta Physiologica Scandinavica* 163:25–32.

Ehrly, A. M. 1991. *Therapeutic Hemorheology*. Berlin: Springer-Verlag.

El Bouhmadi, A. P. Boulot, F. Laffargue, and J. F. Brun. 2000. "Rheological Properties of Fetal Red Cells with Special Reference to Aggregability and Disaggregability Analyzed by Light Transmission and Laser Backscattering Techniques." *Clinical Hemorheology and Microcirculation* 22:79–90.

Elishkevitz, K., R. Fusman, M. Koffler, I. Shapira, D. Zeltser, D. Avitzour, N. Arber, S. Berliner, and R. Rotstein. 2002. "Rheological Determinants of Red Blood Cell Aggregation in Diabetic Patients in Relation to Their Metabolic Control." *Diabetic Medicine* 19:152–156.

El-Sayed, M. S., N. Ali, and Z. E. S. Ali. 2005. "Haemorheology in Exercise and Training." *Sports Medicine* 35:649–670.

Endre, B;, N. Nemeth, F. Kiss, E. Sajtos, T. Hever, L. Matyas, L. Toth, I. Miko, I. Furka. 2010. "Changes of local and systemic hemorheological properties in intestinal ischemia–reperfusion injury in the rat model." Microsurgery. 30:321-326.

Ernst, E. 1987. "Hemorheological Treatment." In *Clinical Hemorheology*, ed. S. Chien, J. Dormandy, E. Ernst, and A. Matrai, 329–373. Dordrecht: Martinus Nijhoff Publishers.

Ernst, E., L. Daburger, and T. Saradeth. 1991. "The Kinetics of Blood Rheology during and after Prolonged Standardized Exercise." *Clinical Hemorheology* 11:429–439.

Ernst, E., A. Matrai, E. Aschenbrenner, V. Will, and C. Schmidlechner. 1985. "Relationship between Fitness and Blood Fluidity." *Clinical Hemorheology* 5: 507–510.

Fadul, J. E. M., T. Linde, B. Sandhagen, B. Wikstrom, and B. G. Danielson. 1997. "Effects of Extracorporeal Hemapheresis Therapy on Blood Rheology." *Journal of Clinical Apheresis* 12, no. 4:183–186.

Falco, C., A. Vaya, J. Iborra, I. Moreno, S. Palanca, and J. Aznar. 2003. "Erythrocyte Aggregability and Disaggregability in Thalassemia Trait Carriers Analyzed by a Laser Backscattering Technique." *Clinical Hemorheology and Microcirculation* 28:245–249.

Feher, G., K. Koltai, G. Kesmarky, L. Szapary, I. Juricskay, and K. Toth. 2006. "Hemorheological Parameters and Aging." *Clinical Hemorheology and Microcirculation* 35:89–98.

Feher, G., K. Koltai, and K. Toth. 2007. "Are Hemorheological Parameters Independent of Aging?" *Clinical Hemorheology and Microcirculation* 36:181–182.

Firsov, N. N., A. Bjelle, T. V. Korotaeva, A. V. Priezzhev, and O. M. Ryaboshapka. 1998. "Clinical Application of the Measurement of Spontaneous Erythrocyte Aggregation and Disaggregation. A Pilot Study." *Clinical Hemorheology and Microcirculation* 18:87–97.

Forconi, S., and T. Gori. 2009. "The Evolution of the Meaning of Blood Hyperviscosity in Cardiovascular Physiopathology: Should We Reinterpret Poiseuille?" *Clinical Hemorheology and Microcirculation* 42:1–6.

Foresto, P., M. D'Arrigo, F. Filippin, R. Gallo, L. Barberena, L. Racca, J. Valverde, and R. J. Rasia. 2005. "Hemorheological Alterations in Hypertensive Patients." *Medicina-Buenos Aires* 65:121–125.

Gaudard, A., E. Varlet, F. Bressolle, J. Mercier, and J-F. Brun. 2004. "Nutrition as a Determinant of Blood Rheology and Fibrinogen in Athletes." *Clinical Hemorheology and Microcirculation* 30:1–8.

Goldin, Y., T. Tulshinski, Y. Arbel, O. Rogowski, R. Ben Ami, J. Serov, P. Halperin, I. Shapira, and S. Berliner. 2007. "Rheological Consequences of Acute Infections: The Rheodifference between Viral and Bacterial Infections." *Clinical Hemorheology and Microcirculation* 36:111–119.

Gordon, A. H., and A. H. Koy. 1985. *The Acute Phase Response to Injury and Infection.* Amsterdam: Elsevier Scientific Publishers.

Gori, T., M. Lisi, and S. Forconi. 2006. "Ischemia and Reperfusion: The Endothelial Perspective. A Radical View." *Clinical Hemorheology and Microcirculation* 35:31–34.

Gruys, E., G. Toussaint, T. A. Niewold, and S. J. Koopman. 2005. "Acute Phase Reaction and Acute Phase Proteins." *Journal of Zhejiang University Science B* 6:1045–1056.

Hacioglu, G., O. Yalcin, M. Bor-Kucukatay, G. Ozkaya, and O. K. Baskurt. 2002. "Red Blood Cell Rheological Properties in Various Rat Hypertension Models." *Clinical Hemorheology and Microcirculation* 26:27–32.

Hadengue, A. L., M. Del-Pino, A. Simon, and J. Levenson. 1998. "Erythrocyte Disaggregation Shear Stress, Sialic Acid, and Cell Aging in Humans." *Hypertension* 32:324–330.

Hager, K., M. Felicetti, G. Seefried, and D. Platt. 1994. "Fibrinogen and Aging." *Aging (Milano)* 6:133–138.

Hammi, H., P. Perrotin, R. Guillet, and M. Boynard. 1994. "Determination of Red Blood Cell Aggregation in Young and Elderly Subjects Evaluated by Ultrasound." *Clinical Hemorheology* 14:117–126.

Hever, T., F. Kiss, L. Sajtos, L. Matyas, and N. Nemeth. 2010. "Are There Arterio-Venous Differences of Blood Micro-Rheological Variables in Laboratory Rats?" *Korea-Australia Rheology Journal* 22:59–64.

Huisman, A., J. G. Aarnoudse, M. Krans, H. J. Huisjes, V. Fidler, and W. G. Zijlstra. 1988. "Red-Cell Aggregation during Normal-Pregnancy." *British Journal of Haematology* 68:121–124.

Ince, C. 2005. "The Microcirculation is the Motor of Sepsis." *Critical Care* 9:S13–S19.

Ji, L. L., and S. Leichtweis. 1997. "Exercise and Oxidative Stress: Sources of Free Radicals and Their Impact on Antioxidant Systems." *Age* 20:91–106.

Justo, D., N. Mashav, Y. Arbel, M. Kinori, A. Steinvil, M. Swartzon, B. Molat, et al. 2009. "Increased Erythrocyte Aggregation in Men with Coronary Artery Disease and Erectile Dysfunction." *International Journal of Impotence Research* 21:192–197.

KaBer, U., G. Altrock, and P. Heimburg. 1987. "Effect of Menstrual Cycle on Viscoelastic Parameters." *Clinical Hemorheology* 7:687–693.

Kameneva, M. V., and J. F. Antaki. 2007. "Mechanical Trauma to Blood." In *Handbook of Hemorheology and Hemodynamics*, ed. O. K. Baskurt, M. R. Hardeman, M. W. Rampling, and H. J. Meiselman, 206–227. Amsterdam, Berlin, Oxford, Tokyo, Washington DC: IOS Press.

Kameneva, M. V., M. J. Watach, and H. S. Borovetz. 1999. "Gender Difference in Rheologic Properties of Blood and Risk of Cardiovascular Diseases." *Clinical Hemorheology and Microcirculation* 21:357–363.

Kayar, E., F. Mat, H. J. Meiselman, and O. K. Baskurt. 2001. "Red Blood Cell Rheological Alterations in a Rat Model of Ischemia–Reperfusion Injury." *Biorheology* 38:405–414.

Khodabandehlou, T., M. R. Boisseau, and C. Le Devehat. 2004. "Blood Rheology as a Marker of Venous Hypertension in Patients with Venous Disease." *Clinical Hemorheology and Microcirculation* 30:307–312.

Khodabandehlou, T., C. Le Devehat, and M. Razavian. 1996. "Impaired Function of Fibrinogen: Consequences on Red Cell Aggregation in Diabetes Mellitus." *Clinical Hemorheology* 16:303–312.

Khodabandehlou, T., H. Zhao, M. Vimeux, F. Aouane, and C. Le Devehat. 1998. "Haemorheological Consequences of Hyperglycaemic Spike in Healthy Volunteers and Insulin-Dependent Diabetics." *Clinical Hemorheology and Microcirculation* 19:105–114.

Kim, A., H. Dadgostar, G. N. Holland, R. Wenby, F. Yu, B. G. Terry, and H. J. Meiselman. 2006. "Hemorheologic Abnormalities Associated with HIV Infection: Altered Erythrocyte Aggregation and Deformability." *Investigative Ophthalmology & Visual Science* 47:3927–3932.

Kirshenbaum, L. A., M. Asis, M. E. Astiz, D. C. Saha, and E. C. Rackow. 2000. "Influence of Rheologic Changes and Platelet-Neutrophil Interaction on Cell Filtration in Sepsis." *American Journal of Respiratory and Critical Care Medicine* 161:1602–1607.

Klingel, R., C. Fassbender, I. Fischer, L. Hattenbach, H. Gumbel, J. Pulido, and F. Koch. 2002. "Rheopheresis for Age-Related Macular Degeneration: A Novel Indication for Therapeutic Apheresis in Ophthalmology." *Therapeutic Apheresis* 6:271–281.

Koppensteiner, R. 1996. "Blood Rheology in Emergency Medicine." *Seminars in Thrombosis and Hemostasis* 22:89–91.

Korotaeva, T. V., N. N. Firsov, A. Bjelle, and M. A. Vishlova. 2007. "Erythrocytes Aggregation in Healthy Donors at Native and Standard Hematocrit: The Influence of Sex, Age, Immunoglobulins and Fibrinogen Concentrations. Standardization of Parameters." *Clinical Hemorheology and Microcirculation* 36:335–343.

Koscielny, J., E. M. Jung, C. Mrowietz, H. Kiesewetter, and R. Latza. 2004. "Blood Fluidity, Fibrinogen, and Cardiovascular Risk Factors of Occlusive Arterial Disease: Results of the Aachen Study." *Clinical Hemorheology and Microcirculation* 31:185–195.

Kumsishvili, T., M. Varazashvili, and G. Mchedlishvili. 2004. "Local Hemorheological Disorders during Chronic Inflammation." *Clinical Hemorheology and Microcirculation* 30:427–429.

Kwaan, H. C., and A. Bongu. 1999. "The Hyperviscosity Syndromes." *Seminars in Thrombosis and Hemostasis.* 25:199–208.

Kwaan, H. C., M. Levin, S. Sakurai, O. Kucuk, M. W. Rooney, L. J. Lis, and J. W. Kauffman. 1997. "Digital Ischemia and Gangrene Due to Red Blood Cell Aggregation Induced by Acquired Dysfibrinogenemia." *Journal of Vascular Surgery* 26:1061–1068.

Le Devehat, C., M. Boisseau, M. Vimeux, G. Bondoux, and A. Bertrand. 1989a. "Hemorheological Factors in the Pathophysiology of Venous Disease." *Clinical Hemorheology* 9:861–870.

Le Devehat, C., T. Khodabandehlou, and M. Vimeux. 2001. "Impaired Hemorheological Properties in Diabetic Patient with Lower Limb Ischemia." *Clinical Hemorheology and Microcirculation* 25:43–48.

Le Devehat, C., M. Vimeux, and G. Bondoux. 1989b. "Hemorheological Effects of Oral Troxerutine Treatment Versus Placebo in Venous Insufficiency of the Lower Limbs." *Clinical Hemorheology* 9:543–552.

Lebensohn, N., A. Re, L. Carrera, L. Barberena, M. D'Arrigo, and P. Foresto. 2009. "Serum Sialic Acid, Cellular Anionic Charge and Erythrocyte Aggregation in Diabetic and Hypertensive Patients." *Medicina-Buenos Aires* 69:331–334.

Lechner, H., R. Schmidt, B. Reinhart, P. Grieshofer, M. Koch, F. Fazekas, K. Niederkorn, et al. 1994. "The Austrian Stroke Prevention Study: Serum Fibrinogen Predicts Carotid Atherosclerosis and White Matter Disease in Neurologically Asymptomatic Individuals." *Clinical Hemorheology* 14:841–846.

Lee, B. K., A. Durairaj, R. B. Wenby, H. J. Meiselman, and T. Alexy. 2008a. "Hemorheological Abnormalities in Stable Angina and Acute Coronary Syndromes." *Clinical Hemorheology and Microcirculation* 39:43–51.

Lee, B. K., A. Durairaj, A. Mehra, R. B. Wenby, H. J. Meiselman, and T. Alexy. 2008b. "Microcirculatory Dysfunction in Cardiac Syndrome X: Role of Abnormal Blood Rheology." *Microcirculation* 15:451–459.

Lee, W. S., and T. Y. Kim. 2009. "Measuring of ESR With TEST 1 Is More Useful Than the Westergren Method in Rheumatoid Arthritis." *American Journal of Clinical Pathology* 132:805.

Leschke, M., W. Motz, H. Blanke, and B. E. Strauer. 1987. "Blood Rheology in Hypertension and Hypertensive Heart Disease." *Journal of Cardiovascular Pharmacology* 10:S103–S110.

Levy, D. E., G. J. del Zoppo, B. M. Demaerschalk, A. M. Demchuk, H. C. Diener, G. Howard, M. Kaste, et al. 2009. "Ancrod in Acute Ischemic Stroke Results of 500 Subjects Beginning Treatment within 6 Hours of Stroke Onset in the Ancrod Stroke Program." *Stroke* 40:3796–3803.

Linderkamp, O. 1996. "Pathological Flow Properties of Blood in the Fetus and Neonate." *Clinical Hemorheology* 16:105–116.

Linderkamp, O., P. Ozanne, P. Y. K. Wu, and H. J. Meiselman. 1984. "Red Blood Cell Aggregation in Preterm and Term Neonates and Adults." *Pediatric Research* 18:1356–1360.

Linderkamp, O., P. Y. Wu, and H. J. Meiselman. 1982. "Deformability of Density Separated Red Blood Cells in Normal Newborn Infants and Adults." *Pediatric Research* 16:964–968.

Lip, G. Y. H. 1995. "Fibrinogen and Cardiovascular Disorders." *Quarterly Journal of Medicine* 88:155–165.

Liu, T. T., W. J. Wong, M. C. Hou, H. C. Lin, F. Y. Chang, and S. D. Lee. 2006. "Hemorheology in Patients with Liver Cirrhosis: Special Emphasis on Its Relation to Severity of Esophageal Variceal Bleeding." *Journal of Gastroenterology and Hepatology* 21:908–913.

Lominadze, D., W. L. Dean, S. C. Tyagi, and A. M. Roberts. 2010. "Mechanisms of Fibrinogen-Induced Microvascular Dysfunction during Cardiovascular Disease." *Acta Physiologica* 198:1–13.

Lominadze, D., I. G. Joshua, and D. A. Schuschke. 1998. "Increased Erythrocyte Aggregation in Spontaneously Hypertensive Rats." *American Journal of Hypertension* 11:784–789.

Longhini, E., R. Agosti, P. Cherubini, A. Clivati, P. Farini, and L. Marazzini. 1986. "Haemorheology in Hypertension." *Clinical Hemorheology* 6:567–576.

Lowe, G. D. O. 1988. *Clinical Blood Rheology*. Boca Raton, FL: CRC Press.

Lowe, G. D. O., F. G. R. Fowkes, J. Dawes, P. T. Donnan, S. E. Lennie, and E. Housley. 1993. "Blood Viscosity, Fibrinogen and Activation of Coagulation and Leukocytes in Peripheral Artery Disease and the Normal Population in the Edinburgh Artery Study." *Circulation* 87:1915–1920.

Luquita, A., L. Urli, M. J. Svetaz, A. M. Gennaro, R. Volpintesta, S. Palatnik, and M. Rasia. 2009. "Erythrocyte Aggregation in Rheumatoid Arthritis: Cell and Plasma Factor's Role." *Clinical Hemorheology and Microcirculation* 41:49–56.

MacPherson, J. L., and D. O. Kasprisin. 2010. *Therapeutic Hemapheresis*. Boca Raton, FL: CRC Press.

Malek, A., S. L. Alper, and S. Izumo. 1999. "Hemodynamic Shear Stress and Its Role in Atherosclerosis." *JAMA* 282:2035–2042.

Manetta, J., I. Aloulou, E. Varlet-Marie, J. Mercier, and J. F. Brun. 2006. "Partially Opposite Hemorheological Effects of Aging and Training at Middle Age." *Clinical Hemorheology and Microcirculation* 35:239–244.

Mantskava, M., N. Momtselidze, N. Pargalava, and G. Mchedlishvili. 2006. "Hemorheological Disorders in Patients with Type 1 or 2 Diabetes Mellitus and Foot Gangrene." *Clinical Hemorheology and Microcirculation* 35:307–310.

Martinez, M., A. Vaya, J. Alvarino, J. L. Barbera, D. Ramos, A. Lopez, and J. Aznar. 1999. "Hemorheological Alterations in Patients with Chronic Renal Failure. Effect of Hemodialysis." *Clinical Hemorheology and Microcirculation* 21:1–6.

Martinez, M., A. Vaya, R. Server, A. Gilsanz, and J. Aznar. 1998. "Alterations in Erythrocyte Aggregability in Diabetics: The Influence of Plasmatic Fibrinogen and Phospholipids of the Red Blood Cell Membrane." *Clinical Hemorheology and Microcirculation* 18:253–258.

Marton, Z., B. Horvath, T. Alexy, G. Kesmarky, Z. Gyevnar, L. Czopf, T. Habon, et al. 2003. "Follow-up of Hemorheological Parameters and Platelet Aggregation in Patients with Acute Coronary Syndromes." *Clinical Hemorheology and Microcirculation* 29:81–94.

Matot, I., and C. L. Sprung. 2001. "Definition of Sepsis." *Intensive Care Medicine* 27 Suppl. 1:3–9.

McCarty, M. F. 2002. "Favorable Impact of a Vegan Diet with Exercise on Hemorheology: Implications for Control of Diabetic Neuropathy." *Medical Hypotheses* 58:476–486.

Mchedlishvili, G., I. Lobjanidze, N. Momtselidze, N. Bolokadze, et al. 2004. "About Spread of Local Cerebral Hemorheological Disorders to Whole Body in Critical Care Patients." *Clinical Hemorheology and Microcirculation* 31:129–138.

Mchedlishvili, G., B. Tsinamdzvrishvili, N. Beritashvili, L. Gobejishvili, and V. Ilencko. 1997. "New Evidence for Involvement of Blood Rheological Disorders in Rise of Peripheral Resistance in Essential Hypertension." *Clinical Hemorheology and Microcirculation* 17:31–39.

McMillan, D. E. 1992. "Hemorheological Therapy to Control Diabetic Vascular Disease." *Clinical Hemorheology* 12:787–796.

Mehta, J., and S. Singhal. 2003. "Hyperviscosity Syndrome in Plasma Cell Dyscrasias." *Seminars in Thrombosis and Hemostasis* 29:467–471.

Meiselman, H. J. 1993. "Red Blood Cell Role in RBC Aggregation: 1963–1993 and Beyond." *Clinical Hemorheology* 13:575–592.

Meiselman, H. J. 1999. "Hemorheologic Alterations in Hypertension: Chicken or Egg?" *Clinical Hemorheology and Microcirculation* 21:195–200.

Meiselman, H. J. 2009. "Red Blood Cell Aggregation: 45 Years Being Curious." *Biorheology* 46:1–19.

Mira, Y., A. Vaya, M. Martinez, P. Villa, M. L. Santaolaria, F. Ferrando, and J. Aznar. 1998. "Hemorheological Alterations and Hypercoagulable State in Deep Vein Thrombosis." *Clinical Hemorheology and Microcirculation* 19:265–270.

Mokken, F. C., C. P. Henny, I. van der Waart, P. T. Goedhart, and A. W. Gelb. 1996. "Difference in Peripheral Arterial and Venous Hemorheologic Parameters." *Annals of Hematology* 73:135–137.

Momtselidze, N., M. Mantskava, and G. Mchedlishvili. 2006. "Hemorheological Disorders during Ischemic Brain Infarcts in Patients with and without Diabetes Mellitus." *Clinical Hemorheology and Microcirculation* 35:261–264.

Monsuez, J-J., J. Dufaux, D. Vittecoq, P. Flaud, and E. Vicaut. 2000. "Hemorheology in Asymptomatic HIV-Infected Patients." *Clinical Hemorheology and Microcirculation* 23:59–66.

Morariu, A. M., Y. J. Gu, R. C. G. G. Huet, W. A. Siemons, G. Rakhorst, and W. von Oeverena. 2004. "Red Blood Cell Aggregation during Cardiopulmonary Bypass: A Pathogenic Cofactor in Endothelial Cell Activation?" *European Journal of Cardio-Thoracic Surgery* 26:939–946.

Murphy, J. R. 1973. "Influence of Temperature and Method of Centrifugation on Separation of Erythrocytes." *Journal of Laboratory and Clinical Medicine* 81:334–341.

Nageswari, K., R. Banerjee, R. V. Gupte, and R. R. Puniyani. 2000. "Effects of Exercise on Rheological and Microcirculatory Parameters." *Clinical Hemorheology and Microcirculation* 23:243–247.

Nash, G. B., R. Wenby, S. O. Sowemimo-Coker, and H. J. Meiselman. 1987. "Influence of Cellular Properties on Red Cell Aggregation." *Clinical Hemorheology* 7:93–108.

Nemeth, N., M. Szokoly, G. Acs, E. Brath, T. Lesznyak, I. Furka, and I. Miko. 2004. "Systemic and Regional Hemorheological Consequences of Warm and Cold Hind Limb Ischemia-Reperfusion in a Canine Model." *Clinical Hemorheology and Microcirculation* 30:133–145.

Neuhaus, D., C. Behn, and E. Ernst. 1992. "Haemorheology and Exercise: Intrinsic Flow Properties of Blood in Marathon Running." *International Journal of Sports Medicine* 13:506–511.

Neuhaus, D., and P. Gaehtgens. 1994. "Haemorheology and Long-Term Exercise." *Sports Medicine* 18:10–21.

Neumann, F-J., H. A. Katus, E. Hoberg, P. Roebruck, M. Braun, H. M. Haupt, H. Tillmanns, and W. Kubler. 1991. "Increased Plasma Viscosity and Erythrocyte Aggregation: Indicators of an Unfavourable Clinical Outcome in Patients with Unstable Angina Pectoris." *British Heart Journal* 66:425–430.

Neumann, F-J., H. Tillmanns, P. Roebruck, R. Zimmermann, H. M. Haupt, and W. Kubler. 1989. "Haemorheological Abnormalities in Unstable Angina Pectoris: A Relation Independent of Risk Factor Profile and Angiographic Severity." *British Heart Journal* 61:421–428.

Nordt, F. J. 1983. "Hemorheology in Cerebrovascular Diseases." *Annals of New York Academy of Sciences* 416:651–661.

Ozanne, P., R. B. Francis, and H. J. Meiselman. 1983a. "Red Blood Cell Aggregation in Nephrotic Syndrome." *Kidney International* 23:519–525.

Ozanne, P., O. Linderkamp, F. C. Miller, and H. J. Meiselman. 1983b. "Erythrocyte aggregation during normal-pregnancy. *American Journal of Obstetrics and Gynecology* 147:576–583.

Paulus, H. E., B. Ramos, W. K. Wong, A. Ahmed, K. Bulpitt, G. Park, M. Sterz, and P. Clements. 1999. "Equivalence of the Acute Phase Reactants C-Reactive Protein, Plasma Viscosity, and Westergren Erythrocyte Sedimentation Rate When Used to Calculate American College of Rheumatology 20% Improvement Criteria or the Disease Activity Score in Patients with Early Rheumatoid Arthritis." *Journal of Rheumatology* 26:2324–2331.

Peto, K., N. Nemeth, E. Brath, I. E. Takacs, O. K. Baskurt, H. J. Meiselman, I. Furka, and I. Miko. 2007. "The Effects of Renal Ischemia-Reperfusion on Hemorheological Factors: Preventive Role of Allopurinol." *Clinical Hemorheology and Microcirculation* 37:347–358.

Petralito, A., L. S. Malatino, and C. E. Fiore. 1985. "Erythrocyte Aggregation in Different Stages of Arterial-Hypertension." *Thrombosis and Haemostasis* 54, no. 2:555.

Pfafferott, C., G. Moessmer, A. M. Ehrly, and R. M. Bauersachs. 1999. "Involvement of Erythrocyte Aggregation and Erythrocyte Resistance to Flow in Acute Coronary Syndromes." *Clinical Hemorheology and Microcirculation* 21:35–43.

Piagnerelli, M., K. Zouaoui Boudjeltia, M. Vanhaeverbeek, and J. L. Vincent. 2003. "Red Blood Cell Rheology in Sepsis." *Intensive Care Medicine* 29:1052–1061.

Pignon, B., D. Jolly, G. Potron, B. Lartigue, J. P. Vilque, P. Nguyen, J. C. Etienne, and J. F. Stoltz. 1994. "Erythrocyte Aggregation—Determination of Normal Values—Influence of Age, Sex, Hormonal State, Oestroprogestative Treatment, Hematological Parameters and Cigarette-Smoking." *Nouvelle Revue Francaise D Hematologie* 36:431–439.

Pirrelli, A. 1999. "Arterial Hypertension and Hemorheology. What Is the Relationship?" *Clinical Hemorheology and Microcirculation* 21:157–160.

Piva, E., M. C. Sanzari, G. Servidio, and M. Plebani. 2001. "Length of Sedimentation Reaction in Undiluted Blood (Erythrocyte Sedimentation Rate): Variations with Sex and Age and Reference Limits." *Clinical Chemistry and Laboratory Medicine* 39:451–454.

Poole, J. C. F., and G. A. C. Summers. 1952. Correction of ESR in Anaemia. Experimental Study Based on Interchange of Cells and Plasma between Normal Anaemic Subjects. *British Medical Journal* 1:353–356.

Popel, A. S., P. C. Johnson, M. V. Kameneva, and M. A. Wild. 1994. "Capacity for red blood cell aggregation is higher in athletic mammalian species than in Sedentary Species." *Journal of Applied Physiology* 77:1790–1794.

Puniyani, R. R., V. S. Agashe, V. Annapurna, and S. R. Daga. 1988. "Haemorheological Profile in Cases of Chronic Infections." *Clinical Hemorheology* 8:595–602.

Puniyani, R. R., V. Annapurna, P. Chaturani, and P. A. Kale. 1987. "Haemorheological Profile in Cases of Hypertension." *Clinical Hemorheology* 7:767–772.

Puniyani, R. R., V. Annapurna, and V. S. Jaiswal. 1988. "Haemorheological Changes in Allergic Skin Disorders." *Clinical Hemorheology* 8:663–667.

Puniyani, R. R., and S. Sonar. 1991. "Hemorheological Changes at Menopause." *Clinical Hemorheology* 11:397–403.

Rampling, M. W. 1988. "Red Cell Aggregation and Yield Stress." In *Clinical Blood Rheology*, ed. G. D. O. Lowe, 45–64. Boca Raton, FL: CRC Press.

Rampling, M. W., H. J. Meiselman, B. Neu, and O. K. Baskurt. 2004. "Influence of Cell-Specific Factors on Red Blood Cell Aggregation." *Biorheology* 41:91–112.

Razavian, S. M., M. Delpino, A. Chabanel, A. Simon, and J. Levenson. 1991. "Increase in Erythrocyte Disaggregation Shear-Stress in Hypertension." *Archives des Maladies du Coeur et des Vaisseaux* 84:1081–1084.

Reinhart, W. H., B. Kayser, A. Singh, U. Waber, O. Oelz, and P. Bartsch. 1991. "Blood Rheology in Acute Mountain Sickness and High-Altitude Pulmonary Edema." *Journal of Applied Physiology* 71:934–938.

Resch, K. L., E. Ernst, A. Matrai, M. Buhl, P. Schlosser, and H. F. Paulsen. 1991. "Can Rheologic Variables Be of Prognostic Relevance in Arteriosclerotic Diseases." *Angiology* 42:963–970.

Ross, R. 1999. "Mechanisms of Disease—Atherosclerosis—An Inflammatory Disease." *New England Journal of Medicine* 340:115–126.

Schooneman, F., C. Claise, and J. F. Stoltz. 1997. "Hemorheology and Therapeutic Hemapheresis." *Transfusion Science* 18:531–540.

Schpitz-Droz, K. R., C. Bucherer, J. Ladjouzi, C. Lacombe, and J. C. Lelievre. 1991. "Are There Two Kinds of Fibrinogen in Diabetic Plasma as Suggested by a Miltivariate Analysis?" *Clinical Hemorheology* 11:387–396.

Schrier, S. L. 1994. "Thalassemia: Pathophysiology of Red Cell Changes." *Annual Review of Medicine* 45:211–218.

Schrier, S. L., and N. Mohandas. 1992. "Globin Chain Specificity of Oxidation-Induced Changes in Red Blood Cell Membrane Properties." *Blood* 79:1586–1592.

Scott, M. D., J. J. van den Berg, T. Repka, P. Rouyer-Fessard, R. P. Hebbel, Y. Beuzard, and B. H. Lubin. 1993. "Effect of Excess Alpha-Hemoglobin Chains on Cellular and Membrane Oxidation in Model Beta-Thalassemic Erythrocytes." *Journal of Clinical Investigation* 91:1706–1712.

Sen, C. K., M. Atalay, and O. Hanninen. 1994. "Exercise-Induced Oxidative Stress: Glutathione Supplementation and Deficiency." *Journal of Applied Physiology* 77:2177–2187.

Senturk, U. K., O. Yalcin, F. Gunduz, O. Kuru, H. J. Meiselman, and O. K. Baskurt. 2005. "Effect of Antioxidant Vitamin Treatment on the Time Course of Hematological and Hemorheological Alterations after an Exhausting Exercise Episode in Human Subjects." *Journal of Applied Physiology* 98, no. 4:1272–1279.

Shapira, I., R. Rotstein, R. Fusman, B. Gluzman, A. Roth, G. Keren, D. Avitzour, N. Arber, and S. Berliner. 2001. "Combined Leukocyte and Erythrocyte Aggregation in Patients with Acute Myocardial Infarction." *Heart Disease: New Trends in Research, Diagnosis and Treatment* 85–88.

Sola, E., A. Vaya, T. Contreras, C. Falco, D. Corella, A. Hernandez, and J. Aznar. 2004. "Rheological Profile in Severe and Morbid Obesity. Preliminary Results." *Clinical Hemorheology and Microcirculation* 30:415–418.

Sola, E., A. Vaya, D. Corella, M. L. Santaolaria, F. Espania, A. Estelles, and A. Hernandez-Mijares. 2007. "Erythrocyte Hyperaggregation in Obesity: Determining Factors and Weight Loss Influence." *Obesity* 15:2128–2134.

Sowemimo-Coker, S. O., P. Whittingstall, L. Pietsch, R. M. Bauersachs, R. B. Wenby, and H. J. Meiselman. 1989. "Effects of Cellular Factors on the Aggregation Behavior of Human, Rat and Bovine Erythrocytes." *Clinical Hemorheology* 9:723–737.

Spengler, M. I., G. M. Goni, G. Mengarelli, M. Bravo Luna, R. Bocanera, R. Tozzini, M. L. Rasia. 2003. "Effect of Hormone Replacement Therapy upon Haemorheological Variables." *Clinical Hemorheology and Microcirculation* 28:13–19.

Spengler, M. I., M. J. Svetaz, M. B. Leroux, M. L. Leiva, and H. M. Bottai. 2004. "Association between Capillaroscopy Haemorheological Variable and Plasma Proteins in Patients Bearing Raynaud's Phenomenon." *Clinical Hemorheology and Microcirculation* 30:17–24.

Stoltz, J. F., M. Singh, and P. Riha. 1999. *Hemorheology in Practice*. Amsterdam: IOS Press.

Stuart, J. 1991. "Rheological Methods for Monitoring the Acute-Phase Response." *Revista Porteguese Hemorreologia* 5:57–62.

Sweetnam, P. M., H. F. Thomas, J. W. G. Yarnell, A. D. Beswick, I. A. Baker, and P. C. Elwood. 1996. "Fibrinogen, Viscosity and the 10-Year Incidence of Ischemic Heart Disease: The Caerphilly and Speedwell Studies." *European Heart Journal* 17:1814–1820.

Szapary, L., B. Horvath, Zs. Marton, T. Alexy, N. Demeter, M. Szots, A. Klabzai, G. Kesmarky, I. Juricskay, V. Gaal, J. Czopf, and K. Toth. 2004. "Hemorheological Disturbances in Patients with Chronic Cerebrovascular Diseases." *Clinical Hemorheology and Microcirculation* 31:1–9.

Tamas, R., N. Nemeth, E. Brath, M. Sasvari, C. Nyakas, B. Debreczeni, I. Miko, and I. Furka. 2010. "Hemorheological, Morphological, and Oxidative Changes during Ischemia-Reperfusion of Latissimus Dorsi Muscle Flaps in a Canine Model." *Microsurgery* 30:282–288.

Tanahashi, N., Y. Fukuuchi, M. Tomita, M. Kobari, H. Takeda, M. Yokoyama, and D. Itoh. 1995. "Effect of Single Intravenous Administration of Batroxobin on Erythrocyte Aggregability in Patients with Acute-Stage Cerebral Infarction." *Clinical Hemorheology* 15:89–96.

Tanahashi, N., M. Tomita, M. Kobari, H. Takeda, M. Yokoyama, M. Takao, and Y. Fukuuchi. 1996. "Platelet Activation and Erythrocyte Aggregation Rate in Patients with Cerebral Infarction." *Clinical Hemorheology* 16:497–505.

Tatarishvili, J., T. Sordia, and G. Mchedlishvili. 2006. "Comparison of Blood Rheological Changes in the Microcirculation during Experimental Hemorrhagic and Traumatic Shock." *Clinical Hemorheology and Microcirculation* 35:217–221.

Taylor, W. F., S. Chen, G. Barshtein, D. E. Hyde, and S. Yedgar. 1998. "Enhanced Aggregability of Human Red Blood Cells by Diving." *Undersea & Hyperbaric Medicine* 25, no. 3:167–170.

Tikhomirova, I. A., A. V. Muravyov, and V. N. Levin. 2002. "Major Alterations in Body Fluid Status and Blood Rheology." *Clinical Hemorheology and Microcirculation* 26:195–198.

Toker, S., O. Rogowski, S. Melamed, A. Shirom, I. Shapira, S. Berliner, and D. Zeltser. 2005. "Association of Components of the Metabolic Syndrome with the Appearance of Aggregated Red Blood Cells in the Peripheral Blood. An Unfavorable Hemorheological Finding." *Diabetes-Metabolism Research and Reviews* 21:197–202.

Toth, K., and G. Kesmarky. 2007. "Clinical Significance of Hemorheological Alterations." In *Handbook of Hemorheology and Hemodynamics*, ed. O. K. Baskurt, M. R. Hardeman, M. W. Rampling, and H. J. Meiselman, 392–432. Amsterdam, Berlin, Oxford, Tokyo, Washington DC: IOS Press.

Tribouilloy, C., M. Peltier, L. Colas, M. Senni, O. Ganry, J. L. Rey, and J. P. Lesbre. 1998. "Fibrinogen Is an Independent Marker for Thoracic Aortic Atherosclerosis." *American Journal of Cardiology* 81:321–326.

Tugral, E., O. Yalcin, and O. K. Baskurt. 2002. "Effect of Donor Age on the Deformability and Aggregability of Density-Separated Red Blood Cells." *Turkish Journal of Hematology* 19:303–308.

Urbach, J., O. Rogowski, I. Shapira, D. Avitzour, D. Branski, S. Schwartz, S. Berliner, and T. Mardi. 2005. "Automatic 3-Dimensional Visualization of Peripheral Blood Slides—A New Approach for the Detection of Infection/Inflammation at the Point of Care." *Archives of Pathology & Laboratory Medicine* 129:645–650.

Uyuklu, M., H. J. Meiselman, and O. K. Baskurt. 2007. "Effect of Decreased Plasma Cholesterol by Atorvastatin Treatment on Erythrocyte Mechanical Properties." *Clinical Hemorheology and Microcirculation* 36:25–33.

van den Broek, N. R., and E. A. Letsky. 2001. "Pregnancy and the Erythrocyte Sedimentation Rate." *British Journal of Obstetrics and Gynaecology* 108:1164–1167.

Varlet-Marie, E., A. Gaudard, J. Mercier, F. Bressolle, and F. Brun. 2003a. "Is the Feeling of Heavy Legs in Overtrained Athletes Related to Impaired Hemorheology?" *Clinical Hemorheology and Microcirculation* 28:151–159.

Varlet-Marie, E., A. Gaudard, J. F. Monnier, J. P. Micallef, J. Mercier, F. Bressolle, and J. F. Brun. 2003b. "Reduction of Red Blood Cell Disaggregability during Submaximal Exercise: Relationship with Fibrinogen Levels." *Clinical Hemorheology and Microcirculation* 28:139–149.

Vaya, A., P. Chorro, D. Juia, C. Falco, L. Ortega, D. Corella, and J. Aznar. 2004. "Menopause, Hormone Replacement Therapy and Hemorheology." *Clinical Hemorheology and Microcirculation* 30:277–281.

Vaya, A., J. Iborra, C. Falco, I. Moreno, P. Bolufer, F. Ferrando, M. L. Perez, and A. Justo. 2003. "Rheological Behaviour of Red Blood Cells in Beta and Deltabeta Thalassemia Trait." *Clinical Hemorheology and Microcirculation* 28:71–78.

Vaya, A., M. Martinez, J. Garcia, and J. Aznar. 1992. "The Hemorheological Profile in Mild Essential Hypertension." *Thrombosis Research* 66:223–229.

Vaya, A., M. Martinez, R. Carmena, and J. Aznar. 1993. "Red-Blood-Cell Aggregation and Primary Hyperlipoproteinemia." *Thrombosis Research* 72:119–126.

Vaya, A., M. Martinez, J. Dalmau, M. Labios, and J. Aznar. 1996. "Hemorheological Profile in Patients with Cardiovascular Risk Factors." *Clinical Hemorheology* 26:1666–1670.

Vaya, A., M. Martinez, J. Dalmau, and J. Aznar. 1997. "Erythrocyte Aggregation and Plasma Viscosity in Children with Polygenic Hypercholesterolemia." *Thrombosis and Haemostasis* 73S:1873.

Vekasi, J., Zs. Marton, G. Kesmarky, A. Cser, R. Russai, B. Horvath. 2001. "Hemorheological Alterations in Patients with Diabetic Retinopathy." *Clinical Hemorheology and Microcirculation* 24:59–64.

Vij, A. G. 2009. "Effect of Prolonged Stay at High Altitude on Platelet Aggregation and Fibrinogen Levels." *Platelets* 20:421–427.

Vin, F., A. Chabanel, A. Taccoen, J. Ducros, J. Gruffaz, B. Hutinel, P. Maillet, and M. Samama. 1994. "Double-Blind Trial of the Efficacy of Troxerutine in Chronic Venous Insufficiency." *Phlebology* 9:71–76.

Vincent, J. L., A. de Mendonca, F. Cantraine, R. Moreno, J. Takala, P. M. Suter, C.L. Sprung, F. Colardyn, and S. Blecher. 1998. "Use of the SOFA Score to Assess the Incidence of Argan Dysfunction/Failure in Intensive Care Units: Results of a Multi-Center Prospective Study. Working Group on 'Sepsis-Related Problems' of the European Society of Intensive Care Medicine." *Critical Care Medicine* 26:1793–1800.

von Tempelhoff, G. F., L. Heilmann, G. Hommel, and K. Pollow. 2003. "Impact of Rheological Variable in Cancer." *Seminars in Thrombosis and Hemostasis* 29:499–513.

Wadworth, A. N., and D. Faulds. 1992. Hydroxyethylrutosides—A Review of Its Pharmacology, and Therapeutic Efficacy in Venous Insufficiency and Related Disorders." *Drugs* 44:1013–1032.

Weng, X. D., G. Cloutier, R. Beaulieu, and G. O. Roederer. 1996. "Influence of Acute-Phase Proteins on Erythrocyte Aggregation." *American Journal of Physiology—Heart and Circulatory Physiology* 271:H2346–H2352.

Whittingstall, P., and H. J. Meiselman. 1991a. "Aggregation Behavior of Neonatal Red Blood Cells." *Clinical Hemorheology* 11:728.

Whittingstall, P., and H. J. Meiselman. 1991b. "Effect of Galactose Incubation on the Aggregation Behavior of Density-Separated Human Erythrocytes." In *Hemorheologie et Aggregation Erythrocytaire*, ed. J. F. Stoltz, M. Donner, and A. L. Copley, 111–122. Paris: Editions Medicales Internationales.

Whittingstall, P., K. Toth, R. B. Wenby, and H. J. Meiselman. 1994. "Cellular Factors in RBC Aggregation: Effects of Autologous Plasma and Various Polymers." In *Hemorheologie et Aggregation Erythrocytaire*, ed. J. F. Stoltz, 21–30. Paris: Editions Medicales Internationales.

Windberger, U., A. Bartholovitsch, G. Wollenek, H. Heinzl, and U. Losert. 1998. "Comparison of Erythrocyte Aggregation and Whole Blood Viscosity of the Fetal Lamb (Third Quarter of Gestation) and the Adult Sheep." *Journal of Experimental Animal Science* 39:34–42.

Wood, S. C., M. P. Doyle, and O. Appenseller. 1991. "Effects of Endurance Training and Long Distance Running on Blood Viscosity." *Medical Science in Sports Exercise* 3:1265–1269.

Yalcin, O., M. Bor-Kucukatay, U. K. Senturk, and O. K. Baskurt. 2000. "Effects of Swimming Exercise on Red Blood Cell Rheology in Trained and Untrained Rats." *Journal of Applied Physiology* 88:2074–2080.

Yalcin, O., A. Erman, S. Muratli, M. Bor-Kucukatay, and O. K. Baskurt. 2003. "Time Course of Hemorheological Alterations after Heavy Anaerobic Exercise in Untrained Human Subjects." *Journal of Applied Physiology* 94, no. 3:997–1002.

Yalcin, O., P. Ulker, U. Yavuzer, H. J. Meiselman, and O. K. Baskurt. 2008. "Nitric Oxide Generation of Endothelial Cells Exposed to Shear Stress in Glass Tubes Perfused With Red Blood Cell Suspensions: Role of Aggregation." *American Journal of Physiology— Heart and Circulatory Physiology* 294:H2098–H2105.

Yang, Z., J. Liu, F. Li, P. Yan, Y. Liu, and F. Sun. 1999. "Effect of Acute Hypoxia and Hypoxic Acclimation on Hemorheological Behavior in Rats with Frostbite." *Clinical Hemorheology and Microcirculation* 20:189–195.

Yang, Z. R., J. Y. Liu, and P. H. Yan. 2003. "Effect of Cold Acclimation on Hemorheological Behavior in Rats with Frostbite." *Clinical Hemorheology and Microcirculation* 29:103–109.

Yarnell, J. W., I. A. Baker, P. M. Sweetnam, D. Bainton, J. R. O-Brien, P. J. Whitehead, and P. C. Elwood. 1991. "Fibrinogen, Viscosity, and White Blood Cell Counts Are Major Risk Factors for Ischemic Heart Disease. The Caerphilly and Speedwell Collaborative Heart Disease Studies." *Circulation* 83:836–844.

Zeltser, D., N. M. Bornstein, R. Rotstein, I. Shapira, and A. S. Berliner. 2001. "The Erythrocyte Adhesiveness/Aggregation Test in the Peripheral Blood of Patients with Ischemic Brain Events." *Acta Neurologica Scandinavica* 103:316–319.

Zhao, H., X. Wang, and J. F. Stoltz. 1999. "Comparison of Three Optical Methods to Study Erythrocyte Aggregation." *Clinical Hemorheology and Microcirculation* 21:297–302.

Zhao, K. S. 2005. "Hemorheologic Events in Severe Shock." *Biorheology* 42:463–477.

Zilberman, L., O. Rogowski, M. Rozenblat, I. Shapira, J. Serov, P. Halpern, I. Dotan, N. Arber, and S. Berliner. 2005. "Inflammation-Related Erythrocyte Aggregation in Patients with Inflammatory Bowel Disease." *Digestive Diseases and Sciences* 50:677–683.

Zilberman-Kravits, D., I. Harman-Boehm, T. Shuster, and N. Meyerstein. 2006. "Increased Red Cell Aggregation Is Correlated with HbA1C and Lipid Levels in Type 1 but Not Type 2 Diabetes." *Clinical Hemorheology and Microcirculation* 35:463–471.

Zorio, E., J. Murado, D. Arizo, J. Rueda, D. Corella, M. Simo, and A. Vaya. 2008. "Haemorheological Parameters in Young Patients with Acute Myocardial Infarction." *Clinical Hemorheology and Microcirculation* 39:33–41.

9 Comparative Aspects of Red Blood Cell Aggregation

9.1 AGGREGATION IS A CHARACTERISTIC OF MAMMALIAN RED BLOOD CELLS

It is well known that the aggregation of red blood cells (RBC) as discussed in this book is a mammalian feature. Mammalian RBC do not contain a nucleus or other organelles, while all nonmammalian species, including avian, reptile, and fish, are characterized by nucleated RBC (see frog RBC in Figure 9.1e as an example). Primarily due to an extensive degree of evolution, RBC in mammalian species are usually biconcave discs (Figure 9.1) with only a few exceptions (e.g., camelids, Figure 9.1c). This special geometry of RBC is common for all sub and infraclasses of the Mammalian class, including placental mammals, marsupials, and monotremes (Baskurt et al. 2010) that represent the first diversification of Mammalia dating back ~200 million years (Grutzner et al. 2003). It has been suggested that this special geometry evolved to optimize blood flow in the circulatory system (Uzoigwe 2006).

Nucleated, nonmammalian RBC do not aggregate, yet observations of the lack of RBC aggregation in nonmammalian species are rarely published. Figure 9.2 presents such an observation for the optical density of RBC suspensions recorded following an abrupt cessation of rapid flow through a transparent tube (Ohta et al. 1992). The records from blood samples for the mammalian species shown exhibited the typical syllectogram curve: a sudden increase in optical density following the stoppage of flow due to recovery of the biconcave shape followed by decreased density due to aggregation of RBC (see Chapter 4, Section 4.1.6). In contrast with the mammalian blood samples, blood obtained from a fowl exhibited a different pattern following stoppage of flow: a slight increment in optical density that is significantly smaller compared to mammalian blood, thus reflecting the lower degree of shape change under flow conditions and thus a smaller change from flow to stasis. This initial period is followed by a further but slower increment in the optical density, probably due to continuing random distribution of the cells. Note that the curve for fowl RBC is similar to those obtained when mammalian RBC are suspended in a nonaggregating medium (e.g., isotonic phosphate-buffered saline).

The biconcave-discoid shape is a prerequisite for RBC aggregation, although this geometry may not always yield measurable RBC aggregation. There are species that have RBC that do not aggregate, as well as those having very high aggregation

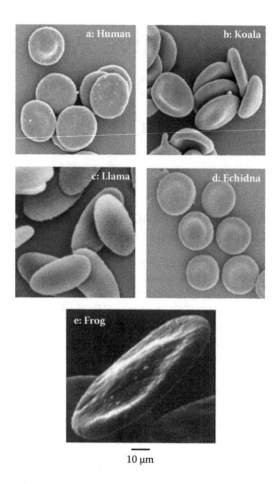

10 μm

FIGURE 9.1 Scanning electron microscope images of red blood cells (RBC) from various species. a. Human (Infraclass: Eutheria, placental mammal); b. Koala (Infraclass: Metatheria, marsupial); c. Llama (Infraclass: Eutheria, placental mammal, camelid); d. Echidna (Subclass Prototheria, monotreme). e. Leopard frog (Rana pipiens, nonmammalian vertebrate). Figure 9.1e is reproduced from Cohen, W. D. 1991. "The Cytoskeletal System of Nucleated Erythrocytes." *International Review of Cytology* 130:37–84, 1991 with permission.

tendency, with a biconcave-discoid geometry being a common feature of both categories. A pattern of RBC aggregation characteristics for species that can be related to phylogenic relationships is not easily recognizable; the two mammalian subclasses (i.e., Theria and Prototheria) and also the two infraclasses of Theria (i.e., Metatheria and Eutheria) exhibit a similar variability in RBC aggregation (Baskurt et al. 2010). Various attempts have been made to determine relationships between obvious characteristics of mammalian species (e.g., athletic capacity) and RBC aggregation properties (see Section 9.4).

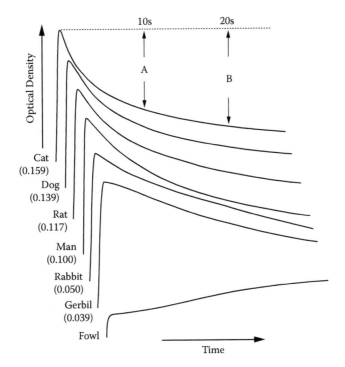

FIGURE 9.2   Time course of optical density of blood from several mammalian species and fowl in a transparent tube following a sudden stoppage of flow. Prior to stoppage, the flow was sufficient for complete disaggregation. The difference in the pattern between mammals and fowl should be noted. Reproduced from Ohta, K., F. Gotoh, M. Tomita, N. Tanahashi, M. Kobari, T. Shinohara, Y. Tereyama, B. Mihara, and H. Takeda. 1992. "Animal Species Differences in Erythrocyte Aggregability." *American Journal of Physiology—Heart and Circulatory Physiology* 262:H1009–H1012 with permission.

9.2   RED BLOOD CELL AGGREGATION CHARACTERISTICS IN VARIOUS MAMMALIAN SPECIES

Robin Fåhraeus (1958) reported his observations on the formation of RBC aggregates and the associated phase separation in bovine, human, and horse blood flowing in a glass tube at a low flow rate. These observations clearly indicated the difference in aggregation between bovine and horse blood: There was almost no visible phase separation for bovine blood, while horse blood exhibited very prominent phase separation under the same flow conditions, indicating very significant differences in RBC aggregation between these two species. Human blood was characterized by an intermediate level of RBC aggregation and phase separation. Erythrocyte sedimentation rate (ESR) reported for bovine and horse blood (0 mm/h and 98 mm/h, respectively) supported his observations in the glass tube (Fåhraeus 1958).

Interspecies comparisons of the shear dependence of blood viscosity published in the 1960s revealed more prominent differences at lower shear rates rather than at high shear (Usami et al. 1969). Usami et al. studied blood viscosity of five species, including elephant, dog, sheep, goat, and human. In measurements with 0.45 l/l hematocrit blood, they observed that the viscosity of sheep and goat blood measured at shear rates below 0.1 s⁻¹ was significantly lower than the other three species, while high-shear blood viscosities were relatively close to each other. Furthermore, they reported similar low-shear viscosities measured in plasma and serum for sheep and goat, whereas the values were higher in plasma than serum for elephant, dog, and human blood (Usami et al. 1969). Given the well-known dependence of low-shear blood viscosity on RBC aggregation (see Chapter 5), these findings strongly suggest that a difference in RBC aggregation properties should be responsible for these results. Usami et al. also demonstrated microscopically that rouleaux formation in sheep and goat blood was significantly less pronounced than in the other three species included in the study (Usami et al. 1969).

Popel et al. (1994) also performed viscometry using a Contraves LS30 viscometer for blood samples from a group of placental mammals and reported contrasting patterns of blood viscosity as a function of shear rate in species with different characteristics. Sheep and cow blood samples exhibited very minor non-Newtonian flow behavior (i.e., shear dependence of viscosity) (Figure 9.3), while there were large differences between low- and high-shear viscosities for other species including horse, dog, and antelope (Popel et al. 1994). They also reported differences in ESR, confirming their conclusion related to differences in RBC aggregation among the species studied.

Johnn et al. (1992) reported blood viscosity measured at low (0.277 s⁻¹) and high (128.5 s⁻¹) shear rates for a wide range of mammals. Figure 9.4 presents the low- and high-shear rate viscosity ranges for 30 species, mostly placental mammals with wallaby representing marsupials. Both low- and high-shear viscosities exhibit

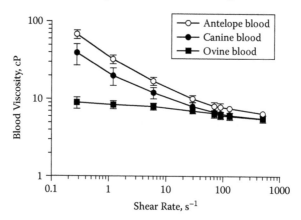

FIGURE 9.3 Blood viscosity measured as a function of shear rate for antelope, dog (canine), and sheep (ovine) blood. The hematocrit in all samples was adjusted to 0.4 l/l. Note the significant difference in non-Newtonian behavior of ovine blood compared to the other two species. Redrawn from Popel, A. S., P. C. Johnson, M. V. Kameneva, and M. A. Wild. 1994. "Capacity for Red Blood Cell Aggregation Is Higher in Athletic Mammalian Species Than in Sedentary Species." *Journal of Applied Physiology* 77:1790–1794 with permission.

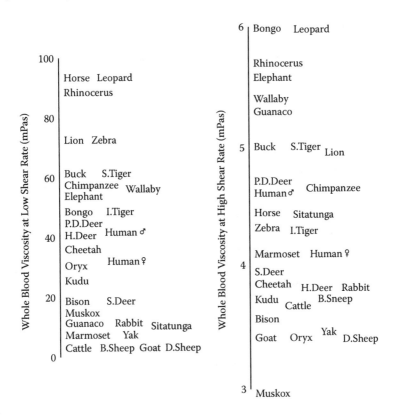

FIGURE 9.4 Whole blood viscosity measured at low (0.277 s⁻¹) and high (128.5 s⁻¹) shear rates for 30 species. Reproduced from Johnn, H., C. Phipps, S. C. Gascoyne, C. Hawkey, and M. W. Rampling. 1992. "A Comparison of the Viscometric Properties of the Blood from a Wide Range of Mammals." *Clinical Hemorheology* 12:639–647 with permission.

interspecies variations, with the range of low- and high-shear viscosities being very different: low-shear viscosity ranges between ~5 mPa.s and ~90 mPa.s (i.e., ~18-fold difference between cattle and horse), while high-shear viscosity ranges only between ~3 mPa.s and ~6 mPa.s (i.e., ~ 2-fold difference between musk ox and bongo, a type of antelope). The relatively small variation of blood viscosity at high shear rate reflects differences in hematocrit and RBC deformability between species; the variation is even smaller after the adjustment of hematocrit to 0.45 l/l for all species (Johnn et al. 1992). The very high variation of blood viscosity at low shear rate indicates differences of RBC aggregation among the species, leading to the conclusion that RBC aggregation is the most variable flow property of mammalian blood.

Direct measurements of RBC aggregation, mostly using photometric methods, have confirmed the large variation in the class Mammalia (Baskurt et al. 1997; Baskurt et al. 2000; Baskurt et al. 2010; Ohta et al. 1992; Spengler et al. 2008; Windberger et al. 2003). Direct comparisons between studies are usually not possible due to the use of different methods, but there is general agreement on categorizing species based on their RBC aggregation properties. It has been suggested that mammals can be classified into three groups as *intense*, *moderate*, and *no aggregation* (Windberger and Baskurt 2007). A fourth group can be added to this classification in order to include species having only a slight degree of RBC aggregation (e.g., rodents); this fourth group, termed the *low aggregation* group, would be positioned between the *moderate* and *no aggregation* groups. Table 9.1 presents this classification method for RBC aggregation measured in autologous plasma.

Detailed analysis of the RBC aggregation process also reveals significant differences among species. These differences can be related to the time course of RBC aggregation as well as to differences of aggregate size in terms of number of RBC per aggregate. Comparisons of many species reveal similar rank orders regardless of the aggregation parameter used (e.g., extent or time course of aggregation), but exceptions may exist for some species. Baskurt et al. (2010) reported similar Myrenne M indexes for echidna and human blood, but the time constant for the fast phase of RBC aggregation was found to be ~100% shorter in human blood (i.e., T_{FAST} for human blood was only ~1/2 of the value for echidna blood, see Chapter 4, Section 4.1.6.2.2.).

Microscopic observations of RBC aggregates of various species also raise interesting questions because while they almost always support the findings of aggregation measurements, they also may present challenges. As an example, Figure 9.5 shows microscopic images for Tasmanian devil and kangaroo RBC suspended in autologous plasma. These two species represent two extremes of aggregation intensity, with the Tasmanian devil exhibiting no measurable aggregation with Myrenne aggregometer while the kangaroo blood exhibits very high values (unpublished observation). Tasmanian devil RBC do approach each other during the several-minute time period of observation, but tight aggregates are not formed, as seen with kangaroos. This observation raises questions about the strength of RBC aggregates even if they are formed, yet unfortunately, studies comparing aggregate strength in various species do not exist.

9.3 CORRELATIONS WITH RED BLOOD CELL PROPERTIES

9.3.1 Plasma Factors versus Cellular Properties

RBC aggregation is determined by both plasmatic and cellular factors (see Chapter 2 for detailed discussion). Interspecies differences in aggregation have been observed for native blood samples (i.e., RBC suspended in autologous plasma) and for RBC suspended in a standard suspending medium (e.g., 0.5% dextran 500 kDa solution). A species with high RBC aggregation in plasma also tends to have high aggregation in a standard aggregating medium (Figure 9.6), suggesting that cellular properties play an important role in the interspecies differences (Baskurt et al. 1997; Baskurt et al. 2000).

Fibrinogen is the most important plasma factor determining the extent of RBC aggregation under physiological conditions (Rampling 1988). Significant correlations have been reported between plasma fibrinogen concentrations and aggregation indices if studied within a given species (e.g., human samples) (Ben Ami et al. 2001; Falco et al. 2005; Rampling 1988; Windberger et al. 2003). However, plasma fibrinogen concentration failed to correlate with the degree of aggregation for various

TABLE 9.1

Classification of Mammals According to Their Red Blood Cell Aggregation Properties

Group	Species	References
Intense	Horse	(Baskurt et al. 1997; Baskurt et al. 2000; Kumaravel and Singh 1995; Popel et al. 1994; Spengler et al. 2008; Windberger et al. 2003)
	Weddell seal	(Castellini et al. 2006)
	Kangaroo	Unpublished observation
	Antelope	(Popel et al. 1994)
	Domestic cat	(Ohta et al. 1992; Windberger et al. 2003)
	Tiger	Unpublished observation
Moderate	Elephant	(Usami et al. 1969)
	Pig	(Windberger et al. 2003)
	Bowhead whale	(Castellini et al. 2006)
	Dog	(Ohta et al. 1992; Popel et al. 1994; Usami et al. 1969; Windberger et al. 2003)
	Koala	(Baskurt et al. 2010)
	Human	(Baskurt et al. 1997; Baskurt et al. 2000; Baskurt et al. 2010; Ohta et al. 1992; Spengler et al. 2008; Usami et al. 1969; Windberger et al. 2003)
	Echidna	(Baskurt et al. 2010)
Low	Rat	(Baskurt et al. 1997;Baskurt et al. 2000;Ohta et al. 1992;Windberger et al. 2003)
	Rabbit	(Baskurt et al. 2000; Kumaravel and Singh 1995; Ohta et al. 1992; Windberger et al. 2003)
	Gerbil	(Ohta et al. 1992)
	Guinea pig	(Baskurt et al. 2000)
None	Cattle	(Kumaravel and Singh 1995; Popel et al. 1994; Spengler et al. 2008; Windberger et al. 2003)
	Sheep	(Popel et al. 1994; Usami et al. 1969; Windberger et al. 2003)
	Goat	(Kumaravel and Singh 1995; Usami et al. 1969)
	Mouse	(Windberger et al. 2003)
	Tasmanian devil	Unpublished observation

Note: The species listed in each category are presented as known examples. The arrow indicates a decreasing level of aggregation measured in autologous plasma. However, the ordering of the species within each category may not be correct due to difficulties in comparing aggregation indexes determined using different methods.

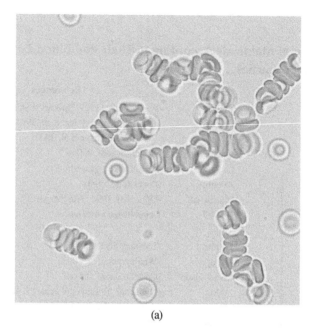

(a)

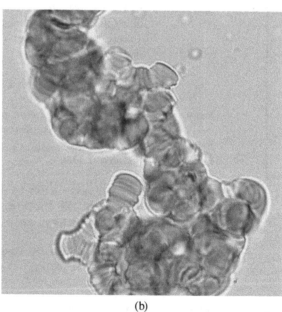

(b)

FIGURE 9.5   Aggregates of Tasmanian devil (a) and kangaroo (b) red blood cells (RBC) suspended in autologous plasma. Note the tight, three-dimensional rouleaux of kangaroo RBC compared to the loose gathering of Tasmanian devil RBC. The time course for the process for Tasmanian devil and kangaroo RBC was also very different, being essentially completed within seconds for kangaroo blood but requiring several minutes for Tasmanian devil blood.

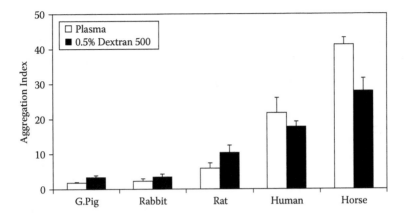

FIGURE 9.6   Aggregation indexes measured by a plate-on-plate photometric aggregometer for RBC in autologous plasma and in 0.5% dextran 500 kDa. The hematocrit was adjusted to 0.4 l/l for both types of suspensions. Data from Baskurt, O. K., M. Bor-Kucukatay, O. Yalcin, and H. J. Meiselman. 2000. "Aggregation Behavior and Electrophoretic Mobility of Red Blood Cells in Various Mammalian Species." *Biorheology* 37:417–428.

species with diverse aggregation properties (Windberger et al. 2003). Table 9.2 presents plasma fibrinogen concentration and degree of aggregation for blood samples from several species. In contrast with expectations, fibrinogen concentration does not positively correlate with the intensity of RBC aggregation: cattle have the highest plasma fibrinogen concentration yet their RBC aggregation is not measurable, while horse blood with the highest degree of RBC aggregation has low plasma fibrinogen concentration. Thus, over a wide range of species, but not within a single species, the data in Table 9.2 indicate a negative correlation between RBC aggregation intensity and plasma fibrinogen concentration.

It has been suggested that the high RBC aggregation for horse may be related to differences in the fibrinogen molecule (Andrews et al. 1992), but unfortunately, there are no data to support such differences for other species. Spengler and Rasia (2001) investigated the role of other plasma proteins in RBC aggregation, but could not identify any specific components of plasma in species with weak (cattle) or intense (horse) RBC aggregation. They concluded that RBC properties are primarily responsible for the differences in aggregation (Spengler and Rasia 2001).

9.3.2   RED BLOOD CELL PROPERTIES AS DETERMINANT OF INTERSPECIES DIFFERENCES

As discussed previously, the biconcave-discoid shape of RBC is a key feature for aggregation, and this feature is very common in the mammalian class with only a few exceptions. Camelids with oval RBC exhibit almost no RBC aggregation (Windberger and Baskurt 2007), as also indicated by the very low blood viscosity at low shear rate for guanaco in Figure 9.4 (Johnn et al. 1992). All other species discussed previously with intense, moderate, low, and no aggregation have biconcave

TABLE 9.2

Plasma Fibrinogen Concentration and Mean Corpuscular Volume (MCV) for Various Species with Diverse Red Blood Cell Aggregation Properties

Species	RBC Aggregation	Fibrinogen (g/l)	MCV (fl)	Reference
Horse	I	1.55	45	(Windberger et al. 2003)
Kangaroo	I	2.20	75	Unpublished observation
Cat	I	1.86	40	(Windberger et al. 2003)
Pig	M	1.72	54	(Windberger et al. 2003)
Dog	M	1.82	61	(Windberger et al. 2003)
Human	M	2.65	90	Normal range
Rat	L	2.33	47	(Windberger et al. 2003)
Rabbit	L	2.86	62	(Windberger et al. 2003)
Cattle	N	3.09	52	(Windberger et al. 2003)
Sheep	N	2.82	34	(Windberger et al. 2003)
Mouse	N	2.83	45	(Windberger et al. 2003)
Tasmanian devil	N	2.00	65	Unpublished observation

Note: The species are presented in the same order as in Table 9.1. The second column indicates the degree of aggregation as intense (I), moderate (M), low (L), and none (N). All data are from Windberger et al. (2003) with the exception of kangaroo and Tasmanian devil; human data are the published normal range.

discoid RBC, including sheep, cattle, horse, human, and so on. Therefore, RBC shape is not a dominant factor when considering cellular properties determining the extent of aggregation.

9.3.2.1 Red Blood Cell Size and Hemoglobin Content

RBC volume is highly variable among mammals, ranging from ~30 fl for goat to ~180 fl for bowhead whale (Castellini et al. 2006; Gascoyne and Hawkey 1992). Table 9.2 presents mean cell volume (MCV) data for many of the species whose RBC aggregation figures are shown in Table 9.1; the species in Table 9.2 are arranged in the same sequence as in Table 9.1. A comparison of the MCV data versus the level of RBC aggregation in different species reveals no relationship: there are species with similar RBC sizes at both ends of the spectrum (e.g., horse with intense aggregation and mouse with no aggregation). Neither negative nor positive correlations could be detected between aggregation indexes and MCV using data sets similar to Table 9.2.

Cytosolic hemoglobin concentration (i.e., mean corpuscular hemoglobin concentration, MCHC) in mammalian RBC is relatively constant among different species, with again only a few exceptions including camelids (Gascoyne and Hawkey 1992). Therefore, RBC hemoglobin content (i.e., mean cell hemoglobin, MCH) is a function of MCV. As with MCV, neither MCH nor MCHC values are useful in explaining interspecies differences in RBC aggregation.

The shape and other properties of camelid RBC represent an evolutionary adaptation to extreme environments. These RBC are characterized by a high hemoglobin concentration to cope with the low-oxygen partial pressure at the altitudes in which they live (Uzoigwe 2006). Their oval shape is an alternative to the biconcave disc and is an adaptation favoring orientation to flow streamlines, thereby aiding *in vivo* flow (Smith et al. 1979). Their RBC also have an exceptional osmotic resistance, which is an advantage under conditions leading to dehydration (Long 2007). However, these adaptations resulted in the loss of aggregation ability and may represent a trade-off in evolutionary history.

9.3.2.2   Red Blood Cell Membrane Structure

The relations between membrane lipid composition and RBC aggregation properties have been investigated in species representing high and low/none aggregators. Plasenzotti et al. (2007) reported a twofold higher unsaturated fatty acid content in sheep RBC membranes compared to horse and pig RBC, both of which aggregate more strongly (Table 9.1). They suggested that higher unsaturated fatty acid content should predict higher membrane fluidity, better RBC deformability, and higher aggregation, and therefore reported their findings as "surprising" (Plasenzotti et al. 2007). Spengler et al. (2008) also reported differences in membrane lipid content for bovine and equine RBC: bovine RBC, with almost no aggregation, are rich in sphingomyelin but have lower phosphatidylcholine content compared to horse RBC with very high aggregation. They also studied membrane fluidity by fluorescence polarization and reported higher membrane fluidity for bovine RBC. Data on membrane lipid composition for a wider range of species are also available in the literature (Roelofsen et al. 1981), yet a general pattern that can be linked to RBC aggregation characteristics is not obvious. Differences in membrane fluidity have been frequently evaluated as a factor contributing to RBC deformability, which may in turn influence RBC aggregation. However, it has been observed that even large changes of membrane lipid composition induced experimentally do not significantly influence RBC deformability (Uyuklu et al. 2007) inasmuch as membrane mechanical properties are mainly determined by the spectrin-based membrane cytoskeletal network (Hochmuth 1981; Mohandas et al. 1983). Differences in membrane protein patterns have also been reported for species with different RBC aggregation characteristics (Baskurt et al. 1997), yet unfortunately, sufficient data do not exist for an interspecies comparison with respect to aggregation characteristics.

9.3.2.3   Red Blood Cell Deformability

Deformability is one of the cellular properties that contributes to the degree of RBC aggregation (Baskurt et al. 2002; Chien 1987; Meiselman et al. 1999; Nash et al. 1987; Sowemimo-Coker et al. 1989). Smith et al. (1979) reported RBC elongation indexes measured by laser diffraction ektacytometry for 20 species with different aggregation characteristics. RBC deformability in most species included in their study remained in a narrow range with only a few species falling outside this range (Smith et al. 1979). The species with exceptionally low RBC deformability were horse, goat, and camelids. Windberger and Baskurt (2007) provided shear stress–elongation index curves for nine species together with shear stress at half-maximal deformation. Horse and elephant RBC required the highest shear stress for half-maximal deformation, while sheep RBC had an exceptionally low maximum elongation index (Windberger and Baskurt 2007). It is interesting to note that the two species with contrasting RBC aggregation characteristics (i.e., ovines and equines) are in the group with exceptional deformability.

It can be suggested that membrane rigidity, rather than the deformability of the entire RBC, is more important in forming parallel surfaces during the aggregation process. Waugh (1992) studied surface rigidity (i.e., membrane shear elastic modulus) of RBC from human, rat, rabbit, opossum, and camel using micropipettes of 1.5–2 μm diameter. Surface rigidity of RBC membranes was reported to be in a narrow range (0.0076–0.0093 dyn/cm) for all species with the exception of camel. Their studies included species with moderate and low RBC aggregation (Table 9.1), but surface rigidity data do not provide any patterns to explain the differences in aggregation.

9.3.2.4 Red Blood Cell Surface Properties

The surface properties of RBC are the most important factors in determining their intrinsic aggregation behavior and hence their aggregability. In general, these properties remain poorly described, although various methods have been applied to reveal the differences in RBC with distinct aggregation characteristics.

Surface charge density has been studied using a cell electrophoresis method for horse, sheep, cow, pig, and human RBC (Eylar et al. 1961). Horse RBC exhibited the highest electrophoretic mobility, suggesting higher surface charge density compared to other species, with mobility lower for sheep and human RBC followed by cattle and dog. Higher surface charge density would predict lower aggregation due to the higher electrostatic repulsion as RBC approach each other (Meiselman 2009; Meiselman 1993; Rampling et al. 2004). Obviously, the highest surface charge density for horse RBC reported by Eylar et al. (1961) does not support the expectation of lower aggregation tendency. Other reports of interspecies differences in surface charge density also do not provide an explanation for differences in aggregation. For example, Baskurt et al. (2000) reported very low electrophoretic mobility for rabbit RBC compared to human, horse, and rat, thus predicting very high aggregation tendency for rabbit, yet rabbit is a low aggregator (Table 9.1). Interestingly, negative correlations between electrophoretic mobility and aggregation indexes were observed for horse, human, and rat RBC if analyzed within a given species, while such correlations were not detected with rabbit and guinea pig RBC that have low aggregation and electrophoretic mobility (Baskurt et al. 2002). Baumler et al. (2001) also reported lower electrophoretic mobility (i.e., lower surface charge density) for cattle compared to horse RBC. In brief, RBC surface charge density differences do not explain the aggregation patterns in various species.

Based on the depletion model for RBC aggregation (see Chapter 3), surface properties including glycocalyx structure are major determinants of the depletion of macromolecules (e.g., fibrinogen, high-molecular-mass dextrans) from the cell surface; increased depletion is expected to increase aggregation. The surface properties of RBC have been investigated by two-phase partitioning as well as by electrophoresis in suspending media containing various polymers (Baumler et al. 2001; Seaman and Uhlenbruck 1963; Walter and Widen 1993; Walter and Widen 1994). Briefly, cell electrophoresis studies conducted in high-molecular-mass polymer solutions revealed that RBC of species with high aggregation tendency (i.e., horse) have a more effective depletion layer compared to species with no RBC aggregation (i.e., cattle) (Baumler et al. 2001). Further investigation of interspecies differences of RBC surface properties and glycocalyx structure may provide valuable information and additional insight into the mechanisms of aggregation. Additionally, such information might be used to understand and model the alterations in RBC aggregation in disease processes and to develop related therapeutic strategies.

9.4   CORRELATION WITH OTHER PROPERTIES OF SPECIES

RBC aggregation has been studied in a wide range of mammals, including placental mammals, marsupials, and monotremes (Baskurt et al. 2010). Comparisons among these animals was deemed of special import since they are main branches of the mammalian kingdom and represent the first major evolutionary diversification dating back ~200 million years. RBC aggregation behavior of these subclasses of mammals with diverse biology was surprisingly similar, suggesting that such behavior is among the fundamental mammalian adaptations (Baskurt et al. 2010). Further studies indicated that a similar degree of variation of RBC aggregation properties (i.e., ranging from none to intense) exist among species in each subclass. Investigations of RBC aggregation have also included marine mammals and revealed a pattern similar to terrestrial mammals (Castellini et al. 2007; Castellini et al. 2010). There have been attempts to correlate various characteristics of species with RBC aggregation properties (see the following text), yet to date there does not seem to be general patterns of RBC aggregation that can be linked to the biological characteristics of various species.

9.4.1   ATHLETIC VERSUS SEDENTARY SPECIES

Various investigators have observed that athletic species are characterized by more intense RBC aggregation compared to those with lower athletic capacity (Johnn et al. 1992; Ohta et al. 1992; Popel et al. 1994). Popel et al. (1994) compared RBC aggregation in several species with their athletic capacity as judged by maximal oxygen consumption (VO_{2max}). RBC aggregation was measured using the ESR and blood viscosity measured at 0.277 s^{-1}. As shown in Figure 9.3, they reported intense RBC aggregation in species with higher athletic capacity (i.e., horse, dog, and antelope) and very low aggregation in species with lower athletic capacity (i.e., cow and sheep). This observation has been extended across the marsupial infraclass: very intense RBC aggregation is found for the eastern gray kangaroo (unpublished observation) and is at a level comparable to athletic placental mammals with high VO_{2max} (Kram and Dawson 1998).

The pattern of intense RBC aggregation in athletic species can be hypothesized to be advantageous for enhancing athletic performance, with this hypothesis based on the *in vivo* effects of RBC aggregation (see Chapter 7). Briefly, intense RBC aggregation may lead to reduced flow resistance in microcirculation due to enhanced axial migration of RBC resulting in lower frictional resistance at the vessel wall and lower microvascular hematocrit. In addition, enhanced RBC aggregation and thus more pronounced non-Newtonian flow behavior seem to be important for moderating changes of capillary blood pressure (Pc) during exercise. This pressure is the most important determinate of the overall relations between filtration and absorption in the microcirculation. In turn, Pc is affected by the ratio of pre- to postcapillary resistance. With exercise there is a decrease of precapillary resistance, increased muscle blood flow, and a tendency for Pc to become higher. However, increased flow also decreases blood viscosity (i.e., non-Newtonian behavior) and postcapillary resistance, thereby restoring the pre-to-post ratio and Pc toward control. In brief, RBC aggregation and the resulting flow behavior of blood affect microcirculatory flow dynamics and tend to minimize increases of Pc and the formation of edema during exercise (Cabel et al. 1997).

An interesting observation that supports the hypothesis related to the role of RBC aggregation in athletic performance was reported by Stoiber et al. (2005). They investigated aggregation indexes of untrained and trained horses before, immediately following, and 30 minutes after submaximal exercise (Table 9.3). Trained horses had a highly significant, 30% increase of aggregation immediately following the exercise episode with aggregation returning to preexercise levels in 30 minutes after exercise. However, the increase of aggregation following submaximal exercise was only 4% in untrained horses and was further increased 30 minutes after the exercise period (Stoiber et al. 2005). Their findings should indicate the contribution of RBC seques-

TABLE 9.3

Aggregation Index (Myrenne M1) Measured in Blood Samples Obtained before, Immediately after, and at 30 Minutes following Submaximal Exercise Performed by Untrained and Trained Horses

		Aggregation Index
Untrained	Before	55.87
	After	58.34
	30 min	65.18
Trained	Before	46.94
	After	61.49
	30 min	51.14

Source: Data from Stoiber, B., C. Zach, B. Izay, and U. Windberger. 2005. "Whole Blood, Plasma Viscosity, and Erythrocyte Aggregation as a Determining Factor of Competitiveness in Standard Bred Trotters." *Clinical Hemorheology and Microcirculation* 32:31–41.

tered in the spleen of horses, recruited during the strenuous exercise and support the role of RBC aggregation in enhanced athletic performance.

9.4.2 Body Size

The intensity of RBC aggregation has a significant but weak correlation ($r = 0.506$, $p < 0.05$) with the logarithm of average body weight (Figure 9.7), and thus resembles various allometric relationships (Li 1996). The most prominent deviations from the main group are related to species with moderate body size but almost zero aggregation (e.g., cattle, sheep, goat, Tasmanian devil) or to those with extremely large body weights (e.g., elephant, whale). Interestingly, the aggregation index versus body weight correlation becomes highly significant ($r = 0.860$, $p < 0.0001$) if the six species with extreme values mentioned previously are excluded.

This relationship of RBC aggregation to the body size of an animal has been suggested to indicate the importance of an evolutionary adaptation with respect to the influence of RBC aggregation on *in vivo* blood flow resistance. Development of RBC axial migration and related hemodynamic effects (e.g., reduced frictional resistance, decreased microvascular hematocrit; see Chapters 6 and 7) occur only if the unbranched length of a blood vessel is sufficient to allow these processes to develop (Windberger and Baskurt 2007). Therefore, RBC aggregation may not have beneficial *in vivo* hemodynamic effects in species with smaller body size and, presumably, shorter vessels. Obviously, this suggestion regarding body size and benefit is not valid in some species (e.g., cattle) and thus requires alternative explanations. Apparently, certain evolutionary or even life style adaptations may surpass the general allometric relationship. Alternatively, this discussion may provide an explanation for the low aggregation intensity for some rodents with small body size, although

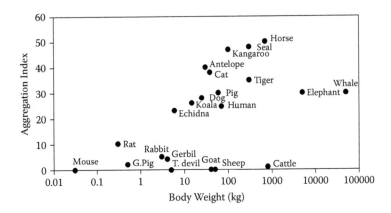

FIGURE 9.7 Relationship between body weight and intensity of red blood cell aggregation. The data were adopted from the references presented in Table 9.1. The correlation coefficient (*r*) for aggregation index versus logarithm of body weight was 0.506 (*p* < 0.05) but increased to *r* = 0.860 (p<0.0001) if species with extreme values (i.e., Tasmanian devil, goat, sheep, cattle, elephant, and whale) are omitted.

they are extremely active animals (i.e., have high athletic capacity) (Windberger and Baskurt 2007).

9.4.3 LIFE STYLE AND NUTRITION

In addition to reports mentioned in Sections 9.4.1 and 9.4.2, comparative studies of RBC aggregation to date have not demonstrated specific patterns that can be attributed to a certain life style. This is obvious by inspection of Table 9.1: species with very different life styles and biological features can be assigned to a group on the basis of RBC aggregation behavior (i.e., intense, moderate, low, or none). For example, the moderate aggregation group includes echidna, an insect-eating animal adapted to survive in a hypoxic underground environment (Nicol 2003); koala, a herbivorous marsupial; dog, a carnivore; bowhead whale, a marine mammal; and human, an omnivore. This assortment of life styles and nutritional patterns also exist for other groups formed according to the intensity of RBC aggregation.

9.4.4 RED BLOOD CELL AGGREGATION IN MARINE MAMMALS

RBC aggregation has been investigated in marine mammals, including various seal species (Castellini et al. 2006; Meiselman et al. 1992; Wickham et al. 1990) and bowhead whales (Castellini et al. 2006). These results indicate a variation of RBC aggregation properties in marine mammals and hence a wide range similar to aggregation behavior for terrestrial mammals.

Seals are diving mammals and encounter long periods of apnea, and thus their RBC aggregation properties have been suggested to provide advantages for this physiological challenge (Castellini et al. 2006). However, different seal species have very different RBC aggregation properties: (1) Ringed seals (Phoca hispida) live in the North Arctic near shelf ice, have an average weight of 80 kg, a 15- to 20-minute diving time and no measurable RBC aggregation (Myrenne M index of zero, ESR of 1–2 mm/hr). (2) Elephant seals (*Mirounga angustirostris*) live near the coast of California and Mexico, have an average weight of 3,000 kg, a 60- to 80-minute diving time, and aggregation somewhat greater than normal human values. (3) Weddell seals (*Leptonychotes weddelli*) live in the Antarctic near shelf ice, have an average weight of 450 kg, a 60- to 80-minute diving time, and aggregation at least twofold greater than human values (Castellini et al. 2010). RBC aggregation for bowhead whales (*Balaena mysticetus*) has also been reported: these whales can reach a weight of 50,000 kg, have a 30- to 60-minute diving time and RBC aggregation 2- to 3-fold greater than humans (Castellini et al. 2006).

The wide diversity of RBC aggregation behavior indicated previously defies easy explanation. All of these marine mammals exhibit the so-called diving response in which, upon diving, blood flow is reduced and the majority of flow perfuses the heart and brain. Upon resurfacing, the low aggregation and blood viscosity as found in ringed seals might be beneficial in promptly reestablishing blood flow in preparation for another dive. However, the merit of this suggestion is questionable in light of the elevated aggregation in elephant seals and the greatly elevated aggregation in Weddell seals; based upon normal values from healthy humans, aggregation in Weddell seals could be considered to be "pathologic," yet these animals are healthy. The diversity in RBC aggregation among these marine mammals must thus reflect evolutionary solutions to the physiological problems encountered during prolonged diving periods, and may indicate a trade-off between vascular resistance and the need to maintain sufficient oxygen storage in the blood (Elsner and Meiselman 1995).

9.5 CONCLUSION

It is clear from the material in this chapter that comparative studies related to RBC aggregation are at the stage of asking questions, rather than at the point of being able to answer them. Obviously, the relationships between aggregation properties and the diverse biology of mammals are very complex and further studies are needed to clarify them. A fuller understanding of the reasons for the large differences in RBC aggregation among mammals would certainly help to improve insight into fundamental physiological mechanisms and should have a beneficial impact on human and veterinary medicine (Baskurt and Meiselman 2010).

LITERATURE CITED

Andrews F. M., N. L. Korenek, W. L. Sanders, and R. L. Hamlin. 1992. "Viscosity and Rheologic Properties of Blood from Clinically Normal Horses." *American Journal of Veterinary Research* 53:966–970.

Baskurt, O. K., M. Bor-Kucukatay, O. Yalcin, and H. J. Meiselman. 2000. "Aggregation Behavior and Electrophoretic Mobility of Red Blood Cells in Various Mammalian Species." *Biorheology* 37:417–428.

Baskurt, O. K., R. A. Farley, and H. J. Meiselman. 1997. "Erythrocyte Aggregation Tendency and Cellular Properties in Horse, Human, and Rat: A Comparative Study." *American Journal of Physiology—Heart and Circulatory Physiology* 273:H2604–H2612.

Baskurt, O. K., S. Marshall-Gradisnik, M. Pyne, M. Brenu E. Simmonds, R. S. Christy, and H. J. Meiselman. 2010. "Assessment of Hemorheological Profile of Koala and Echidna." *Zoology* 113:110–117.

Baskurt, O. K., and H. J. Meiselman. 2010. "Lessons from Comparative Hemorheology Studies." *Clinical Hemorheology and Microcirculation* 41: 105–108.

Baskurt, O. K., E. Tugral, B. Neu, and H. J. Meiselman. 2002. "Particle Electrophoresis as a Tool to Understand the Aggregation Behavior of Red Blood Cells." *Electrophoresis* 23:2103–2109.

Baumler, H., B. Neu, R. Mitlohner, R. Georgieva, H. J. Meiselman, and H. Kiesewetter. 2001. "Electrophoretic and Aggregation Behavior of Bovine, Horse and Human Red Blood Cells in Plasma and in Polymer Solutions." *Biorheology* 38:39–51.

Ben Ami, R., G. Barshtein, D. Zeltser, Y. Goldberg, I. Shapira, A. Roth, G. Keren, et al. 2001. "Parameters of Red Blood Cell Aggregation as Correlates of the Inflammatory State." *American Journal of Physiology—Heart and Circulatory Physiology* 280:H1982–H1988.

Cabel, M., H. J. Meiselman, A. S. Popel, and P. C. Johnson. 1997. "Contribution of Red Blood Cell Aggregation to Venous Vascular Resistance in Skeletal Muscle." *American Journal of Physiology—Heart and Circulatory Physiology* 272:H1020–H1032.

Castellini, M., R. Elsner, O. K. Baskurt, R. B. Wenby, and H. J. Meiselman. 2006. "Blood Rheology of Weddell Seals and Bowhead Whales." *Biorheology* 43:57–69.

Castellini, M. A., O. K. Baskurt, J. M. Castellini, and H. J. Meiselman. 2010. "Blood rheology in marine mammals." *Frontiers in Aquatic Physiology*, 1: 146-1–146-8.

Chien, S. 1987. "Red Cell Deformability and Its Relevance to Blood Flow." *Annual Review of Physiology* 49:177–192.

Cohen, W. D. 1991. "The Cytoskeletal System of Nucleated Erythrocytes." *International Review of Cytology* 130:37–84, 1991.

Elsner, R., and H. J. Meiselman. 1995. "Splenic Oxygen Storage and Blood Viscosity in Seals." *Marine Mammal Science* 11:93–96.

Eylar, E. H., M. A. Madoff, O. V. Brody, and J. L. Oncley. 1961. "The Contribution of Sialic Acid to the Surface Charge of the Erythrocyte." *Journal of Biological Chemistry* 237:1992–2000.

Fåhraeus, R. 1958. "The Influence of the Rouleaux Formation of the Erythrocytes on the Rheology of the Blood." *Acta Medica Scandinavica* 161:151–165.

Falco, C., A. Vaya, M. Simo, T. Contreras, M. Santaolaria, and J. Aznar. 2005. "Influence of Fibrinogen Levels on Erythrocyte Aggregation Determined with the Myrenne Aggregometer and the Sefam Erythro-Aggregometer." *Clinical Hemorheology and Microcirculation* 33:145–151.

Gascoyne, S. C., and C. M. Hawkey. 1992. "Patterns of Variation in Vertebrate Haematology." *Clinical Hemorheology* 12:627–637.

Grutzner, F., J. Deakin, W. Rens, N. El-Mogharbel, and J. A. Marshall Graves. 2003. "The Monotreme Genome: A Patchwork of Reptile, Mammal and Unique Features?" *Comparative Biochemistry and Physiology—Part A: Molecular & Integrative Physiology* 136:867–881.

Hochmuth, R. M. 1981. "Deformability and Viscoelasticity of Human Erythrocyte Membrane." *Scandinavian Journal of Clinical and Laboratory Investigation* 41:63–66.

Johnn, H., C. Phipps, S. C. Gascoyne, C. Hawkey, and M. W. Rampling. 1992. "A Comparison of the Viscometric Properties of the Blood from a Wide Range of Mammals." *Clinical Hemorheology* 12:639–647.

Kram, R., and T. J. Dawson. 1998. "Energetics and Biomechanics of Locomotion by Red Kangaroos (*Macropus Rufus*)." *Comparative Biochemistry and Physiology B* 120:41–49.

Kumaravel, M., and M. Singh. 1995. "Sequential Analysis of Aggregation Process of Erythrocytes of Human, Buffalo, Cow, Horse, Goat and Rabbit." *Clinical Hemorheology* 15:291–304.

Li, J. K-J. 1996. *Comparative cardiovascular dynamics of mammals*. Boca Raton, FL: CRC Press.

Long, C. A. 2007. "Evolution of Function and Form in Camelid Erythrocytes." *Bio'07: Proceedings of the 3Rd WSEAS International Conference on Cellular and Molecular Biology, Biophysics and Bioengineering*,18–24.

Meiselman, H. J. 1993. "Red Blood Cell Role in RBC Aggregation: 1963–1993 and Beyond." *Clinical Hemorheology* 13:575–592.

Meiselman, H. J. 2009. "Red Blood Cell Aggregation: 45 Years Being Curious." *Biorheology* 46:1–19.

Meiselman, H. J., O. K. Baskurt, S. O. Sowemimo-Coker, and R. B. Wenby. 1999. "Cell Electrophoresis Studies Relevant to Red Blood Cell Aggregation." *Biorheology* 36:427–432.

Meiselman, H. J., M. A. Castellini, and R. Elsner. 1992. "Hemorheological Behavior of Seal Blood." *Clinical Hemorheology* 12:657–675.

Mohandas, N., J. A. Chasis, and S. B. Shohet. 1983. "The Influence of Membrane Skeleton on Red Cell Deformability, Membrane Material Properties, and Shape." *Seminars in Hematology* 20:225–242.

Nash, G. B., R. B. Wenby, S. O. Sowemimo-Coker, and H. J. Meiselman. 1987. "Influence of Cellular Properties on Red Cell Aggregation." *Clinical Hemorheology* 7:93–108.

Nicol, S. 2003. "Monotreme Biology." *Comparative Biochemistry and Physiology* A136:795–798.

Ohta, K., F. Gotoh, M. Tomita, N. Tanahashi, M. Kobari, T. Shinohara, Y. Tereyama, B. Mihara, and H. Takeda. 1992. "Animal Species Differences in Erythrocyte Aggregability." *American Journal of Physiology—Heart and Circulatory Physiology* 262:H1009–H1012.

Plasenzotti, R., U. Windberger, F. Ulberth, W. Osterode, and U. Losert. 2007. "Influence of Fatty Acid Composition in Mammalian Erythrocytes on Cellular Aggregation." *Clinical Hemorheology and Microcirculation* 37:237–243.

Popel, A. S., P. C. Johnson, M. V. Kameneva, and M. A. Wild. 1994. "Capacity for Red Blood Cell Aggregation Is Higher in Athletic Mammalian Species Than in Sedentary Species." *Journal of Applied Physiology* 77:1790–1794.

Rampling, M. W. 1988. "Red Cell Aggregation and Yield Stress." In *Clinical blood rheology*, ed. G. D. O. Lowe, 45–64. Boca Raton, FL: CRC Press, Inc.

Rampling, M. W., H. J. Meiselman, B. Neu, and O. K. Baskurt. 2004. "Influence of Cell-Specific Factors on Red Blood Cell Aggregation." *Biorheology* 41:91–112.

Roelofsen, B., G. Van Meer, and J. A. F. Op Den Kamp. 1981. "The Lipids of Red Cell Membranes." *Scandinavian Journal of Clinical and Laboratory Investigation* 41:111–115.

Seaman, G. V. F., and G. Uhlenbruck. 1963. "The Surface Structure of Erythrocytes from Some Animal Sources." *Archives of Biochemistry and Biophysics* 100:493–502.

Smith, J. E., N. Mohandas, and S. B. Shohet. 1979. "Variability in Erythrocyte Deformability among Various Mammals." *American Journal of Physiology—Heart and Circulatory Physiology* 236:H725–H730.

Sowemimo-Coker, S. O., P. Whittingstall, L. Pietsch, R. M. Bauersachs, R. B. Wenby, and H. J. Meiselman. 1989. "Effects of Cellular Factors on the Aggregation Behavior of Human, Rat, and Bovine Erythrocytes." *Clinical Hemorheology* 9:723–737.

Spengler, M. I., S. M. Bertoluzzo, G. Catalani, and M. L. Rasia. 2008. "Study on Membrane Fluidity and Erythrocyte Aggregation in Equine, Bovine and Human Species." *Clinical Hemorheology and Microcirculation* 38:171–176.

Spengler, M. I., and M. Rasia. 2001. "Influence of Plasma Proteins on Erythrocyte Aggregation in Three Mammalian Species." *Veterinary Research Communications* 25:591–599.

Stoiber, B., C. Zach, B. Izay, and U. Windberger. 2005. "Whole Blood, Plasma Viscosity, and Erythrocyte Aggregation as a Determining Factor of Competitiveness in Standard Bred Trotters." *Clinical Hemorheology and Microcirculation* 32:31–41.

Usami, S., S. Chien, and M. I. Gregersen. 1969. "Viscometric Characteristics of Blood of the Elephant, Man, Dog, Sheep and Goat." *American Journal of Physiology* 217:884–890.

Uyuklu, M., H. J. Meiselman, and O. K. Baskurt. 2007. "Effect of Decreased Plasma Cholesterol by Atorvastation Treatment on Erythrocyte Mechanical Properties." *Clinical Hemorheology Microcirculation* 36:25–33.

Uzoigwe, C. 2006. "The Human Erythrocyte Has Developed the Biconcave Disc Shape to Optimize the Flow Properties of the Blood in the Large Vessels." *Medical Hypotheses* 67:1159–1163.

Walter, H., and K. E. Widen. 1993. "Immobilized Metal Ion Affinity Partitioning of Erythrocytes from Different Species in Dextran-Poly(Ethylene Glycol) Aqueous Systems." *Journal of Chromatography* 641:279–289.

Walter, H., and K. E. Widen. 1994. "Cell Partitioning in Two-Polymer Aqueous Phase Systems and Cell Electrophoresis in Aqueous Polymer Solutions. Red Blood Cells from Different Species." *Journal of Chromatography* 668:185–190.

Waugh, R. E. 1992. "Red Cell Deformability in Different Vertebrate Animals." *Clinical Hemorheology* 12:649–656.

Wickham, L. L., R. M. Bauersachs, R. B. Wenby, S. Sowemimo-Coker, H. J. Meiselman, and R. Elsner. 1990. "Red Cell Aggregation and Viscoelasticity of Blood from Seals, Swine and Man." *Biorheology* 27:191–204.

Windberger, U., A. Bartholovitsch, R. Plasenzotti, R. Korbut, and D. D. Heistad. 2003. "Whole Blood Viscosity, Plasma Viscosity and Erythrocyte Aggregation in Nine Mammalian Species: Reference Values and Comparison of Data." *Experimental Physiology* 88:431–440.

Windberger, U., and O. K. Baskurt. 2007. "Comparative Hemorheology." In *Handbook of Hemorheology and Hemodynamics*, ed. O. K. Baskurt, M. R. Hardeman, M. W. Rampling, and H. J. Meiselman, 267–285. Amsterdam, Berlin, Oxford, Tokyo, Washington DC: IOS Press.

Index

9 780367 382315